# Topics in Atmospheric and Oceanic Sciences

Editors: Michael Ghil   Robert Sadourny   Jürgen Sündermann

# The Dawn of Massively Parallel Processing in Meteorology

Proceedings of the 3rd Workshop on Use of Parallel Processors in Meteorology

Edited by
G.-R. Hoffmann and D. K. Maretis

With 126 Figures

Springer-Verlag Berlin Heidelberg New York
London Paris Tokyo Hong Kong

Dipl.-Math. GEERD-R. HOFFMANN
Dr. DIMITRIS K. MARETIS
European Centre for Medium-Range Weather Forecasts
Shinfield Park, Reading Berkshire RG2 9AX, United Kingdom

*Series Editors:*

Prof. Dr. MICHAEL GHIL, Department of Atmospheric Sciences and
Institute of Geophysics and Planetary Physics,
University of California, Los Angeles, CA 90024 / USA

Dr. ROBERT SADOURNY, Laboratoire de Météorologie Dynamique,
Ecole Normale Supérieure,
24 rue Lhomond, 75231 Paris Cedex 05 / France

Dr. JÜRGEN SÜNDERMANN, Universität Hamburg,
Institut für Meereskunde,
Heimhuder Straße 71, 2000 Hamburg 13 / FRG

ISBN-13:978-3-642-84022-7     e-ISBN-13:978-3-642-84020-3
DOI: 10.1007/978-3-642-84020-3

2132/3145-543210

# Contents

# List of Attendees

J. Afifi

University of London Computer Centre, 20 Guilford Street, London WC1N 1DZ, United Kingdom

M. Alestalo

Finnish Meteorological Institute, P.O.B. 503, 00101 Helsinki, Finland

C. van den Berghe

Active Memory Technology, 65 Suttons Park Ave., Reading, RG6 1AZ, United Kingdom

D. Bjørge

Institute of Geophysics, P.O. Box 1022, Blindern, N-0315 Oslo 3, Norway

D. Blaskovich

Environmental Systems Marketing, Cray Research, Inc., 1333 Northland Drive, Mendota Heights, MN 55120, USA

B. Buzbee

National Center for Atmospheric Research, Scientific Computing Division, P.O.B. 3000, Boulder, CO 80307, USA

G. Camara

Instituto de Pesquisas Espaciais, Av. dos Astronautas 1758, Caixa Postal 515, 12201 Sao José dos Campos - SP, Brazil

G. Cats

Royal Netherlands Meteorological Institute, P.O.Box 201, 3730 AE De Bilt, The Netherlands

R. Chamberlain

INTEL Corporation (UK) Ltd., Pipers Way, Swindon SN3 1RJ, United Kingdom

Z. Christidis

IBM Research, T.J. Watson Research Center, P.O.B. 704, Yorktown Heights, N.Y. 10598, USA

P. Cluley

Meteorological Office, London Road, Bracknell/Berks., United Kingdom

S. Dash

Indian Institute of Technology, Centre for Atmospheric and Fluids Sciences, Hauz Khas, New Delhi-110016, India

R. Datta

Dept. of Science & Technology, Technology Bhavan, New Mehrauli Road, New Delhi 110016, India

D. Deaven

NOAA/National Weather Service, Development Division, National Meteorological Center, Washington, DC 20233, USA

A. Dickinson

Meteorological Office, London Road, Bracknell/Berks., United Kingdom

L. Dongxian — State Meteorological Administration, 46 Baishiqiaolu, Beijing, People's Republic of China

I. Duff — Harwell Laboratory, Computer Science and Systems Division, U.K. Atomic Energy Authority, Harwell, OX11 0RA, United Kingdom

H. Eckardt — Siemens AG, ZTI SYS 3, Otto-Hahn-Ring 6, 8000 München 83, F.R. of Germany

A. Foss — Det Norske Meteorologiske Institutt, Postboks 43, Blindern, 0313 Oslo 3, Norway

M. Furtney — Cray Research Inc., 1440 Northland Drive, Mendota Heights, MN 55120, USA

E. Gallopoulos — Center for Supercomputing, Research and Development, University of Illinois, 305 Talbot Laboratory, 104 South Wright Street, Urbana, Illinois 61801-2932, USA

T. Garcia-Meras Jimenez — Instituto Nacional de Meteorologia, Apartado 285, Madrid 3, Spain

P. Garrett — Meiko Ltd., 650 Aztec West, Almondsbury, Bristol, BS12 4SD, United Kingdom

V. Gülzow — Deutsches Klimarechenzentrum, Bundesstr. 55, D-2000 Hamburg, F.R. of Germany

P. Halton — Meteorological Service, Glasnevin Hill, Dublin 9, Ireland

S. Järvenoja — Finnish Meteorological Institute, P.O.B. 503, SF-00101 Helsinki, Finland,

C. Jesshope — University of Southampton, Dept. of Electronics & Comput. Science, Southampton, SO9 5NH, United Kingdom

L. Johnsson — Thinking Machines Corporation, 245 First St., Cambridge, MA 02142-1214, USA

P. Kållberg — The Swedish Meteorological and Hydrological Institute, 60176 Norrköping, Sweden

H. Katayama — NEC, 1 Victoria Road, London, W3 6UL, United Kingdom

D. Majewski — Deutscher Wetterdienst, Postfach 10 04 65, D-6050 Offenbach (Main), F.R. of Germany

D. Martin — Siemens UK Ltd., Windmill Road, Sunbury-on-Thames, Middx., United Kingdom

G. Maxerath — Suprenum GmbH, Hohe Str. 73, D-5300 Bonn 1, F. R. of Germany

G. Mozdzynski — Control Data Ltd., 3 Roundwood Avenue, Stockley Park, Uxbridge, UX11 1AG, United Kingdom

S. Natsuki  FUJITSU LTD., 1015 Kamiodanaka, Nakahara-ku, Kawasaki 211, Japan

M. O'Neill  Cray Research (UK) Ltd., Cray House, London Road, Bracknell/Berks., United Kingdom

B. Owren  Runit Supercomputer Centre, 7034 Trondheim, Norway

S. Pasquini  Servizio Meteorologico, Piazzale degli Archivi 34, 00100 Roma, Italy

S. Potesta  Servizio Meteorologico, Piazzale degli Archivi 34, 00100 Roma, Italy

D. Reynolds  CDC, 60 Garden Ct., Suite 250, Monterey, CA 93940, USA

R. Rohrbach  Siemens AG, Data Systems Division, Systems Engineering Dept., Otto-Hahn-Ring 6, D-8000 München 83, F.R. of Germany

T. Rosmond  Department of the Navy, Naval Environmental Prediction Research Facility, Monterey, CA 93943-5006, USA

S. Saitoh  Information Basic Research Labs., C&C Information Technology Research Labs., NEC Corporation, 1-1 Miyamae-ku, Kawasaki, Kanagawa 213, Japan

D. Salmond  Cray Research (UK) Ltd., Cray House, London Road, Bracknell/Berks., United Kingdom

I. Schmidely  Service Central d'Exploitation de la Météorologie, SCEM/TTI/DEV, 2 Avenue Rapp, 75340 Paris Cedex 07, France

L. Scott  Control Data Limited, 3 Roundwood Avenue, Stockley Park, Uxbridge, Middx. UB11 1AG, United Kingdom

J. Sela  NOAA/National Weather Service, Development Division, National Meteorological Center, Washington, DC 20233, USA

D. Snelling  Department of Computer Studies, University of Leicester, Leicester, United Kingdom

A. Staniforth  Recherche en Prévision Numérique, 2121 Voie de Service Nord, Porte 508, Route Trans-Canadienne, Dorval, Québec, H9P 1J3, Canada

R. Strüfing  Deutscher Wetterdienst, Postfach 10 04 65, D-6050 Offenbach (Main), F.R. of Germany

P. Swarztrauber  National Center for Atmospheric Research, 1850 Table Mesa Drive, Boulder, CO 80307, USA

D. Tanqueray  Floating Point Systems (UK) Ltd., APEX House, London Road, Bracknell/Berks., RG12 2TE, United Kingdom

S. Tett — University of Edinburgh, Dept. of Meteorology & Physics, J.C.M.B., Kings Buildings, West Mains Road, Edinburgh, United Kingdom

C. Thole — Suprenum GmbH, Hohe Str. 73, D-5300 Bonn 1, F.R. of Germany

J. Tuccillo — NOAA/National Weather Service, Automation Division, National Meteorological Center, Washington, DC 20233, USA

J. Tupaz — Commanding Officer, Fleet Numerical Oceanography Center, Monterey, CA. 93943-5005, USA

R. Vanlierde — Institut Royal Météorologique, Avenue Circulaire 3, 1180 Bruxelles, Belgium

R. Wiley — Meteorological Office, London Road, Bracknell/Berks., United Kingdom

H. Xueguang — State Meteorological Administration, 46 Baishiqiaolu, Beijing, People's Republic of China

## E C M W F :

L. Bengtsson — Director
D. Söderman — Head, Operations Department
D. Burridge — Head, Research Department
H. Böttger — Head, Meteorological Division
A. Simmons — Head, Numerical & Dynamical Aspects Section
J. Gibson — Head, Meteorological Applications Section
P. Gray — Head, Computer Operation Section
C. Hilberg — Head, Operating Systems Section
P. Undén — Data Assimilation Section
M. Hamrud — Numerical & Dynamical Aspects Section
D. Dent — Numerical & Dynamical Aspects Section
N. Storer — Operating Systems Section
D. Maretis — User Support
N. Kreitz — User Support
T. Kauranne — Operations Department (Consultant)

## Chairman:

Geerd-R. Hoffmann — Head, Computer Division

# Introduction: The Dawn of Massively Parallel Processing in Meteorology

GEERD-R. HOFFMANN

European Centre for Medium-Range Weather Forecasts, Shinfield Park, Reading, Berkshire RG2 9AX, U.K.

## 1. Background

Since 1984, the beginning of the series of workshops on Use of Parallel Processors in Meteorology (see Hoffmann, Snelling (1988)), the emphasis has always been on a small number of processors, as for example expressed in the statement "it became apparent during the discussions that the smaller the number of processors the better" (see Prior (1988)). However, during the 3rd workshop held in 1988, the proceedings of which are contained in this book, there was a remarkable shift to the investigation of massively parallel processors. The reasons for this change can be attributed to a number of developments outlined below.

## 2. The Case for Massively Parallel Processing

### 2.1 Requirements

As outlined in Bengtsson (1988), numerical weather prediction (NWP) models will require between $10^{12}$ and $10^{16}$ computations to be carried out within 3 to 24 hours. This results in a requirement for a computer to produce up to an average of $10^2$ GFLOPS ($10^9$ Floating Point Operations Per Second), compared with the performance of todays fastest computer of around 1.5 GFLOPS with 8 processors.

### 2.2 Hardware Status

The systems marketed at present as "supercomputers" largely use silicon chips with a system cycle time of down to 3 ns ($10^{-9}$ sec). It is commonly assumed that a system cycle time of 2-1 ns is the maximum which can be achieved without changing the technology to something completely new like optical gates. Therefore, the growth in performance due to faster circuits is limited by a factor of around 3-5. If still higher performance increases are necessary as outlined above, they will have to be delivered by providing for more processors.

### 2.3 Software Status

The use of a small ($\le 16$) number of processors seems to be by now well understood for meteorological applications, provided the system has shared memory access. The use of distributed memory systems, however, is still in its infancy. Since most of the massively parallel systems available today belong to the latter category, their use for meteorological applications cannot be taken for granted yet.

Topics in Atmospheric and Oceanic Sciences
© Springer-Verlag Berlin Heidelberg 1990

The development of system software and supporting applications like compilers for massively parallel systems has been making considerable progress in the last few years, and is now close to the stage where operational implementations of meteorological programs can be envisaged.

## 2.4 Costs

The expected cost benefit of massively parallel systems compared to conventional high performance computers has been borne out. Assuming that ways can be found to harness the inherent capacity of such systems the costs are about a third of those relating to old-style solutions.

## 2.5 Conclusion

For NWP to achieve its goals in the not too distant future, the problems raised by using massively parallel computers have to be aggressively attacked. The time has come to undertake such studies, because both hardware and software developments have progressed sufficiently to allow the implementation of an operational weather forecasting system with some hope of success.

The workshops organised by ECMWF provide a forum to air the problems encountered when embarking on such activities, and to focus the attention of the experts on this field of research.

## 3. Acknowledgements

The author would like to thank all participants of the workshop for their contributions, his colleagues at ECMWF for their support and encouragement and last, but not least, Springer Press for making the papers from the workshop more widely available again.

## 4. References

Bengtsson, L. (1988): Computer Requirements for Atmospheric Modelling. In Hoffmann, G-R., and Snelling, D.F., editors: Multiprocessing in Meteorological Models, Springer Verlag, Berlin Heidelberg New York, 1988, pp. 109-116.

Hoffmann, G.-R. and Snelling, D.F., editors (1988): Multiprocessing in Meteorological Models, Springer Verlag, Berlin Heidelberg New York 1988. ISBN 3-540-18547-0.

Prior, P. (1988): Multiprocessors: Requirements and Consequences. Summary of the Discussion. Ibid. pp. 233-235.

# An Introduction to Parallel Processing in Meteorology

TUOMO KAURANNE[1]

European Centre for Medium-Range Weather Forecasts, Shinfield Park, Reading, Berkshire RG2 9AX, U.K.

## 1. Introduction

The first operational weather model to run on more than one processor was apparently that of the U. S. Fleet Navy Oceanography Center in Monterey, which executed on a CDC multiprocessor already in the late 1960's. The European Centre for Medium Range Weather Forecasts has been in the forefront of multiprocessing since the present spectral model was multitasked on the two processor Cray X-MP/22 in 1985 and on the present four processor Cray X-MP/48 in 1986. The model has now been tested on an eight processor Cray Y-MP with remarkable efficiency, as is reported in David Dent's talk at this workshop [Dent].

The forefront of supercomputing is moving along two lines: the fast processor line of machines with a few very powerful processors, and the massively parallel line in which machines have hundreds, thousands or even more relatively cheap standard processors. Further progress on the former line is impeded by technical difficulties in manufacturing smaller and faster components. The problems on the latter line are mainly due to the difficulty of implementing important applications, like full operational weather forecast models, efficiently on a large number of processors.

The previous two workshops on parallel processing in meteorology held at ECMWF two and four years ago were conceived mainly with using a smallish number of fast processors. The present workshop focuses on the use of massively parallel computers in meteorology. Some recent breakthroughs along this line of research serve to further justify this focus. Even though no current operational weather model runs on more than four processors, several successful attempts to port weather models to massively parallel computers with up to thousands of processors have been made.

Application of massively parallel processors to numerical weather prediction was considered already in the 1960s. Carroll and Wetherald [CarWet] devised a strategy to numerically solve the primitive equations on the SOLOMON II computer which, unfortunately, was never built. A lot of work in this direction was carried out on the Goodyear Aerospace Massively Parallel Processor (MPP) in the early 1980s. The MPP has 16 384 one bit processors. Gallopoulos [Gall] implemented a shallow water equation model on it and also commented extensively on how to implement a model based on the full primitive equations.

1 This work has been supported by TEKES, The Finnish Technology Development Centre, under the program Finsoft III: Parallel Algorithms.

At present massively parallel computers still possess a somewhat exotic reputation in the computer industry, being overshadowed by three decades of hard work on serial von Neumann computers. This wasn't the case in the age before von Neumann, however. In the 1920's the term computer still denoted a person performing manual computations. If extensive calculations were to be performed, the inevitable slowness of this single processor number cruncher caused the turnover times to become intolerably long. 'Parallel computing', using tens of computers was used on many occasions in practice, but the most relevant example for the meteorological community, that which Lewis F. Richardson describes in his book 'Weather Prediction by Numerical Process' in 1922 [Rich], remained a vision.

In Richardson's 'factory' 64 000 human computers were doing operational weather forecasting, calculating the state of the atmosphere on a grid with that same number of grid points. The computers were sitting in a huge circular theatre built to reflect the geometry of the earth. Each computer performed the calculations that belonged to a single grid location. The algorithm resembled present asynchronous or 'chaotic' relaxation algorithms with no stringent synchronization. To prevent the differences in speed between computers leading into spurious instabilities, a soft synchronization scheme was applied by a supreme computer sitting on top of a pillar in the centre of the theatre. He was pointing to areas either behind or ahead of time in their calculations, with lights of different colours to make them adjust their computing speed towards the average. Communication between computers was by message passing (in the form of written notes) in a two dimensional grid topology. All in all, the structure was very similar to a massively parallel asynchronous SIMD computer, with physics calculated apparently in a MIMD mode.

## 2. Levels of parallelism in an atmospheric model

There are many possible levels to produce parallelism in numerical weather prediction. One, albeit nonexhaustive, way to classify them goes as follows:

1. Job level parallelism (operating system)

2. Task level parallelism (macrotasking, subroutine or block level)

3. Code level parallelism (microtasking, parallelizing compilers, SIMD-machines, dataflow machines)

4. Linear algebraic parallelism (parallel vector and matrix algorithms)

5. PDE-numerical parallelism (parallelizing over a set of discrete basis functions)

6. PDE-theoretical parallelism ("microlocal" parallelism over localized functions on the cotangent bundle, rate of decay of Green's functions)

7. Physical parallelism (physical relations between meteorological fields, Monte Carlo molecular dynamics in the extreme)

The above order of levels of parallelism leads from the system software related aspects through the mathematical software and numerical analysis aspects down to the mathematical model and finally reaches the state of our current understanding of atmospheric motion. It also points from more generally valid ideas, of the kind a compiler designer will work with, towards the specific characteristics of modelling the atmosphere. All these levels are relevant in designing a parallel weather model.

A model designer starts his work at level seven and moves up through all the levels to level one. The process of parallelizing a weather model, on the other hand, usually starts at levels three to five (the two first levels are mainly applicable when the entire forecast system is considered for parallelization). However, considerable benefit could ensue from studying parallelization at the highest two levels as well.

In formal logic it is well known that the more general phenomena can be described using a formal language, the fewer statements about any particular phenomenon you can prove in it. The more restrictive the description, the more statements can be proved true or false. A more general language might allow them to be either true or false, and thus unprovable. Formal languages are relevant for all work using computers: they provide the means to communicate with machines.

Hence, when looking at the parallelism inherent in a weather model at the different levels, it will, in principle, be possible to find the most accurate assessment, in terms of provable statements on dependencies or independencies, of parallelism at the physical level. There are features in an atmospheric model which do not generalize to all partial differential equations, certainly not to all matrices or to all DO-loops, but which can be utilized in distributing weather models to many processors.

In particular, high level considerations aid us in picking a discretization technique and a numerical algorithm that respect the parallelism inherent in the physical or mathematical formulation of the problem, rather than just attempting to parallelize a preselected algorithm.

Parallelism at the physical level is understood here as true statements about mutual independence of some atmospheric processes up to some specified accuracy. The complementary statements about mutual dependency are important, too. These statements may be described easily on an abstract level, but they often translate into very complex data dependencies when coded into Fortran. However, if proper attention is paid in programming to keeping consistent with the abstract description, multitasking the code becomes simpler, whether done manually or by the compiler, even when the latter is not particularly sophisticated.


## 3. Some basic concepts in parallel computing

**Computational** *(time)* **complexity** of an algorithm refers to the time required to execute the algorithm. Because the complexity usually grows with the size of the data, even with the same algorithm, it has become customary to talk about complexity as a function of the size of a problem. In the case of a weather model, the size of a problem means the number of spatial degrees of freedom $Cn^3$ multiplied by the number of time steps.

**Parallel complexity** refers to the time required to execute the algorithm on an idealized parallel computer with unlimited parallelism. Serial complexity is proportional to the number of floating point operations required to execute the algorithm. This is not the case with parallel complexity, because in addition to the number of floating point operations, also the degree of parallelism in the algorithm affects the execution time.

Typical complexities are polynomials in $n$. Very difficult problems have an exponential complexity, whereas massively parallel computers with unlimited parallelism render the parallel complexity of many important algorithms to being proportional to $\log n$.

By **speedup** we mean the ratio between the time a program runs on a single processor to its execution time on $p$ processors. Ideally, speedup should remain proportional to p. For any individual code, however, speedup eventually levels off and approaches hyperbolically a constant upper limit. This follows from Amdahl's law, to be discussed in section six below.

**Utilization** is the total CPU time spent by all the processors executing the solution algorithm divided by the elapsed time and the number of processors. Utilization is ideally one, meaning that all processors are performing useful work all the time.

**Granularity** is a concept denoting the size of subproblems assigned to individual processors. A good definition is the number of floating point operations each individual processor can execute without external or internal interruption. This definition refers to what could appropriately be called **absolute granularity**. By **relative granularity** we mean the ratio between the CPU time spent by an individual processor in running the application and the difference between elapsed time and CPU time, that is, the time spent in all overhead operations, serial or parallel, necessary to assign the subtask to the processor and to receive the results from it, I/O waiting times etc. Large relative granularity means lots of effective work gets done compared to the amount of administrative work and queueing needed to organize it.

**Interconnection topology** refers to the topological structure of the parallel computer viewed as a graph with processors as the nodes and communication links as the edges.

**Average** and **maximal connectivity** refer to the average and maximum number of communication links per processor in a parallel computer. A good synonym to connectivity gradually gaining acceptance is **valence**.

**Computing power** is the capacity of the parallel computer to produce results of arithmetic operations with floating point numbers (usually in 64 bit precision) in unit time. Computing power is measured in MFLOPS or GFLOPS, $10^6$ or $10^9$ floating point operations per second respectively.

**Communication bandwidth** denotes the total transfer capacity of all the communication links of the parallel computer counted together. It is convenient to relate this to the computing power.

Commonly investigations on performance of parallel algorithms or architectures put a lot of emphasis on speedup and utilization. This is, however, often not the most appropriate figure to cite. In operational weather forecasting, the limiting factors are the time slot within which the forecast has to be produced and the cost of the computer. Within these constraints, the aim is to maintain maximal spatial and temporal accuracy in the integration of the discrete model, not to fill the computer as completely as possible i. e. by using a solution algorithm with inferior serial complexity but excellent parallelizability, unless its parallel complexity becomes superior to that of the alternative algorithms as a result of this.

A measure related to the last mentioned problem is the **operation efficiency factor**, meaning the ratio between the serial complexity of the best serial algorithm and the serial complexity of the best parallel algorithm. The operation efficiency factor cannot exceed unity.

The **efficiency** of a parallel algorithm is the product of utilization and the operation efficiency factor.

## 4. Classifying massively parallel computers

The following classes are often used to bring some sort of taxonomy into the plethora of massively parallel computers proposed.

A **MIMD** *(Multiple Instruction stream, Multiple Data stream)* **computer** denotes a parallel computer whose individual processors can execute independent programs asynchronously. Some particular problems associated with MIMD computers are *synchronisation overhead* and *load imbalance*. The former arises when there are many stages in an algorithm that have to be processed sequentially. Even though each of these stages may be parallelizable, the tasks assigned to different processors are not equal, and all the processors will have to wait for the slowest one to arrive at the synchronization point, called the *barrier*. This inequality of computational tasks is called load imbalance, and when seen from a purely statistical viewpoint, it grows with the number of processors. The combined effects of load imbalance and synchronisation overhead are a prolonged parallel execution time and a drop in utilization, because most processors are idle most of the time waiting for other processors to complete their tasks.

MIMD computers are difficult to program. Each of the processors has to possess an individual copy of the program, which wastes memory. The exact execution sequence of instructions varies somewhat randomly from one run of the program to another, due to hardware differences or modifications to the program. The latter effect often shows a very delicate interplay between the code, the compiler and the computer, sometimes in a very counterintuitive manner. This, along with a number of other similar difficulties, makes debugging on a MIMD computer sometimes a frustrating experience.

The class in some sense opposite to MIMD computers is called **SIMD** *(Single Instruction stream, Multiple Data stream)* or **data parallel computers**. All the processors of a SIMD computer execute the same instruction simultaneously, applying it to different input data. The structure of processors in a SIMD computer may be considerably simpler, allowing the computers to be 'more massively' parallel. Programming is easier on a SIMD computer than on a MIMD computer, because SIMD programming and debugging are very similar to those on a vector computer. On the other hand, SIMD machines are inherently less flexible when applied to a complicated problem. Some examples of SIMD computers are the ILLIAC IV, The Connection Machine and the AMT DAP.

Parallel computers may have a single **shared memory** accessible to all the processors, or each individual processor may have a local memory. The latter class is called **distributed memory** computers. Hybrid architectures with both kinds of memories are common, possibly in the form of local cache memories.

A shared memory in a MIMD machine is often accessed via a central bus, but this is not necessarily so. A bus based architecture has the problem of *bus contention,* when the number of processors on the bus grows large. The reason is not so much the lack of transport capacity on the bus, but the fact that it can only be used by a single message at any particular instance. The logic required to access the bus and to ensure the safe arrival of a message requires synchronisation and takes time to execute, creating an inherently serial bottleneck.

To avoid other bottlenecks which easily emerge on a shared memory computer when a large number of processors is trying to access the memory simultaneously, the memory is divided into a large number of banks. The most critical and expensive component of a shared memory parallel computer is often the memory switching circuitry which sorts out references between processors and memory banks.

Distributed memory computers on the other hand face the problem of large communication overheads, if the memory reference pattern is not local. Avoiding this calls for the algorithm and the data to be far more carefully distributed among the processors than is necessary in the case of a shared memory parallel computer.

Some of the more exotic types of parallel computers include *dataflow computers, neural networks* and *systolic computers,* which share some important features despite their very different origins.

**Dataflow computers** are parallel computers which operate fully asynchronously. They may have any interconnection topology, which may be modifiable, too. They are programmed by mapping the algorithm, viewed as a directed dependency graph, onto the architecture. Each instruction is executed immediately after it has received all its inputs. After completion, it passes its results forward along the channels assigned for its output edges. The whole process resembles the way a neural pulse passes through the nervous system.

The latter was definitely a model when **neural network computers** were designed. These are an evolution of the perceptrons of the early 1960's used to simulate e.g. adaptive learning of visual patterns. Neural networks consist of huge numbers of simple processors each acting as a simple cellular or finite state automaton. They are most often used to produce images of global patterns in a system where only local rules of behaviour are known beforehand. These local rules may include adaptive, learning or stochastic behaviour. Such complexes of simple automata are able to accomplish many complicated tasks, like understanding speech or recognizing and producing the morphology of languages.

**Systolic computers** consist of a large number of processors as well, but their connection topology is often designed exactly to reflect the communication structure of a particular algorithm, often of numerical origin. Hence the computer resembles a highly automated factory, which receives raw materials at one end and outputs complete products at the other.

The common feature in all these three classes of massively parallel computers is the desire to let the computer adapt to the structure of the algorithm, rather than mapping the algorithm onto the architecture. The relationship between these and the more conventional parallel computers of the earlier paragraphs is somewhat analogous to that between Lagrangian and Eulerian hydrodynamics, with the algorithm as the liquid.

The difficulty of the mapping problem in general varies a lot with the algorithm and the architecture. When understood as generally as with dataflow machines, i.e. mapping the program as an instruction-by-instruction dependency graph onto the architecture at hand, which may have any communication topology, the problem of finding a mapping that, for instance, minimizes the maximum number of communication links onto which any single edge of the dependency graph is mapped, is combinatorial and probably very difficult.

In weather modelling and other physical problems, the geometric consistency is the simplest guide to doing the mapping. If the parallel architecture respects geometricity, e.g. by having a grid embedded in its connectivity graph, and if the algorithm, too, is geometrically consistent, as in the case of many grid point algorithms, the mapping problem is almost trivial.

Unfortunately, however, geometric consistency and optimal accuracy or computational complexity are not always compatible. This holds for spectral transform methods, where we keep switching ungeometrically between the physical space and an abstract transform space. Similar problems might also ensue if the algorithm were viewed exclusively from the linear algebraic point of view.

Indeed, some sparse matrix algorithms used to solve finite element equations resort to graph theoretic techniques to find a permutation of the unknowns that would minimize the fill-in during Gaussian elimination, thus sacrificing geometric consistency. This is not a problem on a serial computer, but will lift the mapping problem on a parallel computer from the trivial geometric case to the difficult combinatorial case. The fundamental reason for this problem is the fact that Gaussian elimination does not preserve geometrical relations. Some of these graph theoretic techniques try to produce a permutation that would restore some of the underlying geometry.

All in all, the mapping problem is a good case for defending the principle of seeking parallelism first in the most abstract and most restrictive setting.

Less intuitive parallel architectures pose mapping problems even in the geometrically obvious case. For the hypercube, Johnsson [John] has made a careful study of mapping the dependency graphs of many numerical algorithms. In most cases he can prove that the maximum number of communication links on which any edge of the dependency graph has to be mapped is at most two, and that these mappings are path disjoint, meaning that the maps of the individual edges of the dependency graph can be communicated in parallel.

## 5. Some common interconnection topologies

The ideal interconnection topology of a parallel computer allows a message to pass from any of the processors to any other in unit time. Such a topology is called the **crossbar switch** topology. The number of links has to grow quadratically with the number of processors, which means, for instance, that a one thousand processor parallel computer requires one million communication links. As a consequence, connectivity grows linearly with the number of processors. These figures explain why the crossbar switch topology does not provide a feasible alternative to a massively parallel computer.

In numerical weather prediction we are solving a problem with a regular geometric structure: that of a hollow ball or a thin spherical shell, which is the space occupied by the atmosphere. Many other important problems in scientific and engineering computation have an equally regular structure, which suggests that the interconnection topology should reflect it. Accordingly, many of the first massively parallel computers, like the ILLIAC IV, were built to have a two dimensional **grid topology**. The modern equivalents should have a three dimensional grid topology, as present massively parallel computers are oriented towards three dimensional simulations. In quantum dynamic calculations an even larger number of dimensions would be desirable. The extreme processors on one side of the grid are usually connected to the extreme processors on the opposite side in a toroidal fashion.

In grid topologies, both maximum and average connectivities remain uniform: four in 2 D, six in 3 D grids, so these architectures are very attractive from the manufacturing point of view. However, many algorithms require some amount of global communication. On a grid, the number of intermediate connections required for communication between two distant processors grows as the square root (2D grid) or the cube root (3D grid) of the number of processors. This quickly results in an unacceptable time penalty for such an algorithm.

A still simpler topology is a **ring**, where the processors sit on a single communication channel like the vertices of a regular polygon with $p$ vertices, $p$ denoting the number of processors. A number of algorithms have been succesfully adapted to a ring, but in general it is a very weak topology, being in fact a one dimensional grid. An example is the ZMOB at the University of Maryland.

To avoid the problems with global communication a number of improved connection topologies have been suggested during the last decade. A **tree** is a structure that appears very commonly in algorithms related to searching or heuristic reasoning. Consequently, tree-like architectures have been suggested for parallel computers intended to run such algorithms. Like grids, trees have uniform connectivity. An example is the Carnegie-Mellon University Tree Machine. Because of lack of geometric regularity, however, trees have met with little success in scientific applications.

A **pyramid** combines a tree with a sequence of grids. The grids have a geometrically decreasing number of processors. The connections between grids are regular in the sense that on a finer grid, for instance, every fourth processor is connected to a processor on the next coarser grid, and so on. This implies that at least half of the processors always reside on the finest grid. Both maximum and average connectivities remain constant, though different (four and six in a two dimensional pyramid, six and eight in a three dimensional one). Because of this, the communication bandwidth grows only linearly with the number of processors. A 2D pyramid is illustrated in Figure 1.

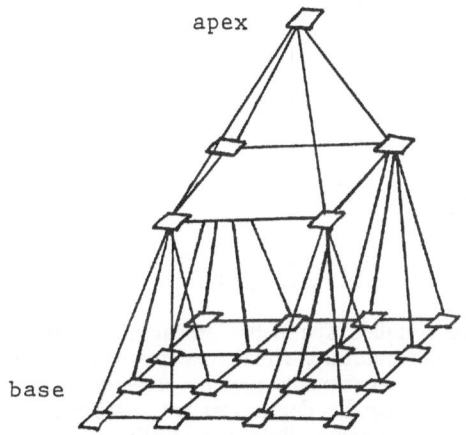

apex

base

Figure 1. A two dimensional pyramid architecture

A **hypercube** is a topology which has gained a lot of popularity among manufacturers during this decade. A $d$ dimensional hypercube has $2^d$ processors in the corners, and the communication links follow the edges of the cube. A four dimensional hypercube can be visualized by thinking of two three dimensional cubes with eight processors each, where each of the processors is connected to the corresponding processor in the other cube. This can again be duplicated by the same principle to form a five dimensional cube and so on. The maximum and average connectivities of a hypercube are the same, namely $d$, and hence grow logarithmically with the number of processors. The communication bandwidth is proportional to $p \log p$. Examples of hypercube machines include the Caltech Cosmic Cube and the Ncube and Intel hypercubes. Some low dimensional hypercubes are illustrated in Figure 2.

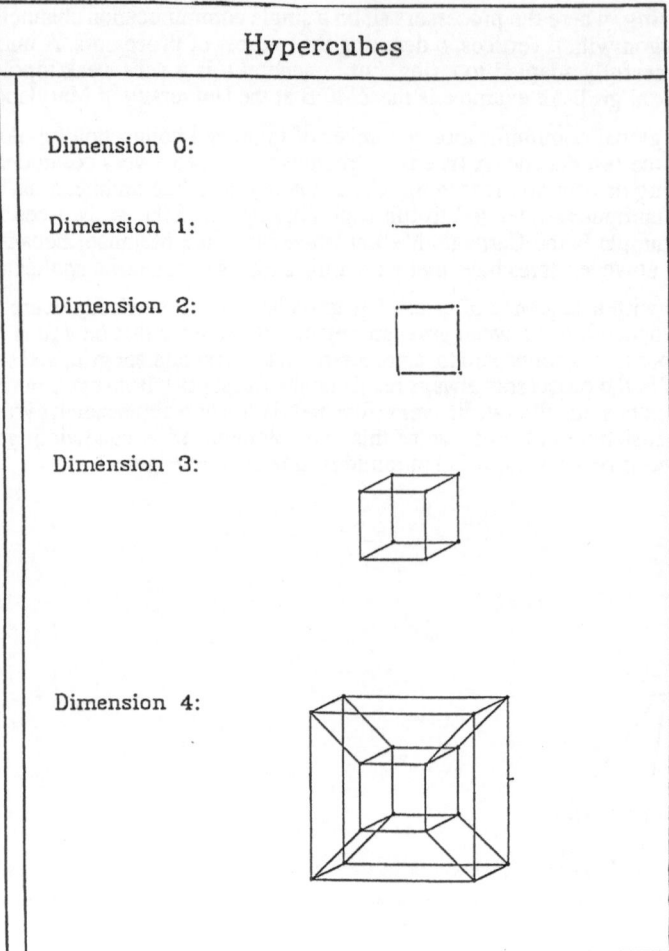

Figure 2

One of the attractive features of a hypercube is its complete symmetry. Because of its rich communication topology, all of the topologies introduced previously, with the exception of the crossbar switch, can be embedded in a hypercube, though this occasionally requires some ingenuity. The biggest problem is the growing connectivity, but to a far smaller degree than with a crossbar switch: a thousand processors only require ten thousand communication links, not a million.

**Perfect shuffle** (or butterfly, or omega network) type interconnection topologies were originally designed to reflect the communication structure of the Fast Fourier Transform algorithm to provide efficient special purpose parallel computers for signal processing applications. A perfect shuffle consists of $\log p$ rows of processors with $p$ processors on each row, with possible nearest neighbour connections between the processors in each row. $p$ is a power of 2, say $p = 2^d$. Connections between the first two rows of processors obey the rule that every processor on the first row is connected to the processor which is exactly $p/2$ processors ahead of or behind it, whichever is possible, on the second row. The second and third rows are connected similarly, but as if both rows had been split in the middle into two separate segments with $p/2$ processors each, and the connection distance having correspondingly been reduced to $p/4$ processors in each segment. The third and fourth rows are connected in the same manner, but as if they had been split into four separate segments with $p/4$ processors in each, and the connection distance now being $p/8$ in each, and so on. The connections consequently become increasingly local when approaching the last rows. The processors in the last two rows are connected as shuffling the two processors of a pair, $p/2$ pairs overall.

The connectivity of a perfect shuffle is uniform, and the communication bandwidth proportional to $p$. There is, however, a collapsed version of the perfect shuffle, which results when corresponding processors on each row are identified. This is the hypercube. A perfect shuffle connected computer can thus be seen as a blown up hypercube with the increase in connectivity replaced by an increase in the number of processors. Consequently, all the good properties of a hypercube also apply to perfect shuffles, with, at most, a logarithmic additional factor.

In some parallel computers a perfect shuffle network switch, also called a banyan switch, is used to link processors and memory banks. This provides for shared memory access, essentially without communication bandwidth problems but with a logarithmically growing access time, when increasing the number of processors. An example is the BBN Butterfly, whose architecture is illustrated in Figure 3.

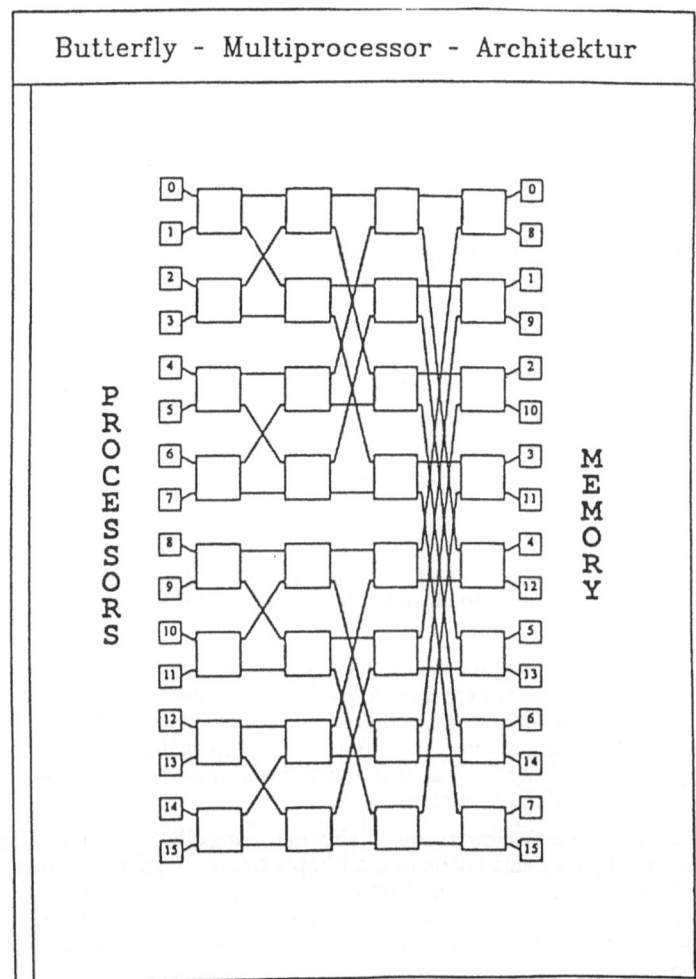

Figure 3

Finally, there are parallel computers with a flexible topology. These are provided with a **reconfigurable switching network** between processors. All the previously introduced topologies have benefits but also drawbacks. The choice of the correct topology for any particular application becomes an exercise in optimization. It should thus make sense to provide a single computer to optimize on. Such a computer would also be applicable to a larger variety of tasks than any computer with a fixed interconnection topology.

The drawback of a reconfigurable switching network is that one has to provide the computer with a crossbar switch, to facilitate a free choice of interconnection topology. Its insurmountable complexity compels the designer to use some combination of switching logic and a hierarchy of buses. This results in a logarithmically (in $p$) growing complexity of any communication or bus contention. Usually, reconfiguring the topology on the fly is extremely slow compelling the user to restrict to a single topology in any single run. Examples of this class include the Meiko Computing Surface and the PARSYS SuperNode.

## 6. Amdahl's law revisited

One of the most frequently cited theorems in parallel and vector computing is the so-called Amdahl's law. In its classical form, Amdahl's law states that any fixed serial component in a code sets an absolute upper limit to the speedup obtainable, namely

$$\text{Maximum Speedup} = \frac{\text{Total CPU time}}{\text{Time consumed in the serial component}} \tag{1}$$

and that the gain from added processors shows a hyperbolic diminishing returns pattern:

$$\text{Speedup} = \frac{1}{S + \frac{P}{p}} \tag{2}$$

where $S$ is the serial fraction and $P$ is the parallelizable fraction of the code, and $p$ is the number of processors. The fractions have been normalized so that $S + P = 1$.

When looking at an atmospheric model from the physical or mathematical point of view, however, there is no identifiable serial portion. The atmosphere works in parallel. Nevertheless, Amdahl's law retains its significance. This is because some other factors affect the obtainable speedup similarly to a serial portion of the code.

The first such factor is any fixed overhead required in, for instance, distributing a new task, however small, to a processor, or the communication startup time, a fixed synchronization overhead, a fixed bus access latency, and so on. Obtaining unlimited speedup on any fixed code when the number of processors is allowed to approach infinity requires that the time spent in each individual subtask must come down towards zero. Clearly, any fixed overhead makes this impossible, and there will be an upper limit to the maximum speedup obtainable.

Another factor having similar effects is an inherent sequentiality in solving elliptic problems. These feature in weather models as implicitly treated parts in semi-implicit time-stepping schemes. The sequentiality arises from the complexity of optimal elliptic solvers, which necessarily grows logaritmically with the accuracy of the discretization. Such a seriality might be measured by the length, in numbers of floating point operations, of a minimal critical path of the directed graph describing such an algorithm.

The inherent assumption in the applicability of Amdahl's law is that we are always solving the same problem, regardless of the total power of the computer available. Clearly, this is not consistent with the quest for maximum accuracy in numerical weather forecasting stated earlier. If we can afford more processors for our parallel computer, we will try to achieve a more accurate representation of the atmosphere.

This was also the point Gustafsson, Montry and Benner made in their now famous paper [GuMoBe], in which they describe their implementation of three real-world problems on a 1024-processor Ncube at Sandia National Laboratories. The problems were a wave diffraction problem, a two dimensional turbulent flow problem and a beam stress analysis. This work brought them the Karp and Bell awards in March 1988.

They introduced the new notion of **scaled speedup**. By this they mean the figure obtained by dividing the time a parallel computer spends in solving a problem that is scaled up to meet the power of the whole computer by the time a single processor would spend in solving the same problem, ignoring the fact that it would possibly not fit into its local memory. This figure grows linearly with the number of processors and they obtain scaled speedups up to 1020 in their sample problems. The serial component of the code appears only in the constant of proportionality between scaled speedup and the number of processors. This same observation was pointed out even earlier by Nicol and Willard [NicWil]:

$$\text{Scaled Speedup} \;=\; S + (1 - S)\, p \tag{3}$$

The serial fraction above now measures the relative time spent in serial computations on the bigger problem that has been scaled up to match the power of the parallel computer. The inherent assumption in the applicability of the 'scaled speedup law' is that the time spent in the serial component remains constant, regardless of the total problem size. Hence the serial fraction above actually decreases with problem size. This is true of the communication startup time, of the synchronization overhead and of fixed overheads in assigning tasks to processors. In general, deterioration of speedup due to fixed overheads can be overcome by keeping the relative granularity of the algorithm high. Figure 4 shows the difference between the classical and scaled speedup curves as a function of the serial fraction.

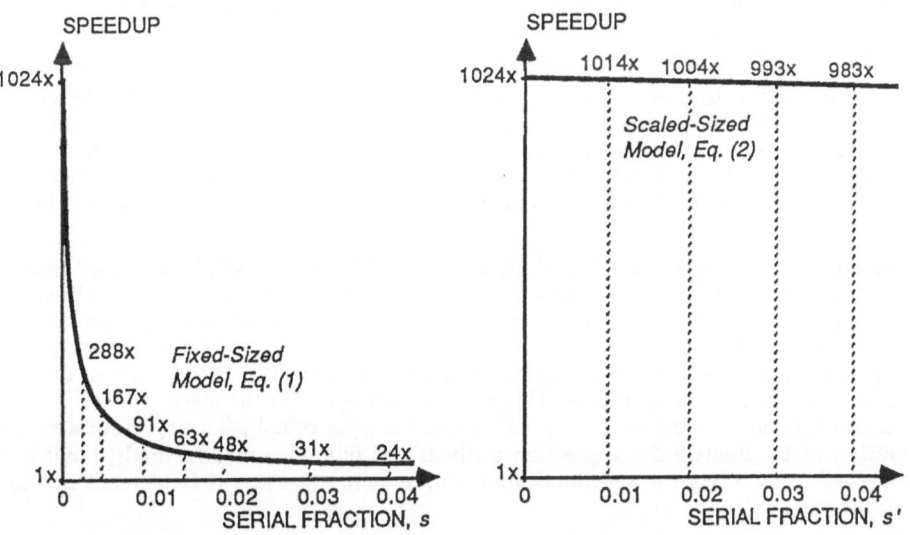

Figure 4. Speedup given by Amdahl's law and by problem scaling

The traffic jams on the bus of a single bus based computer, on the other hand, imply increased communication delays with a growing number of processors. If the bandwidth of the bus does not grow with the number of processors, the increasing amount of communication will jam the bus sooner or later.

For this reason, we do not consider single bus based architectures any further in relation to asymptotic parallelism. In their article [NicWil] Nicol and Willard analyze the gradual jamming of the bus in more detail and they come to the same conclusion on the poor applicability of single bus based architectures to massively parallel computing.

This is not to say, however, that single bus based shared memory computers are useless in large-scale scientific computing. On the contrary, the fastest supercomputers at present and in the near future do and will possess a shared memory. The bottleneck is actually not so much the communication bandwidth, which can be considerably increased by e.g. providing many parallel channels to the memory or employing fibre optic links, but the associated memory reference logic.

With multiple processors and multiple memory banks, the number of potential memory reference patterns grows as the product of the number of processors and the number of banks. Any of the references should still be resolved within just a few machine cycles, and it depends on the ingenuity of memory logic designers whether this will still remain possible with shorter machine cycles and an increasing number of processors and memory banks. The situation is quite analogous to speeding up individual processors. So far, both of these efforts have been fairly successful.

With respect to the overhead incurred in distributing work to a large number of processors, the situation regarding the applicability of the scaled speedup law is somewhere intermediate in between the two different sorts of serial components dealt with above. The same is true of the seriality of elliptic solvers. In both cases, some information will have to be gathered globally and redistributed again. This is optimally done in a divide-and-conquer or tree-like manner. This requires a logarithmically growing number of stages, and consequently, a logarithmic serial complexity, but without any identifiable "serial component". This implies that a typical scaled speedup for elliptic problems would be limited to

$$\text{Scaled Speedup in Elliptic Problems} \quad = \quad S + (1-S)\,p/\log p \qquad (4)$$

The correspondence between conventional and scaled speedups can be seen in figure 5. The amount of total memory available and the detrimental effect of fixed overheads in communication delineate the boundaries of the feasible speedup domain in the problem size - number of processors -plane.

In weather modelling, too, we aim to solve as big a problem as possible within the allowable time slot. This implies that the problem can be solved using a big enough relative granularity to keep each individual processor busy a large portion of the time. As can be seen in figure 5, the communication overhead becomes a problem when relative granularity decreases. Consequently it will not normally be a major problem in weather modelling, where granularity is kept constant by scaling up the problem.

For example, the next operational resolution envisaged for the ECMWF global spectral model approximates the primitive equations by a system of algebraic equations with six million degrees of freedom. Even if we had a hundred thousand processors, this would still leave each processor with a system of sixty equations to be solved each time step. This would take at least a thousand floating point operations, rendering the fixed communication overheads, typically of the order of tens of machine cycles, insignificant. This assumes, though, that the algorithm calls for communication only once during a timestep with a fixed number of neighbouring processors. This is obviously true of domain decomposition methods and block iterative methods; point iterative methods and spectral transform methods, however, require some care in implementation to fulfil this assumption.

JOHN L. GUSTAFSON, GARY R. MONTRY, AND ROBERT E. BENNER

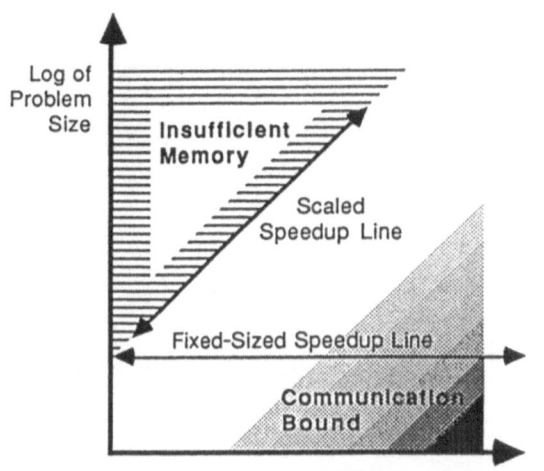

Figure 5. Ensemble computing performance pattern

The effect of fixed communication overheads in parallel computing is somewhat analogous to the effect of round off error in floating point computation. The art of writing parallel programs tells us how to keep granularity high enough for the effect of fixed overheads not to hamper efficiency beyond a tolerance limit. This is analogous to the way the art of numerical analysis tells us how to write stable algorithms where the round off error does not pollute the results beyond a certain tolerance limit. In both domains, there exist ill-behaved problems or algorithms, where adverse effects are inherently unavoidable because of fixed overheads or round off error.

As it is often wise to use double precision in calculations, it is certainly advisable to try to minimize all potential fixed overheads when designing a novel parallel architecture, often even at the cost of lower sustained performance in computation or communication. With respect to Amdahl's law, this minimization plays the same role as maximizing the scalar speed of a vector computer, hence making the parallel computer less sensitive to small granularity. Figures 6 and 7 are essentially surface representations of Figure 5, demonstrating the effect of fixed overheads on speedup. They originate in a combined theoretical and experimental analysis of implementing various domain decomposition algorithms for elliptic problems on a parallel computer by Gropp and Keyes [GroKey]. Figure 6 shows the speedup on a parallel computer with large overheads, whereas Figure 7 displays the corresponding surface on a more balanced parallel architecture. In the latter case, the effect of the communication bottleneck only shows up when the absolute granularity has dropped very low.

The seriality of the elliptic solvers, however, will become a problem before long. It causes the synchronisation overhead to grow logarithmically with the number of processors, thus causing the parallel complexity of each time step to grow logarithmically, too. This implies that, eventually, if the time slot available for running the model cannot be extended, there will be a point beyond which no further accuracy can be obtained from parallelism, as long as semi-implicit methods are used. This defect is, however, not as severe as the corresponding problem with vector supercomputers. Even with a fixed time slot, the number of processors that can be utilized grows exponentially with the speed of individual processors. Hence, the potential total computational power of a massively parallel computer grows by a factor $p$ when we furnish the computer with two times faster processors, instead of just doubling.

The other alternative would be to take recourse to explicit time-stepping schemes. They fit the scaled speedup paradigm, and consequently do not suffer any loss of speedup. In an explicit solver, a single time-step has a constant parallel complexity regardless of spatial accuracy, but a more

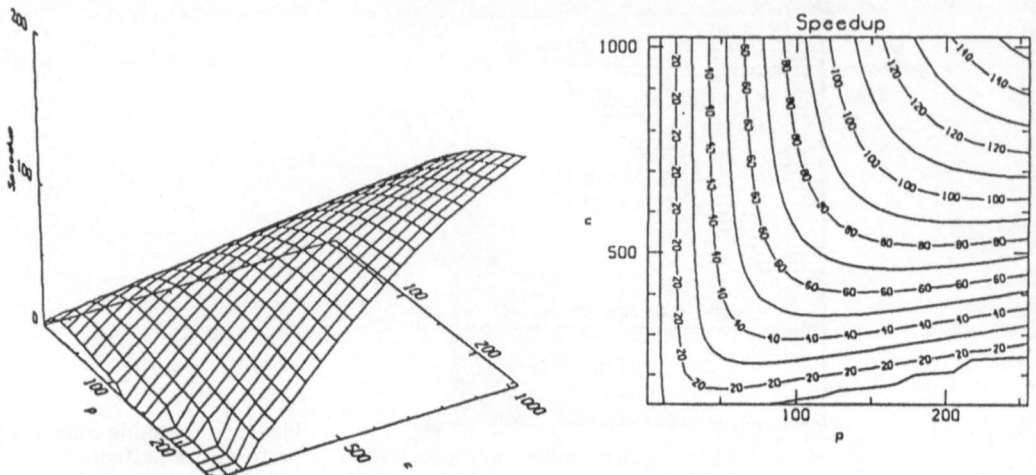

Figure 6. Speedup as a function of problem size and the number of processors on a parallel computer with large fixed overheads

W. D. GROPP AND D. E. KEYES

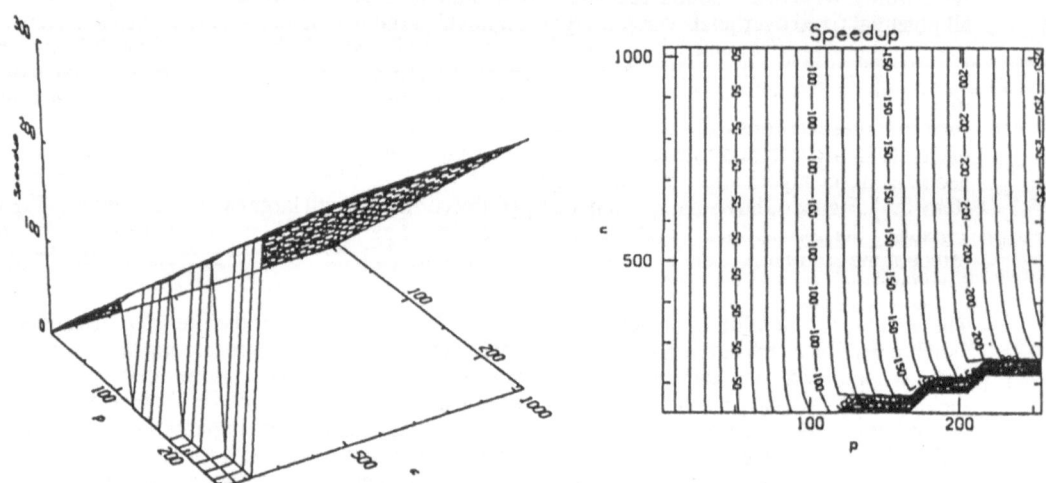

Figure 7. Speedup as a function of problem size and the number of processors on a parallel computer with small fixed overheads

serious stability restriction forces an explicit method to use more time-steps than a semi-implicit method. Because of the nonlinear advection of momentum, however, even the semi-implicit methods have to abide by a modified Courant-Friedrichs-Lewy stability condition. As a consequence, the asymptotic number of time-steps needed in a semi-implicit method differs at most by a constant factor from the number needed in an explicit method. Hence there will be a break-even point in the number of processors after which explicit methods become more efficient compared to linear semi-implicit methods.

The only algorithm that can ignore the CFL condition is a fully implicit, nonlinear time-stepping scheme. So far, these have been far too expensive as a nonlinear, partially elliptic equation has to be solved every time-step, involving in the process several linear solutions. With new and efficient nonlinear elliptic solvers, like the multigrid Full Approximation Scheme, now available it might, however, be worthwhile to take another look at them.

Gustafsson, Montry and Benner noticed in their tests that, when running algorithms on a massively parallel computer with a thousand or more processors, some rather unexpected causes of load imbalance become the dominant source of loss of efficiency, when algorithms are tuned to be in perfect load balance. Such a thing is, for example, data dependent execution rates of floating point operations.

Another increasingly dominant factor is subtle hardware deficiencies, causing the processor in question to occupy itself frequently in retrying unsuccessful communications, performing dynamic RAM refreshing, or correcting errors in the memory. This effect is getting increasingly prominent as parallel computers become more and more massively parallel. It might be averted to some extent by providing the computer with a number of processors in reserve.

These reserve processors could reside in a separate pool, sitting on a very fast control bus connected to every processor. Consequently, any of them could replace a defective processor anywhere. If the bus were used only for diagnostics and for the traffic to the reserve processors, it would not necessarily jam until the number of defective processors had grown fairly large. This would allow for performing maintenance of the computer at regular intervals. The necessity of replacing a processor whenever one becomes defective will become very burdensome when the number of processors grows large.


## 7. Meteorological supercomputing in the 1990's: a vision

The introduction described a vision of a supercomputing system for an operational weather forecasting centre of the 1920's. This last section attempts to do the same for the 1990's. We assume here that the functions of such centres remain largely unaltered. That is, they will, in addition to producing operational forecasts, continue to provide large-scale number crunching capacity both to local researchers at each centre, and to an increasing number of remote researchers accessing the resources via international data communication networks.

Many trends in scientific computing point to the increasing functional specialisation of components in a supercomputing system. The era of the general purpose computer seems to be passing away. The trends include the increased accessibility of many different computers to any individual user through computer networks. Though threatened by some viral diseases, the trend towards further integration seems to continue.

The user's interface to the outside world of computing resources will be his **workstation**. It provides the user with all basic editing, compilation, local file storage, file transfer, interactive graphics and hardcopy services. The workstation is likely to be based on a RISC architecture, run Unix, communicate using ISO/OSI protocols, have multiple windows for multiple sessions and be mouse driven. Its most important task will be to hide the differences between the various different computing resources on the network from the user. This will call for a lot of standardisation in operating systems and data communication.

Another equally important function of the workstation is to reduce the frequency of interaction between the user and remote computing resources. Asynchronous character-based interaction, with lots of short messages with long headers, is particularly detrimental to communication efficiency, indeed displaying the 'Amdahl effect' in this context. Big chunk messages are far more efficient to communicate.

The future demand for processing power on the workstation depends very much on the development of high speed data communication. If new standards, like FDDI and HSC, get widely implemented in the near future, compatibility rather than power will be the major factor in each workstation's commercial success. If not, processing power will continue to play a role, as more of the graphical post-processing will have to be done locally.

Data communication in general will be rather invisible to the user, and even to the supercomputing system designer. This is because increasingly abstract layers of message representation are going to be embedded in the physical equipment itself. The equipment will contain considerable processing power for example to route various protocols, but the supercomputing system will only see a plug which can be fed with messages coded in a high level protocol. This will also relieve the computers on the network from the burden of routing messages.

The supercomputing system proper is likely to consist of at least four separate functional components, as described below.

Philip Chen pointed out in his talk [Chen] at the graphics workshop at ECMWF in December 1988 that the production of graphical output ('visualisation') is likely to move from the supercomputer to a dedicated **graphics engine** attached to it via a very fast communication link. Graphics also calls for a lot of computing power, but of a rather different kind than the actual solution of numerical models. The graphics subsystem would also feature a facility for animation analogous to hard-copying: video recording. This is likely to become a standard feature on the workstations later on.

At present, the best graphics engines would appear to be the minisupercomputers, being both powerful and compatible. They are also replacing superminis as front-ends to supercomputers. As long as many users still depend on a terminal rather than a workstation, minisupers will also provide these users with the basic user interface.

When the need for the latter function diminishes, however, graphics processing might well provide an opportunity for massively parallel SIMD computers. Many graphical algorithms are very suitable for parallelization and, being local in nature, treat the computational domain uniformly, as required by SIMD type computing. Some promising examples of such implementations already exist on, for instance, the Connection Machine and the AMT DAP.

It should be borne in mind, however, that the above scenario bears considerable uncertainty. It is clear that visualisation of scientific results will be increasingly fought over between 'minisuper-like' and 'superworkstation-like' solutions. Superworkstations will keep adopting functions from the dedicated front-end graphics engines, the more powerful they become, and the survival of the latter then depends on their ability to offer new, attractive and computing intensive visualisation options that are too heavy for the workstations to execute.

The second component of a supercomputing system, particularly important for operational weather forecasting systems, is the **data base engine**. This will be, and indeed in many centres already is, a computer that administers the pool of disks common to all the computers in the supercomputing system. It also provides for long term archival of files and old forecasts, probably on optical disks.

The actual access to disks is likely to be hidden from the user, except when efficiency calls for explicit disk control. With bigger virtual and real memories and RAM-disk type devices, it seems likely that programmers in the future will be able to dispense with I/O statements altogether, and rely instead upon automatic storage and retrieval of data structures, just as virtual memory has relieved us of the burden of writing explicit overlays, except when this is dictated by the demand for efficiency. The above would particularly apply to experimental research codes, making coding new algorithms far more straightforward.

Coming to the final two components of the supercomputing system, even the heart of such a system, the supercomputer itself, is likely to be split into two. When thinking about a large number of users with a large number of computing intensive programs, each still only consuming a tiny fraction of the total power of the supercomputer, Amdahl's law is likely to remain very relevant to most of them. The task of writing automatic parallelizing compilers for general Fortran programs to be executed on massively parallel computers seems to be a formidable one. Hence, for the near future, only programs that are explicitly hand-tuned for massively parallel computers are likely to benefit from their superb computing power. The majority of programmers will find the speedups on massively parallel computers rather disappointing.

Consequently, there seems to remain a niche also for the present vector supercomputer with a single or, at most, a few extremely fast processors, though it might more appropriately be called a **fast scalar/vector computer**. Development of operating systems, parallelizing compilers and high-level tools like parallelized subroutine libraries may eventually make this single processor engine obsolete, as the corresponding tools for vector computers have made separate front-end computers less necessary for present supercomputers, but this may not happen until perhaps the next millenium.

The title of the fastest computer in the world, however, is likely to pass to a massively parallel computer during the next decade, unless some revolutionary developments take place in chip manufacturing, facilitating the manufacture of ultrafast but extremely expensive individual processors. (This could take place as a consequence of developments e.g. in high temperature superconductivity, or optical computing.)

This **massively parallel supercomputer** may be attached to the fast scalar/vector computer, perhaps in the way attached processors of the early 1980's were attached to their hosts. Thus, the fast scalar/vector computer may assume the role of a front-end computer to the massively parallel supercomputer. Or both might be integrated as subsystems of a single computer, as are vector processors in present mainframes.

Operational global weather forecast models in particular will, with all likelihood, be executing on thousands of processors in the late 1990's, continuing to be among the most demanding computational tasks in the world.

## 8. References

[CarWet]    Carroll, A. B., Wetherald, R. T.: Application of Parallel Processing to Numerical Weather Prediction. J. ACM. 14(1967)3, pp. 591-614.

[Chen]      Chen, P. C.: Computer Graphics Systems. Talk at the Workshop on Graphics in Meteorology. ECMWF, Reading, December 1988.

[Dent]      Dent, D.: A Modestly Parallel Model. These proceedings.

[Gall]      Gallopoulos, E.: Processor Arrays for Problems in Computational Physics. PhD Thesis, Department of Computer Science, University of Illinois at Urbana-Champaign, 1985.

[GroKey]    Gropp, W. D. and Keyes, D. E.: Complexity of Parallel Implementations of Domain Decomposition Techniques for Elliptic Partial Differential Equations. SIAM J. Sci. Stat. Comput. 9 (1988) 2, pp. 312-326.

[GuMoBe]    Gustafsson, J. L., Montry, G. R. and Benner, R. E.: Development of Parallel Methods for a 1024-processor Hypercube. SIAM J. Sci. Stat. Comput. 9(1988)4, pp. 609-638.

[John]      Johnsson, S. L.: Communication Efficient Basic Linear Algebra Computations on Hypercube Architectures. J. Par. Distr. Comput. 4(1987), pp. 133-172.

[NicWil]    Nicol, D. M. and Willard, F. H.: Problem Size, Parallel Architecture, and Optimal Speedup. J. Par. Distr. Comput. 5(1988), pp. 404-420.

[Rich]      Richardson, L. F.: Weather Prediction by Numerical Process. Cambridge University Press, Cambridge 1922.

## Sources of figures

Figure 1.   Reproduced with permission from Stout, Q.F.: Sorting, Merging, Selecting and Filtering on Tree and Pyramid Machines. Siegel, H.J. and Siegel, L. (eds.): Proceedings of the 1983 International Conference on Parallel Processing. IEEE Computer Science Press, Silver Spring, 1983. Copyright by IEEE. All rights reserved.

Figure 2.   Reproduced with permission from Schwichtenberg, G.: Parallelrechner. A set of overheads from the Computer Centre of the University of Dortmund, 1988. All rights reserved.

Figure 3.   Same as Figure 2.

Figure 4.   Reproduced with permission from Gustafsson, J.L., Montry, G.R., and Benner, R.E., Development of Parallel Methods for a 1024-Processor Hypercube, SIAM J. Sci. Statist. Comput., 9 (1988), pp. 609-638. Copyright 1988 by the Society for Industrial and Applied Mathematics. All rights reserved.

Figure 5.   Same as Figure 4.

Figure 6.   Reproduced with permission from Gropp, W.D., and Keyes, D.E., Complexity for Parallel Implementations of Domain Decomposition Techniques for Elliptic Partial Differential Equations, SIAM J. Sci. Statist. Comput., 9 (1988), pp. 312-326. Copyright 1988 by the Society for Industrial and Applied Mathematics. All rights reserved.

Figure 7.   Same as Figure 6.

# A Modestly Parallel Model

DAVID DENT

European Centre for Medium-Range Weather Forecasts, Shinfield Park, Reading, Berkshire RG2 9AX, U.K.

## 1.    INTRODUCTION

ECMWF has run a parallel spectral model in daily production since 1985. Initially, this executed on a Cray X-MP/22 and later on a Cray X-MP/48. The degree of high-level parallelism so far exploited is low because it is sufficient to provide the basic need, that is minimum wall clock execution time in a production environment. However, low-level parallelism in the form of vectorised Fortran code is very substantial. 99% of all floating point operations are performed in vector mode.

As a means of gaining insight into how such models might behave on more massively parallel architectures, this paper will look at the present implementation to see how far the parallelism can be reasonably extended.

The basic structure of the model and the multitasking strategy have been described in previous ECMWF workshops [1,2] but will be briefly included here for completeness.

## 2.    BASIC STRUCTURE

Figure 1 shows schematically the relations between workfiles, memory and the scan structure of the code. The memory SPEC represents a relatively large area of shared memory which contains fields in spectral form. This area is present for most of the model execution. An important aspect is the 'out of core' nature of the model since this imposes restrictions on vectorising and multitasking. It also demands efficient random I/O on a fast secondary memory (such as SSD).

Scan 1 involves translation between Fourier space, grid-point space and spectral space. It includes all the physical parameterisations present in the model. Scan 2 consists of an inverse Legendre transform from spectral space into Fourier space. Between the scans are some computations in spectral space involved with semi-implicit time stepping and horizontal diffusion.

Topics in Atmospheric and Oceanic Sciences
© Springer-Verlag Berlin Heidelberg 1990

22

WORK FILES

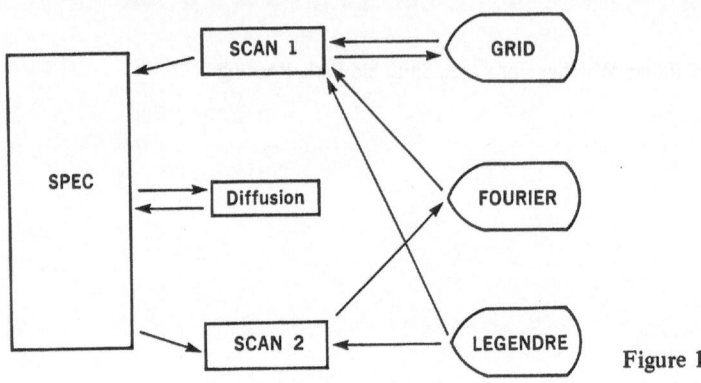

Figure 1

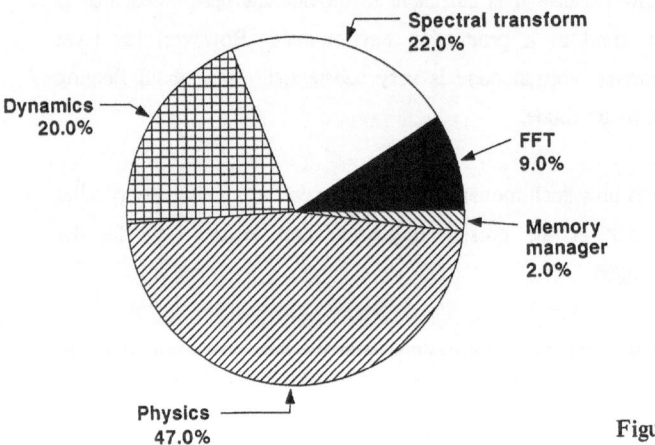

Figure 2

The relative cost of various significant parts of the model code is shown in figure 2. The parts differ substantially in terms of data organisation and therefore in terms of multitasking strategies. The cost of the spectral transform is of particular interest since it increases as the cube of the resolution. Therefore, it must be computed efficiently as it will dominate the model computation at higher resolutions.

## 3.    MULTITASKING STRATEGIES

For most of the model execution (process 1 or P1), computations on a latitude row are independent of computations on other rows. Hence, it is natural to split the work in this fashion and compute lines of latitude in separate tasks. However, because of some characteristics of the Legendre transform, it is economic to compute in North-South pairs so that the end product is a contribution to the Legendre transform (P2) from each pair of lines. This can be done either

(a)    by computing North and South in parallel using different tasks in which case synchronisation is necessary after P1 (figure 3), or

(b)    by computing North followed sequentially by South in which case the data is then ready for P2 and no synchronisation is necessary (figure 4).

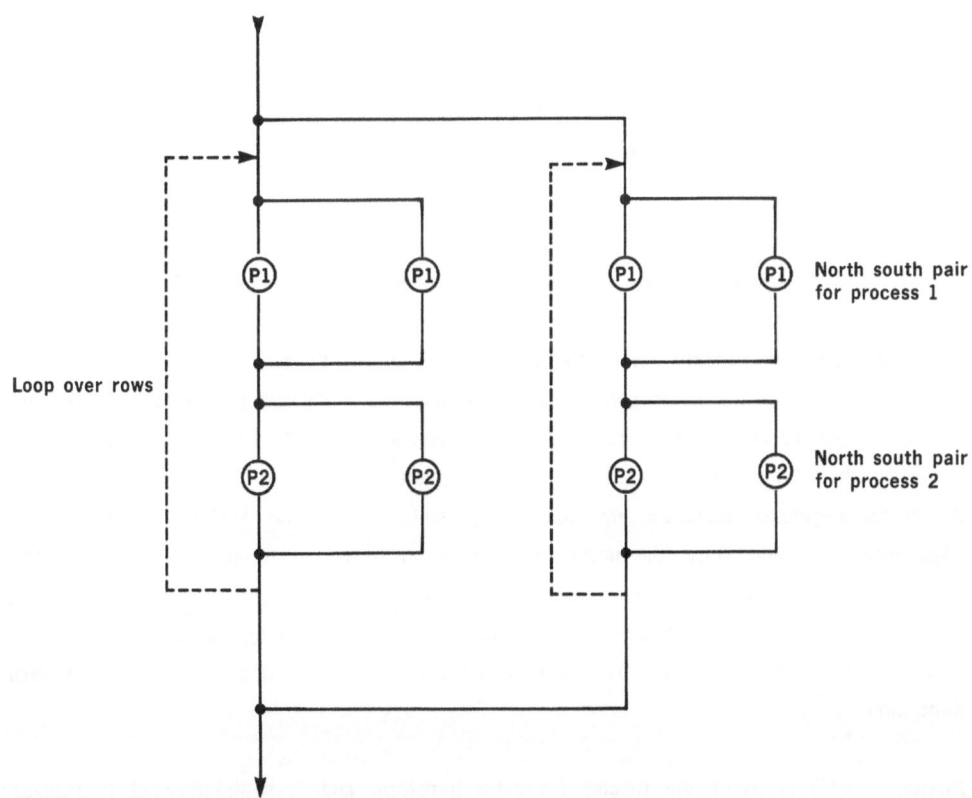

Schematic of multitasking strategy for scan 1 of 4 processor model          Figure 3

24

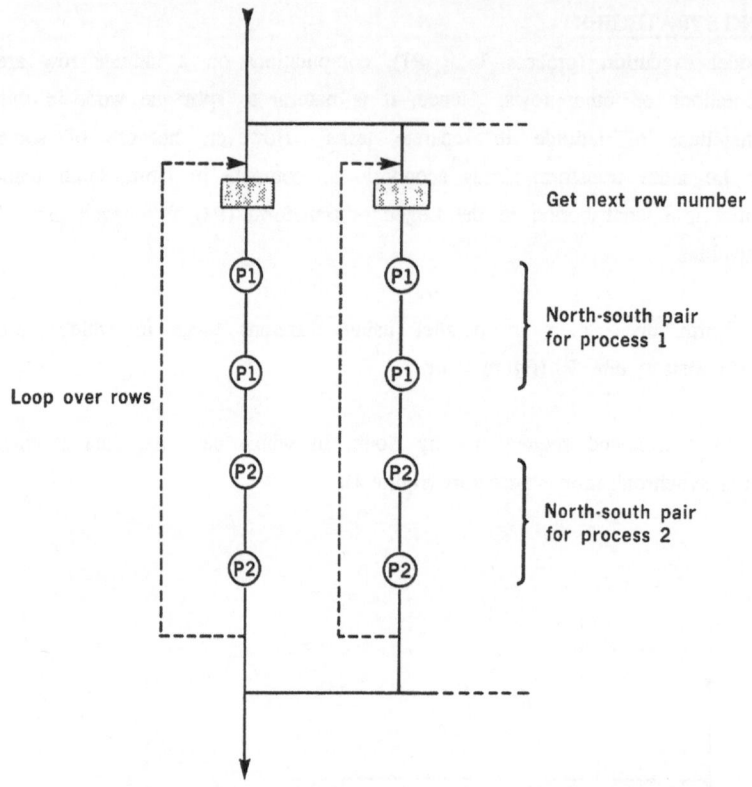

Loop over rows

Get next row number

North-south pair
for process 1

North-south pair
for process 2

Dynamic multitasking strategy for scan 1 any number of processors          Figure 4

Process 2 involves updating the spectral fields held in shared memory (SPEC in figure 1). Since this can take place in a non-deterministic order, there is a small non-reproducibility introduced due to rounding error (figure 5). We choose to remove this by forcing the execution of P2 in a predetermined (serial) order. The inefficiencies introduced by this mechanism appear as a startup and rundown expense as shown in figure 6. They are small providing that the number of tasks is small. Alternative methods for the computation of the direct Legendre transform are certainly possible [3,4]. The availability of sufficient main memory would allow the order of computation to be modified and the multitasking strategy to be such that tasks are truly independent of each other. Additionally, vectorization could be applied over a different dimension giving greater cp efficiency through longer vector lengths. The current implementation vectorises P2 over the vertical dimension.

Process 3 (P3) performs the inverse Legendre transform and is straightforward to organise in such a way that the tasks are independent.

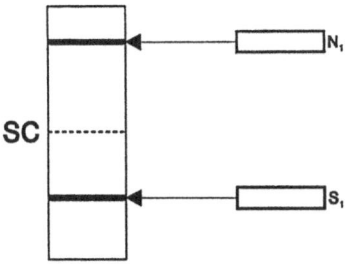

$$SC(l,^*) = SC(l,^*) + CINC(^*)$$    Figure 5

REPRODUCABILITY WASTE

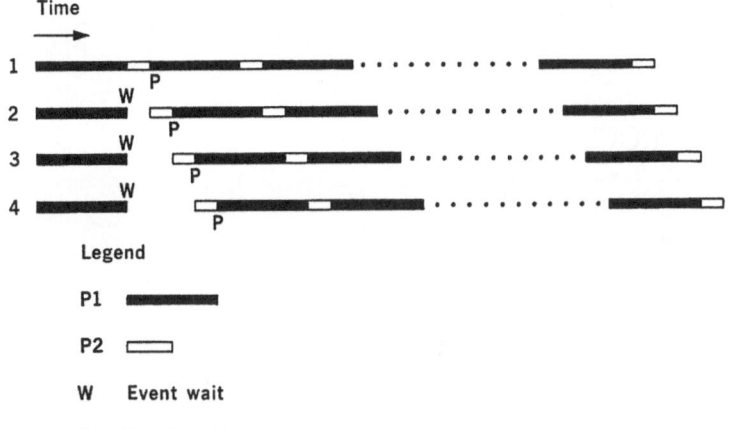

Figure 6

Process sizes and variability are important in assessing the efficiency of a multitasking implementation. Figure 7 shows average sizes and variability. Only P1 exhibits a noticable variability. In practice, because of the way the code is organised, this is due almost entirely to convective activity which differs substantially between polar and equatorial rows as shown in figure 8. The effect of this variability is minimised when North South pairs are executed sequentially since out of balance effects only become noticable when synchronization takes place at the end of the scan. The maximum out of balance is of the order of 10% of the process size, i.e. about 10ms.

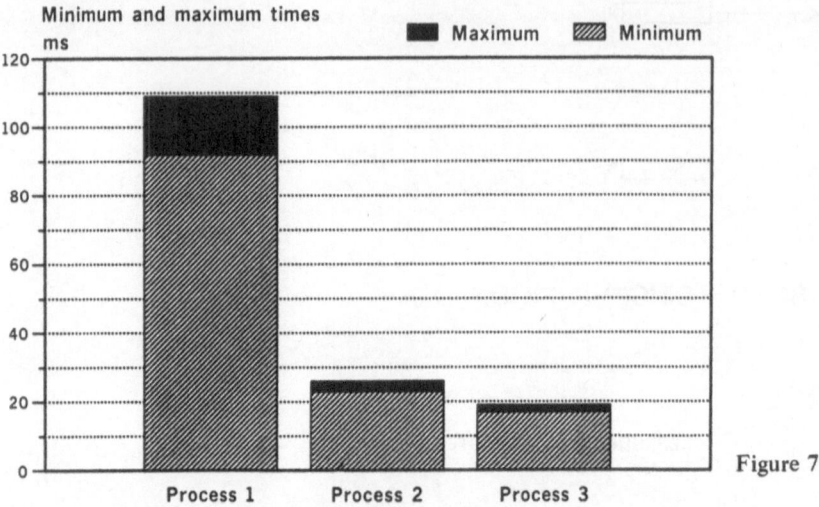

### PROCESS SIZES

Minimum and maximum times

Figure 7

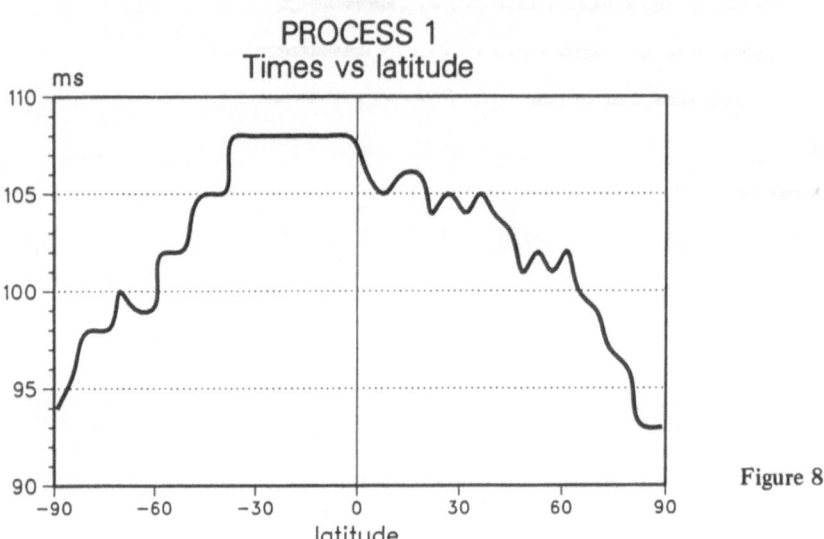

### PROCESS 1
### Times vs latitude

Figure 8

Another method for reducing the effect of out of balance costs has been investigated. This depends on the ability to mix low level parallelism in the form of microtasking with the existing high level parallelism (macrotasking). The microtasking mechanism is such that on most occasions no additional processors are available (macrotasking is utilising all available processors). Microtasking is then inoperative and suffers no noticable overhead. But if

OUT OF BALANCE WASTE

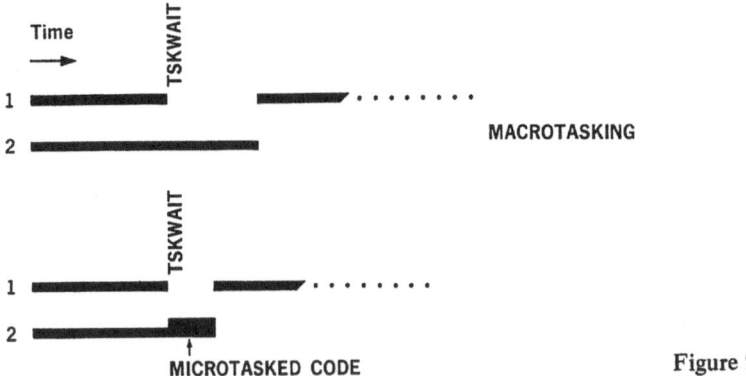

Figure 9

at least one processor is idle, the microtasking parallelism can become active and utilise the spare cp capacity (figure 9). This is particularly attractive in process P2 (direct Legendre transform) which is the last cp intensive code to be executed for each line of latitude and is easily microtasked. It also helps to reduce reproducability waste (see figures 10 and 11).

## 4.    FURTHER PARALLELISM

At the present operational resolution of T106L19, there are 80 row pairs in P1, 80 completely independent tasks in P3, and 80 tasks in P2 forced to execute in a fixed order. On the Cray X-MP/48, this leads to figures of 93% for the multitasked portion of the code (utilising all processors) with a speedup over one processor of 3.7 (see figures 10 and 12). A speedup of 7.3 has been measured on a Cray Y-MP/832.

If the implementation is kept constant and we estimate the performance as more processors (of the same kind) become available, then as the limit of 80 is approached the out of balance effect in P1 grows to about 5%. P3 is well balanced and should not show any degredation. However, the P2 reproducability mechanism would have to be abandoned beyond 8 or 16 processors due to inefficencies. If however, the main memory were also enlarged in proportion to the increasing number of processors, alternative methods of computing the direct Legendre transform could be introduced.

Since we will certainly be running a higher resolution model on future generations of supercomputer, it is interesting to consider potential increased parallelism. In parts, additional parallelism can be achieved without great effort by microtasking at the outer loop

## MULTI-TASKING EFFICIENCY
Macro-tasked model

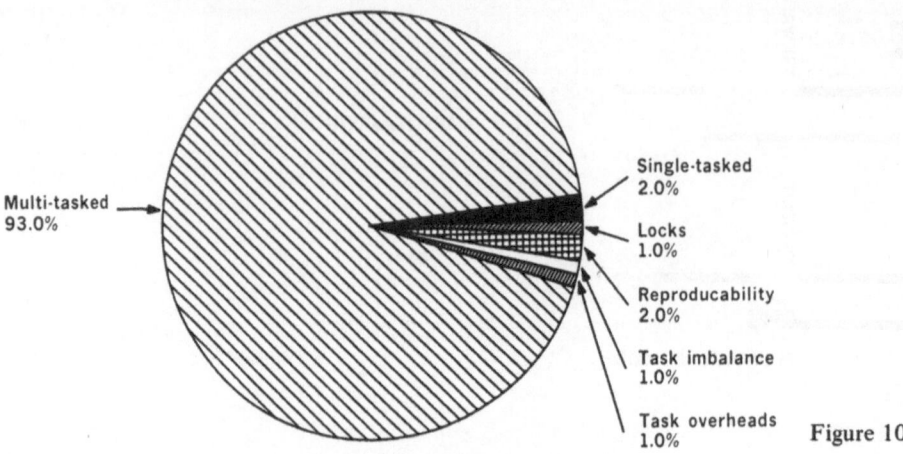

Multi-tasked
93.0%

Single-tasked
2.0%

Locks
1.0%

Reproducability
2.0%

Task imbalance
1.0%

Task overheads
1.0%

Figure 10

## MULTI-TASKING EFFICIENCY
Macro + micro tasked model

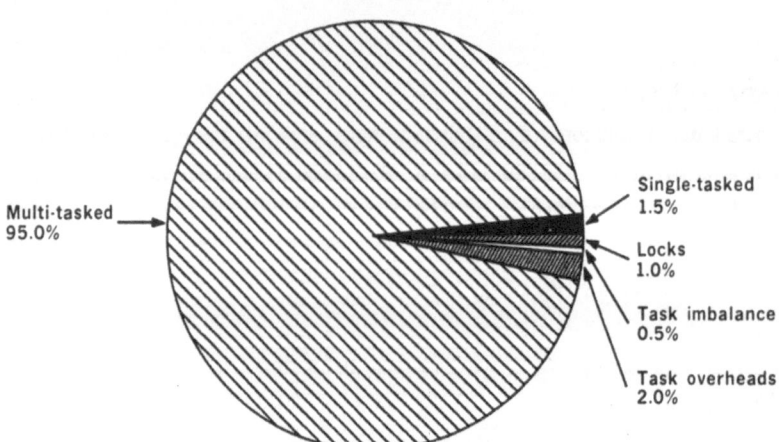

Multi-tasked
95.0%

Single-tasked
1.5%

Locks
1.0%

Task imbalance
0.5%

Task overheads
2.0%

Figure 11

level and without disturbing the vector efficiency. Considering each process separately, P3 is the easiest and consists of an outer loop running up to the value of T (currently 106). The work content in each pass varies with the value of T, though these variations could be minimised by dealing with appropriate pairs in each process. On average the additional parallelism would be T/2 (order 100).

MULTI—TASKING EFFICIENCY

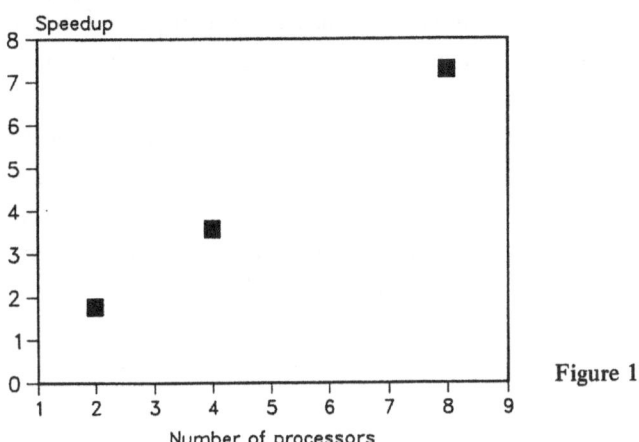

Figure 12

P2 would have to be reorganised (given sufficient memory) so that each macrotask updates a unique subarray, e.g. legendre polynomial index M (figure 13). This implies vectorising in a different direction e.g. over latitude rows thus leaving the vertical dimension free for microtasking. Additionally, the different meteorological fields can be processed in separate tasks since they are independent over this transform. This leads to a total additional parallelism of order 400. If vectorisation is no longer relevant, the vertical levels can also be handled independently, increasing the parallelism to order 10000.

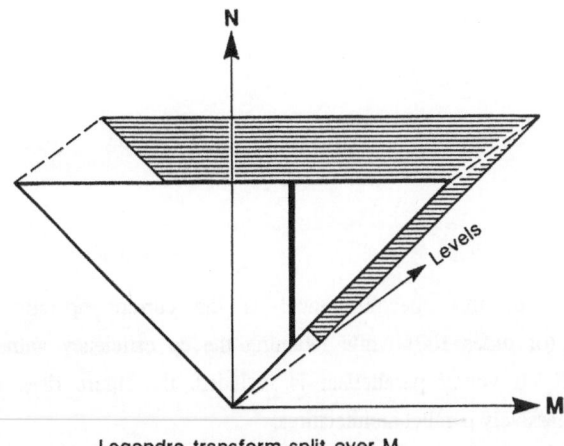

Legendre transform split over M                    Figure 13

P1 contains the largest quantity of code and uses the most cp time. There is no easy way to modify the code in its present form. However, the parallelism inherent in the system is large since for the vast majority of the code only data belonging to a vertical column is related (figure 14). Assuming that vectorisation is no longer relevant, currently nested loops must be inverted so that the contents of each pass over the inner loop (computation over a vertical column) are computed independently on separate processors. This leads to parallelism of order 5000. There will be some variability in work content due for example to convection calculations. However, for a distributed memory environment the quantity of data needed per column is quite small (order of 1000 words).

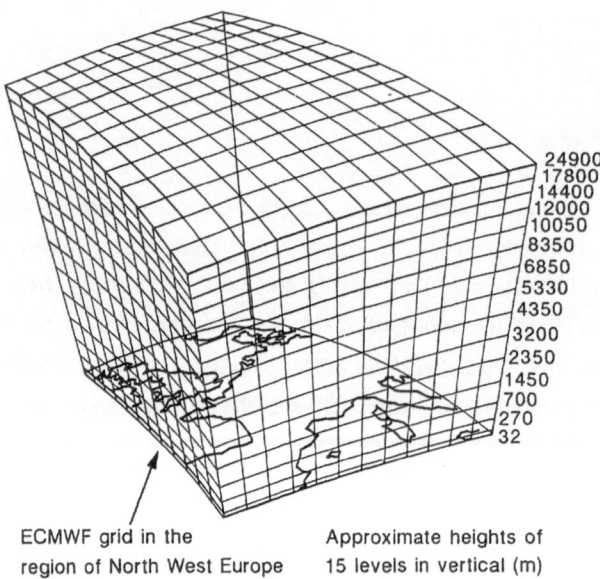

24900
17800
14400
12000
10050
8350
6850
5330
4350
3200
2350
1450
700
270
32

ECMWF grid in the          Approximate heights of
region of North West Europe   15 levels in vertical (m)          Figure 14

## 5.    SUMMARY

In summary, the potential parallelism of this spectral model at the current operational resolution can be seen to be modest (of order 100) while retaining the cp efficiency gained by a highly vectorised implementation. If vector parallelism is included, the figure rises to order 10000 - 50000, more suitable for massively parallel architectures.

References

[1]  Dent, D.W., The Multitasking Spectral Model at ECMWF, in: G.-R. Hoffmann and D.F. Snelling, eds., Multiprocessing in Meteorological Models (Springer Verlag, Berlin, 1988) 203-213.

[2]  Dent, D.W., The ECMWF model: Past, Present and Future, in: G.-R. Hoffmann and D.F. Snelling, eds., Multiprocessing in Meteorological Models (Springer Verlag, Berlin, 1988) 369-381.

[3]  Hoffman, G.-R. and D.F. Snelling, A Comparative Study of the ECMWF Weather Model on several Multiprocessor Architectures, in: G.-R. Hoffmann and D.F. Snelling, eds., Multiprocessing in Meteorological Models (Springer Verlag, Berlin, 1988) 419-432.

[4]  Carver, G., A Spectral Meteorological Model on the ICL DAP, Parallel Comput. $\underline{8}$, (1988) 121-126.

# Implementation of Atmospheric Models on Large Multi-Processor Surfaces

SIMON F. B. TETT[1], R. S. HARWOOD[2], R. D. KENWAY[3]

[1] Departments of Physics and Meteorology
[2] Department of Meteorology
[3] Department of Physics, University of Edinburgh, The King's Buildings,
West Mains Road, Edinburgh, EH9 3JZ, U.K.

## Abstract

In this paper we discuss an implementation of a spherical shallow water model and the U.K. Meteorological Office rainfall benchmark code on the *Edinburgh Concurrent Supercomputer* consisting of approximately 200 Inmos T800 transputers.

Multi-tasking in atmospheric models can be divided into two distinct problems:

Dynamics — time evolution using an approximation to the Navier-Stokes and thermodynamics equations. An implementation of a simple two dimensional model, which is representative of the dynamics code used by the Meteorological Office, has been completed successfully. Linear speedup has been demonstrated for the model on a plane but the spherical version suffers scaling problems.

Subgrid processes — a series of parameterisation schemes for the physical processes happening in the atmosphere. A naive implementation of the subgrid processes may lead to load balancing problems depending on the state of the atmosphere. This paper discusses this issue using the U.K. Meteorological Office's rainfall benchmark code as an example.

# 1 Introduction

The traditional way that multiprocessing has been used in the meteorological community has been to use small numbers of very complex and powerful processing units (for example the Cray-XMP and the ETA-10). Here we consider a different approach, an architecture based on large numbers of identical microprocessors. The micro-processor is the Inmos transputer and networks of these micro-processors have a natural two dimensional topology.

Topics in Atmospheric and Oceanic Sciences
© Springer-Verlag Berlin Heidelberg 1990

Such micro-processors are expensive to design but cheap to mass produce. Consequently, architectures which utilize large numbers of identical processors to achieve high-performance may be more cost-effective than complex processing units built of many different special-purpose chips. However the problems involved in designing algorithms to utilise the potential power from such architectures are by no means simple.

To gain experience and understanding of the issues involved we have implemented a spherical shallow water model and some parts of the U.K. Meteorological Office's benchmark code. Other authors have carried out work on similar architectures based on the FPS-T series [3], [5].

First we describe briefly the transputer and the Edinburgh Concurrent Supercomputer. Next we discuss the implementation of the shallow water equations, the extension to a full spherical model and the problems associated with this. Finally we describe the work in progress implementing the U.K. Meteorological Office's rainfall benchmark code.

# 2 Transputers at Edinburgh

## 2.1 The Transputer

Generically, a transputer consists of a central processing unit, on-chip memory, an external memory interface and four communication links, or external links, which provide bidirectional communications with other transputers.

There are several varieties of transputer; at Edinburgh we use the T800. On chip it has a floating-point unit, in addition to the integer processor, and four kilobytes of memory. In our configuration each transputer has four megabytes of external memory. This computing resource is capable of sustaining approximately $0.8 \times 10^6$ floating-point calculations per second per transputer. To get this kind of performance the data must be stored in the on-chip memory rather than the slower external memory. Each external link is capable of communicating at a rate of $0.8 \times 10^6$ bytes per second each way.

Each processor can simulate the running of several parallel processes by timeslicing them. If a process is doing nothing, for example if it is waiting for a communication to proceed, then it takes up only a very small part of the processor's time. It is also possible to do communications and calculations concurrently. This is real concurrency, the communications being handled by link controllers leaving the arithmetic unit to proceed with calculations, but it causes roughly a one percent reduction in the calculation speed.

## 2.2 The Edinburgh Concurrent Supercomputer (ECS)

The ECS is an electronically reconfigurable array of transputers built by Meiko Ltd. Its computational engine currently comprises 200 T800 transputers. The number of processors and hence the performance can increase with additional funding. The machine is divided into several 'domains'. Each domain is a single-user resource ranging in size from one to 131 processors. The processors within each domain can be reconfigured by the user into most two dimensional topologies; reconfiguration of the domains is determined by the system manager.

The computer is fileserved from a microVAX running VMS and three 800 megabyte discs which run a very simplified 'Unix'[1]-like operating system. The machine can be programmed in any one, or mixtures of three languages: Occam, C and Fortran 77.

# 3    Parallelising a Finite Difference Scheme

As a first step towards mounting a three dimensional dynamics code and to gain experience, we have implemented a shallow water model using a similar finite difference grid and time integration scheme to that used by the U.K. Meteorological Office [1]. The main feature of this model is that it uses the Arkawara 'B' grid on a regular longitude-latitude grid. The scheme is 'split' in the time integration; three adjustment timesteps are followed by one advection timestep. Velocities are stored at gridpoints labelled by $(i+\frac{1}{2}, j+\frac{1}{2})$ and geopotentials are stored at points labeled by $(i, j)$. See appendix C for a description of the difference equations used.

## 3.1    Geometrical Decomposition

In geometrical decomposition each processor has the same code but different data. For efficient finite difference schemes, data should be divided up such that neighbouring processors, have neighbouring data points as in figure 1.

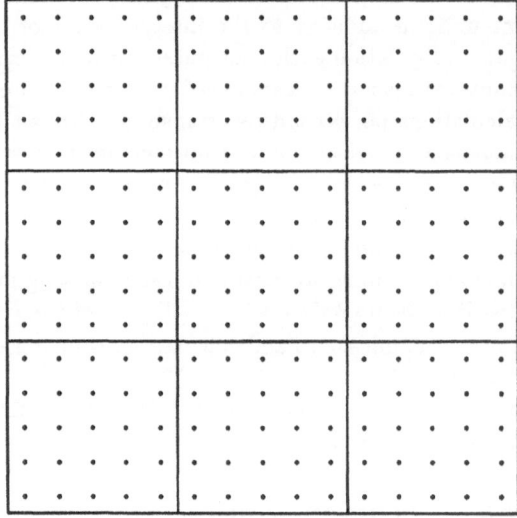

Figure 1. Processors with grid overlayed. Processors are shown as thick boxes and grid points by dots. Each processor has 5 × 5 grid points.

---

[1]Unix is a trademark of AT&T

It is useful to introduce the concept of *temporal decomposition*, in which first some variables are communicated, then all the calculations that are possible upon these variables are carried out. Algorithms naturally decompose into *blocks* where each block consists of communications of some variables followed by some calculations using these variables. In general, the blocks are ordered in time. For example, the finite difference form of the geopotential evolution equation is

$$\phi^{n+1} = \phi^n - \delta t \{ \overline{\delta_x u \overline{\phi}^{xy}}^y + \overline{\delta_y v \overline{\phi}^{xy}}^x \} \tag{1}$$

where

$$\overline{f(i,j)}^x = \frac{f(i+\frac{1}{2},j) + f(i-\frac{1}{2},j)}{2}$$

and

$$\delta_x f(i,j) = \frac{f(i+\frac{1}{2},j) - f(i-\frac{1}{2},j)}{\delta x}$$

The values of $\phi$ are communicated first then the function $\overline{\phi}^{xy}$ is calculated (which requires values of $\phi$ on adjacent grid points). Next the calculated values of $\overline{\phi}^{xy}$ are communicated. Then the values of $u\overline{\phi}^{xy}$ and $v\overline{\phi}^{xy}$ are calculated, and so on. This means that some code written for serial machines may need rewriting.

Another useful idea is that of *interaction range*. This characterises the neighbourhood size of the algorithm. Consider a set of variables defined on a grid of points. For example in the shallow water model, some of these variables would be $\phi$, $u$, and $v$ (where the usual meaning is assumed). Define a vector $\mathbf{X}^0(i,j)$ at each grid point where $X_0^0 = \phi$, $X_1^0 = u$, $X_2^0 = v$, and so on for any other variables. Then, given some new vector $\mathbf{X}^1(i,j) = f(\{\mathbf{X}^0(n,m)\})$ which is a function of a subset of the old values, labeled by $m,n \in$ a subset of all the grid points, define $I_R$ (the interaction range) as the smallest integer such that

$$\forall n,m \quad |n-i| \le I_R(i,j), |m-j| \le I_R(i,j) \tag{2}$$

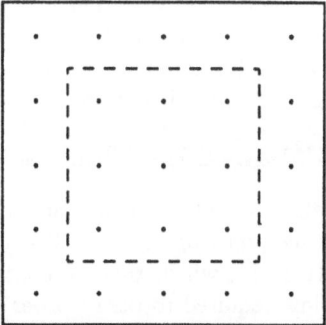

Figure 2. How the grid points on a processor splits up into two regions for an interaction range of one; an outer region (outside the dashed lines) and an inner region (inside the dashed lines).

Note that the set of all grid points $\{(n,m)\}$ for any particular $(i,j)$ define the neighbourhood of the point $(i,j)$ for the algorithm and $I_R$ can take on a different values at different grid points. The interaction range is best defined for a block of the algorithm and so different parts of the algorithm may have different interaction ranges. Consider the algorithmic block $f(\phi) = \delta_x \overline{\phi}^y$. From our definition of interaction range and the way that we store variables on the grid we can see for this case that the interaction range is one for all grid points. The concept of interaction range is particularly useful for finite difference grid point methods. For semi-lagrangian methods the interaction range may well be different at each grid point and may not be such a useful concept.

Using the interaction range it is possible to split the subgrid on each processor into two different regions: an inner region where data to carry out the calculations is already on the processor and an outer region where some data needs to be exchanged with neighbouring processors. The size of the outer region depends only on the interaction range; see figure 2.

To get the maximum performance, we must overlap communications and calculations. We have just discussed how the subgrid on each processor can be split into two regions. If each processor sends data from its outer region to neighbouring processors while simultaneously calculating in the inner region we can effect a substantial overlap of calculations and communications. When both of these processes have completed, every processor calculates in the outer region. There are two effects from this: first communications are at a very low level in the code and second a minimum size for the subgrid on each processor is imposed, if we wish to get the maximum speed from the processor. The corresponding segment of code will look like this

```
SEQ
  PAR
    ... Do communications for outer region
    ... Do calculations for inner region
  ... Do calculations for outer region
```

For a brief explanation of Occam see appendix A. The constraint on array size is to ensure that there are enough calculations to overlap communications (otherwise the processor will waste time waiting for communications to finish). For a finite difference scheme, if the subgrid size is $n^2$ this requires:

$$4(I_R n + 2I_R^2)\tau_{comm} \leq 2(n - 2I_R)^2 \tau_{calc} \tag{3}$$

where $\tau_{comm}$ is the time to transfer one byte between the two processors and $\tau_{calc}$ is the time carry out one floating point operation. The above is obtained by considering the simplest operation, $\overline{f}^y$, which takes 2 floating-point operations. The factor of $2I_R^2$ is the extra time required to pass 'corner' data points between diagonally adjacent neighbouring processors.

For the T800, the bandwidth is $0.8 \times 10^6$ bytes per second per link, and the calculation speed (coincidently) is $0.8 \times 10^6$ floating point operations per second for 32 bit arithmetic. We can then write equation (3) as

$$4(I_R n + 2I_R^2) \leq 2(n - 2I_R)^2 \qquad (4)$$

which has the solution

$$n \geq 6I_R \qquad (5)$$

For $I_R = 1$ and $I_R = 2$ this gives $n \geq 6$ and 12 respectively. Considering the FPS-T series where $\tau_{comm} \approx 12\tau_{calc}$ [5], the corresponding constraint is that $n \geq 17$ and $n \geq 34$ for $I_R = 1$, 2. These minima of $n$ depend on the difference scheme used and in particular on the least computationally intensive block.

The effect of this is to give a constraint on the maximum number of processors that can be used for a grid of given size. Considering the FPS-T series, we have just shown that the minimum number of points on each processor should be $34 \times 34$ and at present the scheme [1] uses $\approx 200 \times 100$ points. This means that for this scheme the full processing power can only be obtained from $\approx 18$ processors for each horizontal slice compared to $\approx 150$ processors using the ECS.

Of course, reducing the array size on each processor enables more processors to be used in the calculations and gives, overall, a greater rate of computation even though each processor is doing less work. In the case where each processor is at least 50% efficient (i.e communications take less than twice as long as calculations) we find that the size of the subgrid on each transputer is $n \geq 4I_R$, allowing us $\approx 325$ processors. Exactly how much computational power we will get out of these processors depends on a detailed analysis of the algorithm.

If we were to consider a three dimensional model with $h$ vertical levels on the same network topology, where each processor would have a $n \times n \times h$ volume of the atmosphere, then the same argument as above will apply, giving exactly the same values for the number of processors. We could consider having more complex decompositions where the atmosphere is divided into $n \times n \times k$ volumes. Each computational node would need two transputers to give it the required connectivity, i.e. six links (east, west, north, south, up and down).

Next we consider the model on a spherical geometry. Due to the way that the scheme handles the poles it is necessary to average various quantities over the northernmost and southernmost rows. This might be inefficient because it involves long-distance communications. However only one number per processor need be sent around the ring of processors in each latitude band, and so the actual time wasted on communications is small compared to the total time taken by the program. Figure 3 shows a diagram of the processor topology used to map a sphere.

## 3.2 Fourier Transforms

In finite difference models on a regular longitude-latitude grid one way to deal with the instabilities near the poles is to transform to Fourier space, damp each mode by its amplification factor and transform back again to grid space.

However, we are constrained by the processor geometry imposed by geometrical decomposition, so we cannot use the optimal topology for fast Fourier transforms (FFT). An FFT needs global communications and these tend to be inefficient for transputer architectures, without using additional processors to enhance the connectivity. It is impossible to carry out the techniques we described earlier of

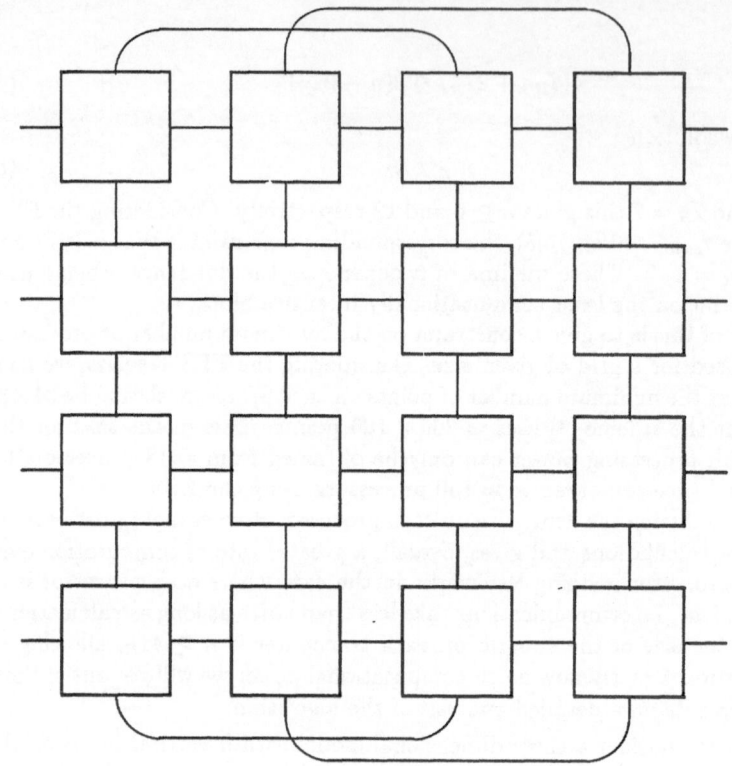

Figure 3. Processor topology used to map a sphere onto, in this case, four by four processors. The links shown at the east and west sides wrap around. The links shown at the north and south edges are connected to processors $\frac{M}{2}$ $(= 2)$ away, to provide the 'over the pole' communications.

overlapping calculations and communications, due to the structure of the algorithm. So in order that communications do not dominate, we require that the time to carry out calculations must be of the same order as that required to carry out communications:

$$\frac{N \ln N}{M} \tau_{calcs} \sim N \tau_{comms} \tag{6}$$

where $N$ is the total number of points and $M$ is the total number of processors in the ring. This implies $M \sim \ln N$. So doubling the number of points will only let us have one more processor if we require that communications do not dominate. This gives bad scaling behaviour. Similar problems occur if spherical harmonic methods are implemented using this type of network topology, as these methods also need global communications. These problems may be reduced or even avoided by using some extra processors to increase the connectivity of the network.

Since only a subset of the processors need to carry out Fourier transforms, if we want to get the maximum efficiency from the transputers we might attempt to share this computational load among the remaining processors which would otherwise be idle. We have not investigated this possibility yet.

# 4  Getting Data Out of the Network

As forecasters are interested in how the state of the atmosphere changes with time it is necessary to either store data at regular intervals on each processor and then write it to store after the model has terminated, or to write data out while the model is running. We chose to write data out while the model is running, as in a larger and more complex model there may not be enough memory on each processor to store the data.

To get data out of the processor network we map a 'snake' onto the network and send data out along it. A snake is just a chain through all the processors where one processor outputs to store and each processor has one input and one output. Figure 4 shows such a network, the only interprocessor links shown are those which are active during output.

As some calculations are more expensive compared to their associated communications, there will be some spare capacity in the communication links. In order to make use of this, we shall now discuss the three processes required to get data out of the system without interfering with the communications necessary to do the calculations discussed earlier.

Two of these processes are fairly similar, they are what we call switches.

An *in.switch*, or demultiplexer, process can take in a message from an external link and then pass it out on any one of its internal channels. Which internal channel is chosen is determined by the message.

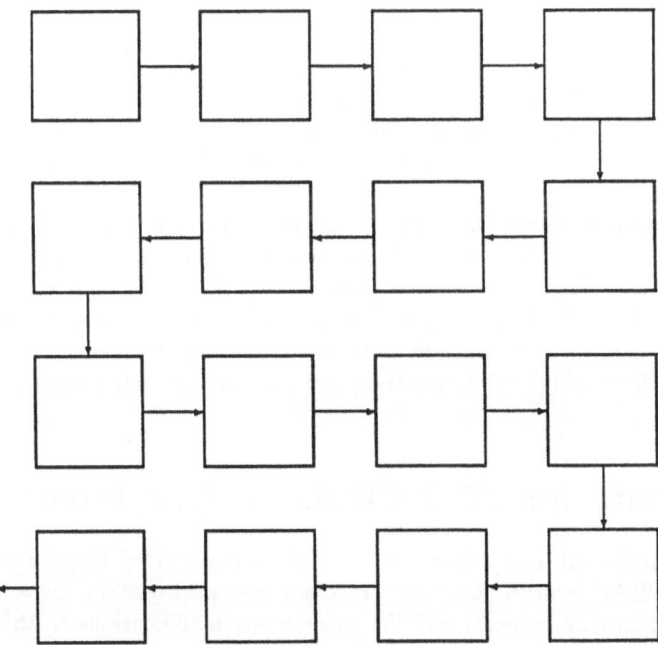

Figure 4. The snake; only active links are shown.

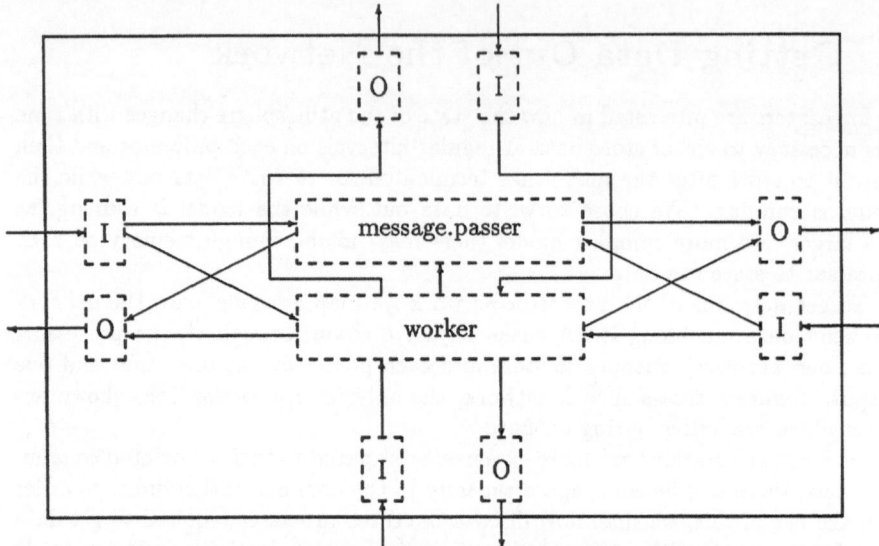

Figure 5. The separate processes are all running in parallel and are connected together by channels. Processes are shown by dashed boxes with their name inside ( I is an in.switch, O is an out.switch), channels are shown as lines and the processor by a box.

An *out.switch*, or multiplexer process, can take in a message from any one of the internal channels and pass it out again on the external channels. This process gives priority to messages from the *calculations* process, in the sense that if two messages are both waiting the *out.switch* process will take the one from the *calculations* process first.

Switches are placed on external links in pairs: an *in.switch* and an *out.switch*.

A *message.passer* process first gets data from the *calculations* process, passes it out, then gets data in from an upstream processor in the snake and passes that out. It repeats this until it gets a message from the end processor in the snake, whereupon it passes that out and gets some more data from the *calculations* process. See figure 5 for a diagram of how the processes are connected together by channels.

# 5    Implementation of the Rainfall Benchmark

The Rainfall Benchmark code was written in FORTRAN and ported directly to the ECS. The FORTRAN 'worker' on each processor was packaged by Occam processes which handled communications. We made some modifications to this benchmark in order to efficiently distribute it over several processors, and bring out any difficulties that might occur in a full model. As the benchmark rainfall scheme has no interactions with horizontal neighbours, but in a full model there would be communication, so we implemented a synchronising communication after every timestep.

The main problem that we were concerned with in the implementation was load balancing. We expected that as only some areas of the globe would have rainfall at any one time, we would find some processors would be doing no work while others would carry out the majority of the computational load. Provisional results seem to indicate that this may not be the case. See section 6. This result is probably an artifact of the data set used, a full model (one that has both dynamics and rainfall processes) may be unbalanced.

# 6    Results

Figure 6 shows the variation of time for one iteration against the fraction of iterations displayed for the model on the plane. In the case when one in a hundred timesteps are displayed, linear speedup is obtained in the range 1 to 64 processors.

Shown in figure 7 is the model on a sphere, for small numbers of processors the FFT's do not change the timings much, but at 64 processors the FFT's are having a significant effect. In both cases there is a saturation effect. This is because in our model every processor has the same patch size, and so for large processor networks there will be much more data to move out from the network to the outside world and all this data must go through one link.

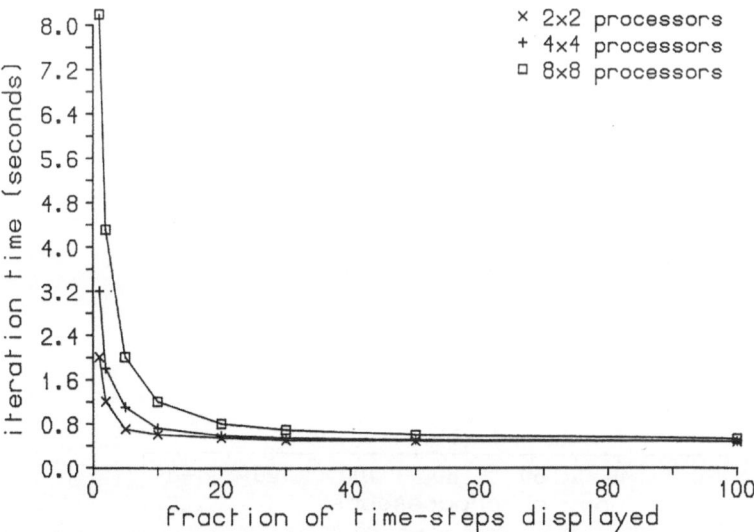

Figure 6. Iteration time versus fraction of timesteps displayed for model on a plane

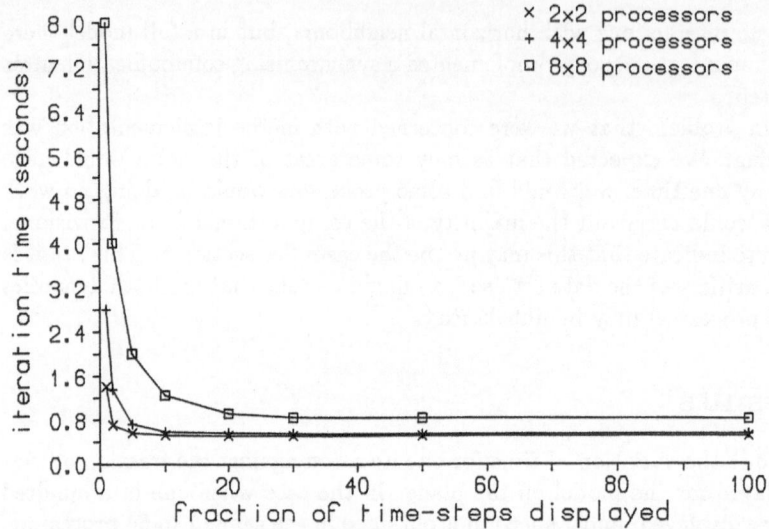

Figure 7. As figure 6 for a sphere

Figure 8 shows a graph of the number of processors against time for 160 iterations multiplied by the number of processors. The results indicate a linear speed-up. This is something of a surprise because of the anticipated load balancing problems due to the subgrid processes. The explanation appears to be that the dataset used was not sufficiently inhomogeneous to expose any significant problem.

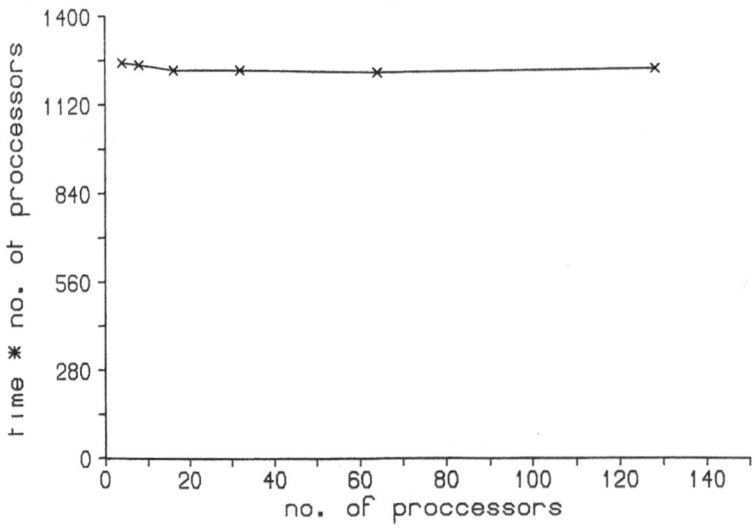

Figure 8. Linear speed-up with the number of processors.

# 7 Conclusion

We have obtained, for a finite difference scheme similar to that used by the U.K. Meteorological Office, linear speedup with the number of processors in a transputer array in the range 4 to 64 processors. We believe that this result is generalisable to most finite difference schemes used by meteorologists. We have also obtained a minimum patch size for the scheme considered. This minimum patch size depends on the interaction range of the scheme and on the ratio of calculation speed to communication speed for the processor.

Next we considered a version of the model on a sphere. Here it is necessary to carry out Fourier transforms. We showed that these are inefficient for the processor topology we considered. We therefore believe that future work should be on algorithms that don't need these transforms or (more likely) on network topologies which have better global communications. It should be noted that filtering operators with a large neighbourhood may also be inefficient due to the communications overhead.

Our preliminary results on the Rainfall Benchmark code did not indicate any load balancing problems from the modelling of subgrid processes. This we believe was due to homogeneities in our data set, indicating the need for care in choosing benchmark data for parallel machines.

# A  Occam

Occam is a process language. A program is build up from simpler processes. A process is an independent computation. Processes, in Occam, can be linked together by channels. Channels have two ends, an input and an output. Communication is synchronised and only occurs on a channel when an input and output are both ready. Processes can be named via a PROC statement. This is similar to a subroutine in FORTRAN and, like FORTRAN, parameters can be passed into the process. All variables and channels in Occam can be arrays. Array subscripting runs from 0 . In writing Occam

```
... stuff
```

is used a lot. This just means that there is some Occam code 'folded away' inside.

```
chan ? var
```

means input var on the channel called chan.

```
chan ! var
```

means output var on chan.

```
PAR
    ... stuff1
    ... stuff2
```

This means do stuff1 and stuff2 in PARallel. The compound process will terminate when both stuff1 and stuff2 have terminated.

```
SEQ
  ... stuff1
  ... stuff2
```

All this means is do **stuff1** and **stuff2** in SEQuence.

```
ALT
  in1 ? var
    ... stuff1
  in2 ? var
    ... stuff2
```

This process will wait until one of the input channels is ready for input. The first one ready will activate and then do the resulting process. In this case if **in1** becomes ready it will then do **stuff1**.

```
IF
  cond1
    ... stuff1
  cond2
    ... stuff2
```

This process is very similar to the FORTRAN IF, ELSEIF , ENDIF though the first condition ( in textual order ) that is TRUE will be taken. The PAR, SEQ, ALT and IF processes can all be 'replicated'. This is similar to the FORTRAN DO loop so a replicated SEQ is logically the same as a DO loop. A replicated PAR is a PAR DO. For more details on Occam see [2], and [4].

# B  Occam Code for Switch Processes

```
PROC out.switch(CHAN out, []CHAN in.choice)
  ... variable decs and initialisation
  WHILE TRUE
    ALT i=0 FOR no.of.in.chans
      in.choice[i] ? message
        out ! message
:
```

```
PROC in.switch(CHAN in, []CHAN out.choice)
  ... variable decs and initialisation
  WHILE TRUE
    SEQ
      in ? message
      IF  i=0 FOR no.of.codes
        code=which.code[i]
          out.choice[which.choice[i]] ! message
:
```

In both cases **message** consists of an integer which is the length of the message, a code integer and then the actual data itself.

# C   Finite Difference Scheme Used

## C.1   Adjustment Step

Each adjustment step consists of:

$$\phi^{n+1} = \phi^n - \delta t\{\overline{\delta_x u\overline{\phi}^{xy}}^y + \overline{\delta_y v\overline{\phi}^{xy}}^x\} \tag{7}$$

followed by

$$u^{n+1} = w_1 u^n + w_2 v^n - w_3\overline{\delta_x\phi^{n+1}}^y - w_4\overline{\delta_y\phi^{n+1}}^x \tag{8}$$
$$v^{n+1} = w_1 v^n - w_2 u^n - w_3\overline{\delta_x\phi^{n+1}}^x - w_4\overline{\delta_y\phi^{n+1}}^y \tag{9}$$

where

$$\delta_x f \equiv \frac{f(x+\frac{1}{2},y) - f(x-\frac{1}{2},y)}{\Delta x}$$
$$\overline{f}^x \equiv \frac{(f(x+\frac{1}{2},y) + f(x-\frac{1}{2},y))}{2}$$
$$w_1 \equiv \frac{1 - f^2\delta t^2/4}{1 + f^2\delta t/4}$$
$$w_2 \equiv \frac{f\delta t}{1 + f^2\delta t^2/4}$$
$$w_3 \equiv \frac{\delta t}{1 + f^2\delta t^2/4}$$
$$w_4 \equiv \frac{f\delta t^2/2}{1 + f^2\delta t^2/4}$$

This adjustment step is carried out three times.

## C.2   Advection Scheme

The advection step consists of a modified Lax-Wendroff scheme
$$u^{n+\frac{1}{2}} = [\overline{u}^{xy} - \Delta t\{\overline{u}^{xy}\delta_x\overline{u}^y + \overline{v}^{xy}\delta_y\overline{u}^x\}]^n \tag{10}$$
$$v^{n+\frac{1}{2}} = [\overline{v}^{xy} - \Delta t\{\overline{u}^{xy}\delta_x\overline{v}^y + \overline{v}^{xy}\delta_y\overline{v}^x\}]^n$$

where $\Delta t = 3\delta t$, followed by
$$u^{n+1} = u^n - \Delta t\{\overline{u}^{xy}[(1+c)\delta_x\overline{u}^y - c\{2/3\delta_{3x}\overline{u}^y + 1/3\delta_{3x}\overline{u}^{3y}\}] \tag{11}$$
$$+\overline{v}^{xy}[(1+c)\delta_y\overline{u}^x - c\{2/3\delta_{3y}\overline{u}^x + 1/3\delta_{3y}\overline{u}^{3x}\}]\}^{n+\frac{1}{2}}$$

$$v^{n+1} = v^n - \Delta t\{\overline{u}^{xy}[(1+c)\delta_x\overline{v}^y - c\{2/3\delta_{3x}\overline{v}^y + 1/3\delta_{3x}\overline{v}^{3y}\}]$$
$$+\overline{v}^{xy}[(1+c)\delta_y\overline{v}^x - c\{2/3\delta_{3y}\overline{v}^x + 1/3\delta_{3y}\overline{v}^{3x}\}]\}^{n+\frac{1}{2}}$$

where
$$c = 3/4(1-\mu^2)$$
$$\mu = (\tfrac{u^2\Delta t^2}{\Delta x^2} + \tfrac{u^2\Delta t^2}{\Delta y^2})^{\frac{1}{2}}$$

Note that all the blocks in equation (10) have an $I_R$ of 1 while some of the blocks in equation (11) have an $I_R$ of 2.

# References

[1] R.S. Bell and A. Dickinson. The meteorological office operational numerical prediction system. Scientific Paper No.41, U.K. Meteorological Office, Her Majesty's Stationary Office , London, 1987.

[2] K.C. Bowler, R.D. Kenway, G.S. Pawley, and D.Roweth. *An Introduction to Occam 2 Programming*. Chartwell-Bratt, Sweden, 1987.

[3] G.-R.Hoffmann and D.F. Snelling. *Multiprocessing in Meteorological Models*. Springer-Verlag, Berlin, 1988.

[4] D. Pountain and David May. *A Tutorial Introduction to Occam Programming*. Blackwell Scentific Publications/McGraw Hill, 1988.

[5] David F. Snelling and David A. Tanqueray. Performance modelling of the shallow water equations on the FPS-T series. In *CONPAR 88 B*, pages 156–159. British Computer Society - Parallel Processing Specialist Group, 1988.

# A Meteorological Model on a Transputer Network

GERARD CATS[1], HANS MIDDELKOOP[2], DICK STREEFLAND and
DOAITSE SWIERSTRA[2]

[1] Royal Netherlands Meteorological Institute, P.O. Box 201, 3730 AE,
De Bilt, The Netherlands
[2] Department of Computer Science, University of Utrecht, The Netherlands

This document is a report of a study into the feasibility of parallellising numerical weather prediction programs for execution on a grid of transputers.

Our investigations concern programs, in which a set of partial differential equations, describing the behaviour of the atmosphere, are numerically solved.

Two different implementation models are presented, together with a criterion for choosing the fastest method. Expressions are derived by which speedup for both models can be calculated.

Applied to HIRLAM, a specific weather prediction program, these results show that both internal and external communication time are negligible with respect to calculation time. As a consequence, for this program, speedup is almost linear in the number of transputers used.

## 1    INTRODUCTION

Finite difference solution methods to the Navier-Stokes equations are among the finest examples of problems that allow parallel processing. Richardson applied (human) parallel processors to a weather prediction problem as early as 1922. The first electronic computers were very much based on the concept of sequential processing. Therefore all software developed for weather prediction models became inherently sequential - culminating in software for the vector hardware of modern supercomputers. The capacity of a supercomputer can be equalled by a large number of small computers, each working on a small part of the grid, interconnected by a network. The performance/price ratio for small computers is, in general, better than for supercomputers. If the overhead due to inter-processor communication is limited, a network of small computers could provide the performance of a much bigger computer at lower costs.

The *transputer* microprocessor is especially well suited for network applications. It has been designed to make parallel processes and inter-processor communication easy to use and fast. Moreover, transputers are relatively cheap and very little extra hardware is necessary to build a network of them. Therefore, networks of low cost are easy to build using transputers.

Each transputer in a transputer-network can possibly run a different program. This gives such a system a great flexibility, which could for instance be used to perform different calculations at the border of a grid. With vector machines such exceptions are clumsy and inefficient to implement.

Computations with embedded conditions also cause problems for vector computers, but not for a transputer-network. The conditional expressions at different gridpoints possibly evaluate to different truth values. A vector-computer has to perform both branches of a conditional statement for each gridpoint and select the desired one afterwards. This inefficiency is avoided with a distributed system of transputers.

Topics in Atmospheric and Oceanic Sciences
© Springer-Verlag Berlin Heidelberg 1990

Another advantage of such a network is the flexibility to increase the performance gradually by adding more computers.

In this paper we report on an investigation as to how and in what sense weather prediction can benefit from the use of a multiprocessor system and how performance predictions can be made. In order to be more specific we will consider the following issues:

**Partitioning** Is there an obvious way to partition the computations in numerical weather prediction into a number of tasks, each task to be assigned to one processor, so that the load for each processor will be approximately equal?

**Communication** The need for communication results from dependencies between tasks that are assigned to different processors. For most multiprocessor systems without shared memory, communication via links is rather time consuming compared to computation. Care should be taken that partitioning of the computation process does not lead to excessive communication requirements. It is therefore important to know what dependencies can occur in numerical weather prediction programs.

**Distribution** In a multiprocessor system it is in general not the case that each processor is directly connected to every other processor. This implies that sending some data from one processor to another might require the data passing through several other processors, thus increasing overhead. In the multiprocessor system it is therefore important to assign communicating tasks to processors that are close to each other.

**Finite number of processors** It might be that an obvious partition of the computation leads to a number of tasks that exceeds the number of processors. Thus several tasks must be assigned to the same processor or a new partition must be made. On the other hand, when the number of processors exceeds the number of tasks, a different partition is even necessary, which generally causes extra communication.

**Implementation model** If several tasks are assigned to the same processor, these tasks have to communicate. There are several ways to implement this "internal" communication.

**Adding processors** One advantage of a multiprocessor system is the ease with which processors can be added, thus increasing the performance of the system. It is important to be able to predict how the addition of processors affects the overall computation time of a fixed numerical weather prediction program.

The weather prediction model component of the HIRLAM system[1] was kindly made available to us for experimentation. The HIRLAM model is based on modern numerical techniques, which make it an appropriate program for testing the suitability of a parallel transputer system for numerical weather prediction.

The HIRLAM program uses a 3-dimensional grid to model the atmosphere. This 3-dimensional structure is reflected in the computation. Basically, the computation consists of a series of updates of variables associated with gridpoints. Therefore, the most obvious way to partition computations is by taking the calculations associated with one gridpoint as a separate task. When the number of these tasks exceeds the number of available processors, each processor should process a number of tasks.

---

[1]The HIRLAM system was developed by the HIRLAM-project group, a cooperative project of Denmark, Finland, Iceland, The Netherlands, Norway and Sweden.

In numerical weather prediction, many numerical methods can be used for the solution of the partial differential equations. In explicit finite-difference schemes, the unknown quantities at a certain gridpoint depend on known values at other gridpoints. From numerical viewpoint, it suffices to use values at neighbouring gridpoints only (local dependencies). In implicit finite-difference schemes, however, the unknowns at a certain gridpoint also depend on unknown values at other gridpoints. Thus there is interdependency of variables at many gridpoints (global dependencies).

The disadvantage of an implicit scheme is that it requires communication between many tasks, and therefore between tasks assigned to processors that are not close to each other in the multiprocessor system. Explicit methods do not have this disadvantage. However, a smaller time step in the integration procedure is necessary because of stability requirements. This loss in computation time can be easily compensated for in a multiprocessor environment. We have restricted ourselves to explicit computations.

In the HIRLAM model the number of tasks exceeds the number of transputers and there are more dependencies in the vertical direction than in the two horizontal directions. Therefore the tasks associated with a vertical column of gridpoints should be executed on the same transputer. Typically, even the number of columns exceeds the number of transputers, so a cluster of columns should be assigned to the same transputer. To lower dependencies between transputers, and therefore communication overhead, such a cluster has to consist of neighbouring columns. Inspection of the communication requirements with respect to the distribution of tasks among transputers, reveals how the shape of these clusters should be chosen in order to minimise the amount of communication between transputers.

There are different ways to implement the computations associated with a cluster of neighbouring columns. Two different implementation models are presented, as well as a criterion to determine in advance which implementation model will be faster.

For both implementation models expressions for the execution time were derived, and these were used to derive expressions for the speedup, in case the number of transputers would be increased. These formulas contain a number of machine-dependent constants and a number of application-dependent constants.

The machine-dependent constants were measured by running benchmarks on the transputer system. The application-dependent constants of the HIRLAM program were obtained by close inspection of the program text, and by using estimations based on the running time of the HIRLAM program on a Harris HCX-9 computer.

Substitution of these constants in the formulas for execution time and speedup shows that the time for communication is negligible with respect to calculation time. As a consequence, for the HIRLAM program, speedup is almost linear in the number of transputers used.

This report is set up as follows: In Section 2 a short description of the meteorological model is given. Some properties of the transputer microprocessor are discussed in Section 3.

In Section 4 the implementation of the computations on a staggered grid using explicit finite-difference schemes for approximating partial differential equations is investigated. It is argued that the most appropriate network topology for the computations is a 2-dimensional grid of transputers, each one calculating a subgrid of the discrete model. The two different implementations for performing the calculations for such a subgrid on one transputer are presented, together with some possible optimalisations. We present expressions for the speedup that is to be expected when more processors are added in both implementations.

In Section 5 the results of Section 4 are applied to the HIRLAM program. Figures with the expected running time for the HIRLAM program on a grid of T414, respectively T800 transputers, are shown.

Finally, in Section 6 conclusions are drawn and some suggestions for future work are presented.

## 2    The Meteorological Model

### 2.1    The HIRLAM System

For the investigations described in this paper, we used a version of the HIRLAM (HIgh Resolution Limited Area Model) system. In principle, the HIRLAM system consists of three parts: An analysis scheme, an initialisation scheme and the proper forecast model; this forecast model will here be simply referred to as the (meteorological) model. To give an impression of the relative use of computer resources of the three components of the full system, the forecast model takes 32 minutes of CP time on a Cray-XMP (single processor) (48-hours forecast, 13000 gridpoints in the horizontal), the analysis 10 minutes and the initialisation 2. The initialisation scheme is organised much along the same lines as the model, and therefore the results presented here concerning parallellisation are straightforwardly extendable to that scheme. The analysis scheme, however, requires a different approach, that we did not investigate at all.

### 2.2    The HIRLAM Model

The model version that was made available to us (see HIRLAM Documentation Manual, 1988) had 34 x 34 gridpoints, and 9 levels. It was coded in standard Fortran-77. It is optimised for a vector machine, allowing very long loops, with lengths up to the number of gridpoints (if sufficient memory is available). The bulk of the calculations within the HIRLAM model is in subroutine DYN (the so-called "dynamical" part) (approximately 25% of the central-processor time) and in the group of subroutines collectively called "physics", and initiated from subroutine PHCALL (approximately 60% of CP time). The program structure is shown in Fig. 1.

* read start data
* read boundary data 1
* read boundary data 2
* initialization :
    * BDINIT
    * MAPFAC
    * INIPHY
* loop for each time-step :
    * DYN (local communication)
    * TSTEP
    * PHCALL:
        * HYBRID
        * RADIA
        * VDIFFX(1) (local communication)
        * VDIFF
        * VDIFFX(2) (local communication)
        * KUO
        * COND
        * QNEGAT
    * HDIFF (local communication)
    * SICALL (global communication) (only in semi-implicit scheme)
    * BDMAST
    * STATIS (global communication) (only used for diagnostics)
    * array copying
    * PRSTAT
    * 6 hourly input of new boundary data
    * output results STATIS

Figure 1: Structure of the HIRLAM Program


## 3    Hardware

### 3.1    The Transputer

#### 3.1.1  What is a Transputer?

The British chip manufacturer INMOS recently developed a new range of microprocessors (INMOS, 1986). These microprocessors are especially designed for building parallel computers. These chips were named *transputers*, because they are, according to INMOS, similar to transistors in the sense that both are elementary building blocks for complex systems. The first transputers were produced in 1985.

At present there are three different types of transputers available. The transputer T212 is a 16-bit processor, which is mainly meant to be used in controller-applications. Two other members of the transputer family are the T414 and the T800; they are both 32-bit processors. Transputers have a RISC-architecture (*Reduced Instruction Set Computer*) and are available with performances of 5 to 10 MIPS.

A transputer is in fact a complete computer on one chip. It contains a processing unit, 4K of very fast (50 ns) static RAM (2K for T414), 2 timers, 4 high speed serial links with DMA (*Direct Memory Access*) capability, operating bidirectionally at 10 or 20 Mbit/s and a memory interface for controlling up to 4 Gbyte of external memory. As an extra the T800 has an integrated, very fast (1.5 Mflops for a 20 MHz device) floating point unit.

### 3.1.2 Properties of Transputers

What makes a transputer so well suited for building distributed systems are the four integrated serial links. These links enable high-speed communication and synchronisation between different transputers. Because the links are serial, just a simple cable is needed to connect two transputers.

Communication via these links is completely controlled by hardware. After a process has set up a link-communication, the communicating process is suspended until the communication is completed. The hardware takes care of reading and writing data from and to memory by DMA. This DMA does hardly decrease processing speed, even when all four links operate at the same time.

Another nice feature is the on-chip RAM. This memory is fast enough to keep up with the processing unit, whereas external memory references will slow down the processor. For some applications, it might not even be necessary to use external memory. In that case, the cost for building a network of transputers is greatly reduced. The required interconnections for such a network are minimal; only the power supply, clock and reset signal and of course the four links have to be interconnected. No extra components are necessary in this case.

A transputer can boot from a ROM or from a link. For members of a transputer network, the latter method has the advantage of not requiring a boot ROM for every transputer. The desired boot method is selected by wiring a pin to logic "0" or "1". When booting from a link is selected and the transputer is reset, the first block of data the transputer receives via one of its links is loaded into the internal RAM and executed. To start up a whole network, one must first boot one transputer which has a connection with the host. After that, this transputer is instructed to send a boot-program to its neighbours, etc.

The transputer has the built-in capability to manipulate processes. There are instructions to start and stop processes. These instructions manipulate a linked list which contains all the processes which are ready to execute. A simple form of time-sharing is also built-in; processes that are running longer than some fixed time-interval are moved to the end of the list and the next process is picked from the beginning of the list for execution. When a process has to wait for a timer or for a communication, it is temporarily removed from the list of processes and the next one is started. A context-switch between two processes takes, because it is built-in, less than 1 μs. It follows that the use of processes is very simple and also does not lead to much overhead.

Communication between processes is also very simple. With a single instruction one can send or receive a message, using a *channel*. Communication via a channel takes place unidirectionally, according to the so-called *rendez-vous* principle: the sender and receiver should simultaneously engage in the communication. If one of them is not ready to communicate, the other must wait. A waiting process will automatically be removed from the list of ready processes. Communication is the only way to synchronise processes. Because an identification of the suspended process is stored in a memory word associated to the channel, it is not possible for more than one process at either side to use a channel at the same time.

This form of asynchronous communication gives rise to data-driven execution. Although asynchronous communication is not as efficient as synchronous communication between processes which are synchronous in time, it is a convenient programming tool, which enables the programmer to abstract from instruction timings.

There are two types of channels: *internal* channels for communication between processes on the same transputer; *external* channels for communication between processes on different transputers. External channels are mapped onto a link; internal channels are mapped onto an arbitrary memory word. The same communication instructions can be used for internal and external channels; for the program, the difference is transparent.

Because of the hardware restrictions mentioned above, at most one external channel in each direction can be mapped onto a link. In order to get more "external" channels, it is necessary to implement a number of "virtual" channels on one link. Virtual channels can be implemented by using "multiplexer" and "demultiplexer" processes on each side of the link. A multiplexer process communicates at one side with a number of processes via internal channels and at the other side with a demultiplexer on another transputer via the link.

The communication between a multiplexer and the corresponding demultiplexer at the other side of a link, can be done in two different ways. The first approach is to send an identification of the virtual channel with each value. Another approach is to agree on sending values in a fixed order. Of course, the latter method can only be used if it is known beforehand how the virtual channels will be used.

## 3.2   Working Environment

The software development was done on a Harris HCX-9 mini-computer with Fortran-77 and C compilers. This machine served as a host to the transputer network: One of the transputers on the edge of the network was connected to it. On the HCX-9, a C compiler, supporting parallel structures in C, to the transputer system was available (UNICOM *TCC version 1.10a*).

## 4   Grid-Based Computations on a Network of Transputers

### 4.1   Choosing a Network Topology

When designing a network for parallel computation, one should keep in mind two important objectives:

1.   It should be possible to divide the computation in such a way that each processor has to do approximately the same amount of work.

2.   The need for communications between different processors, and in particular the waiting time caused by these communications should be minimised.

In the case of a computation on a grid, the first objective is achieved easily. The calculation times at the gridpoints will in many cases be approximately the same, so the gridpoints can be distributed evenly across the available processors.

The second objective might cause more problems, because it is sometimes possible to avoid communications by modifying the algorithm. Communications are then replaced by some extra calculations. Thus there is a tradeoff between communication and computation. In order to achieve maximum performance, one must know precisely what the communication costs and the cost of the extra calculations are, and how they relate to each other. Moreover, one must take into account that communication between two processors which are not directly connected requires the processors in between to involve in the communication. This places an extra burden on these processors.

Many algorithms on grids, including the finite differencing methods used for numerical weather prediction require only local information on each gridpoint; that is information stored at that gridpoint, or information from one of its direct "neighbours". For these cases the obvious network architecture is a grid of processors, where each processor will do the calculation for some rectangular area. The optimal shape of these rectangular areas depends on the relative amount of vertical and horizontal communications.

Because a transputer has only four links, it is not possible to form a grid of transputers of dimension 3 or more. A more-dimensional grid however, can always be projected onto a 2-dimensional grid, which can then be distributed over a 2-dimensional grid of transputers.

We will use only orthogonal projections. The choice which dimension(s) to project depends on a number of properties of the program, such as:

- The amount of communication in each direction. The dimension in which most communication takes place, is a good candidate for projecting, because this gives the greatest saving of communications between transputers. External communication is always slower than internal communication, so lowering the number of these communications has a positive effect on the speedup.

- The shape of the space of gridpoints. By choosing the dimensions with the least number of gridpoints for projecting, the resulting 2-dimensional grid has a maximum number of gridpoints. As a result, more transputers can be used without being forced to split gridpoints computations over different transputers, which possibly results in a large number of extra communications.

- The use of global data. To avoid data-duplication, gridpoints using the same global data should preferably reside on the same transputer.

In the remainder of this chapter, we will consider grid-based computations on a network of transputers, in which only local communications take place. The grid may be 1- or 2-dimensional, but we will concentrate on 2-dimensional grids. The total number of gridpoints will be denoted by $N$, the total number of transputers by $P$. $P$ will never be greater than $N$.

Each transputer will do the computations for some rectangular subgrid of the total grid. We will call these gridpoints the *local grid* of a transputer. The actions needed for calculating the new state of one gridpoint consist of a sequence of communication and calculation steps. The communication steps are needed if the next calculation step requests some value from a "neighbour" gridpoint. Because the actions for each gridpoint are equal, this implies that at the same time the neighbour on the opposite side also needs a value from this gridpoint. Thus one communication generally involves a send and a receive action with two opposite neighbours.

In the next sections we will first treat the impact of staggering on projecting a physical grid on a grid of transputers. After that, we will look at different ways of calculating the gridpoints of a local grid on one transputer. In the last sections we will investigate the effect of the number of transputers on the total execution time.

## 4.2 Implementation of Staggering on a Grid of Transputers

The usual implementation of staggering (in which the staggered gridpoints are stored in computer memory as if they coincide with the unstaggered gridpoints) yields an efficient program for both vector computers and a grid of transputers. When calculating a space centered (with respect to a staggered variable) finite-difference scheme, only one neighbour variable is needed. Thus besides a gain of a factor two in computation time by removing the computational mode, the proposed programming model reduces both internal and external communication by a factor two compared to the program that implements staggering directly.

In this section we will use the term "gridpoint" to denote a *logical* gridpoint, that is, the superposition of a staggered and an unstaggered gridpoint.

## 4.3 Two Computational Models

Each transputer has to perform the calculations for the gridpoints of its local grid. In general, the calculations at neighbouring gridpoints are mutually dependent, so they must be performed quasi parallel, with interleaved communication steps.

The most obvious way to implement these calculations on a transputer is to create a separate process for each gridpoint. The exchange of values between neighbour gridpoints can be accomplished by using internal "channels" for inter-process communication. This model will be called the parallel model.

The other possibility is to use one process for all gridpoints on a transputer, by storing the gridpoint data in arrays indexed by the relative position of the gridpoint. Each calculation step will typically be a loop which iterates over the grid and communications are realised simply by using array subscripts. We will refer to this second model as the sequential model. In the following subsections, we will discuss these two different models in detail.

### 4.3.1 The Parallel Model

The parallel model is a conceptual simple one. One writes a program to be executed at one gridpoint using only local variables. Values from neighbouring gridpoints are requested using communication via channels.

For each gridpoint a process executing this program has to be created and run in parallel with the others. Using the built in timesharing facilities of a transputer this is easy to do. There is no need to control or synchronise these processes, they synchronise automatically through their communications.

Figure 2 shows as an example a parallel implementation of a 2-dimensional grid-based computation.

```
FOR t := 1 TO Nsteps
  BEGIN
    BEGINPAR
      BEGIN
        SEND U TO east;
        SEND U TO west;
        SEND U TO north;
        SEND U TO south
      END;
      BEGIN
        RECEIVE U_west FROM west;
        RECEIVE U_east FROM east;
        RECEIVE U_south FROM south
        RECEIVE U_north FROM north;
      END
    ENDPAR;
    U := f(U, U_west, U_east, U_south, U_north)
  END
```

Figure 2   Parallel implementation of a 2-dim. grid-based computation

The gridpoint processes at the boundary of a local grid, have to communicate with their counterparts on a neighbour transputer. Because there is only one communication link at each side, it is necessary to implement a number of "virtual channels" on one physical link by using multiplexers (see section 3.1.2). Each gridpoint process at the boundary will then communicate with its corresponding gridpoint process on another transputer via a virtual channel.

Multiplexers at the boundary of the global grid must be special. They should only simulate a multiplexer process and should not communicate over their link. These dummy multiplexers should ignore values they receive from the gridpoint processes and send values corresponding to the boundary conditions to the gridpoint processes requesting a value.

Communications deserve special attention. As noted before, each communication step consists of a send action and a receive action from the opposite direction. Consider what happens when these two actions are performed sequentially. When for example each process starts with a send to their right neighbour, they all have to wait until their right neighbour performs a receive, which results in a deadlock situation. This situation can be prevented by executing the send and the receive actions concurrently.

Another approach is to run two different variants of the program for alternating gridpoints, the only difference between these two being that the first always starts with send actions, while the second starts with the receive actions.

When using the first approach, it is possible to create the two processes for the send and receive actions afresh for each communication, but this causes some overhead. The alternative is to create a send and a receive process once, and let them do all communications.

The problem with the latter approach is the synchronisation needed between the calculations and the two communication processes. The only way to do this is using communications via a channel. Thus the second alternative saves two process creations per communication step, but at the cost of two extra channel communications.

A disadvantage of the parallel model is the data duplication caused by the communications. Each communication of a value in a particular direction requires an extra variable at each gridpoint. In section 4.4.1, the different variants of the parallel model are compared.

### 4.3.2 The Sequential Model

In the sequential model there is only one process per transputer and the data for all local gridpoints are stored in arrays. Each calculation step is performed for all local gridpoints, typically by iterating over the arrays. Internal communication steps are not necessary, because "neighbour values" can be found in the arrays. Only neighbour values that reside on another transputer deserve special attention. The simplest solution is to extend all arrays by one column and one row on every side. A communication step then consists of a copy of a complete row or column from a neighbour transputer to this extended border.

An important issue is the order of the iteration over the arrays. When for instance the calculation of some variable depends on the previous value of this variable at the gridpoint to the left, then one clearly should not calculate new values left-to-right but right-to-left. If the calculation depends on both the previous left and right values, we are forced to use a temporary array for the results of one row and copy the contents of this array back to the original array when the row is completely calculated.

Compared with the parallel model, there is some overhead caused by array subscripting and loop control. The overhead can be reduced by copying values that are multiply accessed into plain variables, which can then be used instead of the original array element. Other possible optimisations are for instance the use of arrays of arrays instead of multidimensional arrays. In section 4.4.2, different optimisations of the sequential model are compared.

Figure 3 shows a sequential implementation of a 2-dimensional grid-based computation.

```
FOR t := 1 TO Nsteps
  FOR x := 1 TO GridX
    FOR y := 1 TO GridY
      BEGIN
        U_west := U[x-1][y];
        U_east := U[x+1][y];
        U_south := U[x][y-1];
        U_north := U[x][y+1];
        U[x][y] := f(U[x][y], U_west, U_east, U_south, U_north)
      END
```

Figure 3 Sequential implementation of a 2-dim. grid-based computation

#### 4.4 Comparing the two Models

In order to compare the execution speed of the two models, we have constructed test programs which perform the same computations on the same grid size, but using the two different computational models.

In these test programs, we have forced all variables into the external RAM, to exclude variations due to the much higher access speed of the internal RAM.

Because the amount of link communication is the same for both models, it is possible to compare the two models using only one transputer. The external communications using links will therefore be omitted, but the parallel test program will include the four multiplexer processes.

Internal communications will take more time in the parallel model than in the sequential model. On the other hand, variable accesses in the sequential model will be slower. The decision which model to choose for a particular problem therefore depends on the amount of communications versus the amount of variable references. We will define $Q$ to be the ratio of the number of references to local variables, and the number of communications.

The test programs will perform calculations on a 2-dimensional grid of size 10 x 10. Each test program consists of a communication step followed by a calculation step, which are performed 10 times in order to get more reliable results. The communication step consists of communications in all four directions. The calculation step is a loop with a calculation using four local variables. $Q$ will be a parameter of the test programs; it will control the number of iterations of this loop. The total execution time is plotted as a function of $Q$.

The resulting graph will consist of a straight line, where the slope of this line gives information about the time needed to perform one iteration of the calculation loop. The vertical offset is the time needed to perform the communications (plus a little loop overhead).

As said before, a number of optimisations are possible for both models. For a comparison of the models, we should choose the fastest variants of the two models. This will be done in the next two subsections.

#### 4.4.1 The Optimal Parallel Model

The three different variants of the parallel model were timed as a function of $Q$. For the first two variants, the send and receive actions are performed by separate processes. There are two different ways to achieve this. The first method is to create these processes anew for every communication step; the second method is to create them outside the outer loop. The latter method has to use two extra channels per gridpoint to synchronise these processes with the calculation process. The third variant uses explicit alternating communication, caused by executing different code on adjacent gridpoints.

The reported times are net times; the constant startup time is subtracted from each measurement, so only the time for 10 loop-iterations is counted. The results in Fig. 4 show that for all values of $Q$ , the function Par3 performs slightly better. The reason why the slopes of the plotted lines are not exactly parallel, is the fact that the variables in the three functions do not have the same position in the function's "workspace". References to variables outside the first 16 words of the workspace are slower because they need an extra "prefix" instruction.

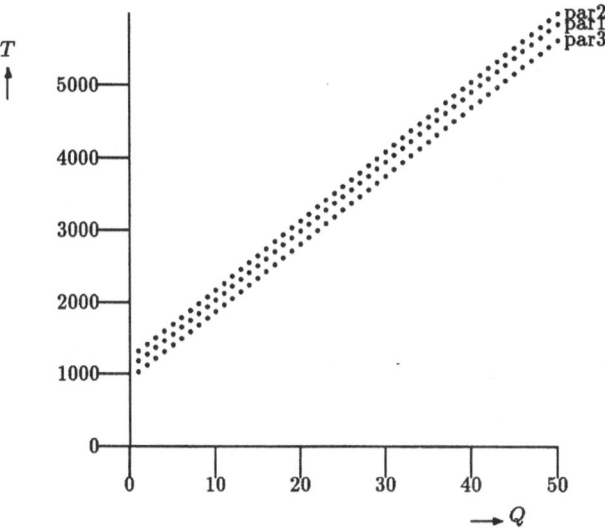

Figure 4  Results of Parallel Program

### 4.4.2  The Optimal Sequential Model

For the sequential model, a number of optimisations are possible. We compared three different approaches. The first approach uses normal 2-dimensional array subscripts to access a variable. The second approach uses a 1-dimensional array which is aliased to a column of the 2-dimensional array at the start of each column. The functions seq3a and seq3b implement yet another optimisation. In these functions no array subscripts are used, but pointers instead. These pointers are initially set to point to the first array element and are incremented each iteration.

When the same array subscript or pointer dereference is used a number of times, it might be attractive to make a copy of the value into a plain variable, and to use this variable instead. The number of occurrences of a variable, above which this approach results in a gain in efficiency, depends strongly on the compiler used and on the question whether the value is modified during the computations, in which case it has to be copied back.

The minimum number of occurrences of a variable, for which this optimisation is advantageous has been determined for the compiler we had available (UNICOM *TCC version 1.10a*). Separate values are determined for 2-dimensional arrays, 1-dimensional arrays and pointers. Both the case that a variable is only read, and the case that a variable is modified and has to be copied back are covered. The next table shows the results:

|          | 2-dim | 1-dim | pointer |
|----------|-------|-------|---------|
| read-only | 3 | 3 | 4 |
| read/write | 4 | 4 | 6 |

In the remainder of this subsection, we will assume that the number of times the same variable is used, is too small for these optimisations.

There are two versions of optimisation using pointers, for the number of times a variable is used has a great impact on performance. This difference is caused by the increment operations which are needed for every variable used. The two functions therefore differ in the number of times a variable is used. The function seq3a simulates a computation where each variable is used only once; the function seq3b does the same for two references to each variable, so the number of increment operations is halved.

The results of the four different methods used in this program are plotted in Fig. 5. The function seq3b is only slightly faster than seq2, and seq3a slightly slower. Because the performance of seq2 is independent of the number of times each variable is used, this seems to be the most appropriate variant of the sequential implementation model.

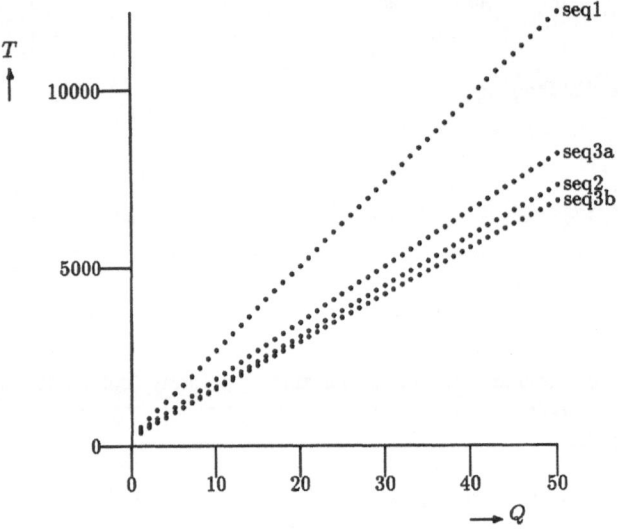

Figure 5 Results of Sequential Program

### 4.4.3 Comparison of the two Optimal Models

When we compare the methods par3 and seq2 by drawing their results together in Fig. 6, we see that there is no absolute "winner". Function seq2 performs better for $Q < 15$ and function par 3 performs better for $Q > 15$.

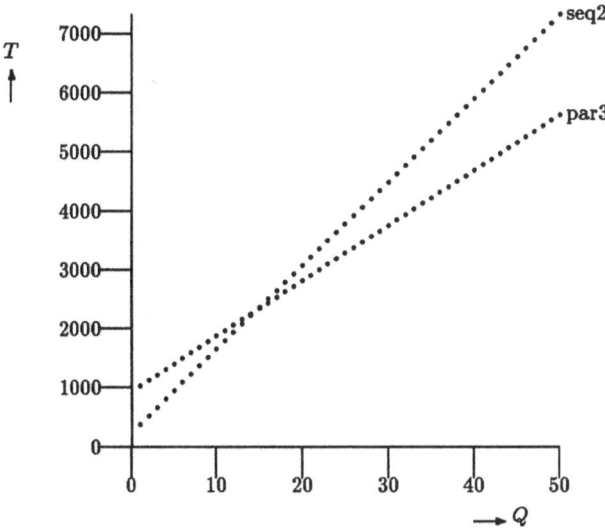

Figure 6 Comparison between parallel and sequential model

The Figure also shows that the communication costs for par3 are approximately three times as high as for seq2, but this is compensated by faster variable references for larger values of $Q$.

It is good to note however, that these results depend strongly on the compiler used. Because the slopes of the two lines in the Figure are so close, the intersection point will rapidly move if another compiler generates different code.

**4.5  Using More Transputers**

So far, we only looked at the organisation of the calculations for the local grid of one transputer. Now we will take a look at the behaviour of a network of transputers, together performing the computations for the whole grid.

First we need a few definitions, which are listed in table 1. We will assume that $N_x$ is a multiple of $P_x$, and $N_y$ is a multiple of $P_y$. For grid sizes where this is not the case, it is always possible to extend the grid at the boundary with dummy gridpoints.

Table 1  Definitions for Transputer Grids

| | |
|---|---|
| $N_x$ | number of gridpoints in horizontal direction |
| $N_y$ | number of gridpoints in vertical direction |
| $N$ | total number of gridpoints; equal to $N_x . N_y$ |
| $P_x$ | number of transputers in horizontal direction |
| $P_y$ | number of transputers in vertical direction |
| $P$ | total number of transputers; equal to $P_x . P_y$ |
| $x$ | size of the local grid in horizontal direction |
| $y$ | size of the local grid in vertical direction |

Because all transputers perform the same calculations, they will run almost synchronously. Therefore, it suffices to use the execution time of one processor as a measure of the execution time of the complete transputer network. It is possible to express this time for the optimal variants of the two implementation models from the previous section as a function of the grid sizes x and y and a number of other parameters. These parameters apply to the local grid of size x x y of one transputer; they are listed in table 2.

Table 2  Definitions for Execution Time Calculations

### Application Dependent Constants

| | |
|---|---|
| $C$ | total number of calculation steps |
| $C_x$ | number of communications in horizontal direction per gridpoint |
| $C_y$ | number of communications in vertical direction per gridpoint |
| $C_{xy}$ | total number of communications per gridpoint (abbreviation for $C_x + C_y$) |
| $C_{ext}$ | total number of external communications (abbreviation for $C_x \cdot y + C_y \cdot x$) |
| $V$ | number of subscripts of 1-dimensional arrays, needed only in the sequential model, per gridpoint ($V = Q \cdot C_{xy}$) |
| $F$ | fraction of link-communication time which is not overlapped with internal processing ($0 \leq F \leq 1$) |
| $T_{calc}$ | total calculation time for one gridpoint |

### Application Independent Constants

| | |
|---|---|
| $T_{subs}$ | additional costs for referencing a 1-dimensional array instead of a plain variable |
| $T_{array}$ | time needed to access a "neighbour value" by doing a subscript in a 2-dimensional array |
| $T_{chan}$ | time for performing one internal communication (a send plus the corresponding receive) |
| $T_{loop}$ | time for performing one iteration of a loop over the local grid |
| $T_{link}$ | additional costs for doing a link-communication instead of an internal communication |

We can now give a formula for the time needed to perform the computations for the optimal variant of the parallel model (par 3):

$$
\begin{aligned}
T^{par} &= T_{calc} \cdot x \cdot y \\
&\quad + C_x \cdot T_{chan} \cdot x \cdot y \\
&\quad + C_y \cdot T_{chan} \cdot x \cdot y \\
&\quad + (C_x \cdot y + C_y \cdot x) \, (T_{chan} + F \cdot T_{link}) \quad\quad\quad (1) \\
&= x \, y \, (T_{calc} + C_{xy} \, T_{chan}) + C_{ext} \, (T_{chan} + F \, T_{link}) \quad (2)
\end{aligned}
$$

The first term of (1) represents the total calculation time for all local gridpoints. The second and third term give the time needed to do the internal communications in horizontal respectively vertical direction, including the communications with the multiplexer processes.

The last term denotes the time needed for external communications in horizontal and vertical direction, which are done by the (de)multiplexer processes. The parameter F has a value between 0 and 1. The value of $T_{link}$ is approximately the time the hardware needs to send and receive a value over the links. Because the link-hardware works concurrently with the processor, link-communication is partly overlapped with processing. If $F = 0$, link-communication is completely overlapped; if $F = 1$, there is no overlapping at all.

If $x = 1$ or $y = 1$, the vertical respectively the horizontal multiplexer processes can be eliminated. In that case $y$ can be replaced by $y - 1$ in the third term, respectively $x$ can be replaced by $x - 1$ in the second term.

For the chosen sequential model (seq2), a similar formula can be derived:

$$
\begin{aligned}
T_{seq} &= (T_{calc} + V \cdot T_{subs} + C \cdot T_{loop}) \cdot x \cdot y \\
&+ (C_x + C_y) \cdot T_{array} \cdot x \cdot y \\
&+ (C_x \cdot y + C_y \cdot x) \cdot (T_{chan} + T_{link}) \quad\quad\quad (3) \\
&= x\,y\,(T_{calc} + V\,T_{subs} + C\,T_{loop} + C_{xy}\,T_{array}) + C_{ext}\,(T_{chan} + T_{link}) \quad\quad (4)
\end{aligned}
$$

As with $T^{par}$, the first term of (3) gives the calculation time. Additional time compared with the parallel model is needed for subscripting (as a result of the optimisation in an 1-dimensional array), and for iterating. The number of loops over the grid equals the number of calculation steps C, which also equals the number of communication steps.

The second term represents internal communication which is done by subscripting in a 2-dimensional array. The third term is like the last term of (1), except that the parameter F has disappeared. This is because in the sequential model communication is done sequentially, by the same process which does the calculations, so communication and calculation do not overlap. Depending on the application it might be possible to overlap some calculations with external communication. However, this is the responsibility of the programmer, whereas in the parallel model it is inherent.

As can be seen in the formulas (2) and (4), the time needed for external communications is proportional to $C_x\,y + C_y\,x$, while the time for internal processing is proportional to $x \cdot y$. To minimise the overhead caused by external communications, the shape of a local grid should be chosen in such a way that $C_x\,y + C_y\,x$ is minimised with respect to $x \cdot y$. So for the case that $C_x \approx C_y$, $x$ and $y$ should be chosen approximately equal.

## 4.6   Speedup Calculations

To see what the effect on the processing time of adding extra transputers is, we will compare the processing times for different numbers of transputers, using the same global grid in all cases. When the number of transputers in the horizontal direction ($P_x$) is increased by a factor p and the number of transputers in the vertical direction ($P_y$) is increased by a factor q, the speedup for the parallel model is given by:

$$S^{par}(p,q)$$

$$= \frac{T^{par}(px, qy)}{T^{par}(x, y)}$$

$$= \frac{pqxy(T_{calc} + C_{xy}T_{chan}) + (C_x qy + C_y px)(T_{chan} + FT_{link})}{xy(T_{calc} + C_{xy}T_{chan}) + (C_x y + C_y x)(T_{chan} + FT_{link})}$$

$$= pq - \frac{((pq - q)C_x y + (pq - p)C_y x)(T_{chan} + FT_{link})}{xy(T_{calc} + C_{xy}T_{chan}) + (C_x y + C_y x)(T_{chan} + FT_{link})} \qquad (5)$$

The speedup calculation for the sequential model is analogous:

$$S^{seq}(p,q)$$

$$= \frac{T^{seq}(px, py)}{T^{seq}(x, y)}$$

$$= \frac{pqxy(T_{calc} + VT_{subs} + CT_{loop} + C_{xy}T_{array}) + (C_x qy + C_y px)(T_{chan} + T_{link})}{xy(T_{calc} + VT_{subs} + CT_{loop} + C_{xy}T_{array}) + (C_x y + C_y x)(T_{chan} + T_{link})}$$

$$= pq - \frac{((pq - q)C_x y + (pq - p)C_y x)(T_{chan} + T_{link})}{xy(T_{calc} + VT_{subs} + CT_{loop} + C_{xy}T_{array}) + (C_x y + C_y x)(T_{chan} + T_{link})}$$

$$(6)$$

As the number of transputers increases, the size of the local grid decreases. For small local grids, the time devoted to external communications will increase with respect to processing time. When the amount of external communication dominates the total processing time, $C_{ext}$ $(T_{chan} + T_{link})$ respectively $C_{ext}$ $(T_{chan} + FT_{link})$ will be much greater than the other terms in the denominators. In this case, the speedup for both models reduces to:

$$S(p,q) \sim pq - \frac{(pq - q)C_x y + (pq - p)C_y x}{C_x y + C_y x} \qquad (7)$$

This formula gives a lowerbound for the speedup attainable. If we choose p and q equal, this reduces to:

$$S(p,p) \sim p \qquad (8)$$

For the case that $C_x \approx C_y$, x and y should be chosen approximately equal, so that (7) reduces to:

$$S(p,q) \approx \frac{p + q}{2} \qquad (9)$$

## 4.7    Variables for Speedup Calculation

A number of variables in the speedup formulas do not depend on the type of computation performed, but they only depend on the hardware and the compiler used. These variables are: $T_{subs}$, $T_{array}$, $T_{chan}$, $T_{loop}$ and $T_{link}$. The first four variables have been measured using an appropriate benchmark program running on a single transputer. To determine $T_{link}$, a network of transputers is needed.

The next table shows the results of the benchmarks:

| | | |
|---|---|---|
| $T_{subs}$ | = | 0.61 μseconds |
| $T_{array}$ | = | 3.36 μseconds |
| $T_{chan}$ | = | 6.03 μseconds |
| $T_{loop}$ | = | 0.57 μseconds |
| $T_{link}$ | = | 5.15 μseconds |

## 4.8    Conclusion

Two different implementation models for grid-based computations were introduced, together with a criterion by which the fastest model for a given application can be selected, on a basis of the ratio of local variable usage and the number of communications.

Formulas were derived to calculate the execution time for each model, using a number of program dependent constants, and a number of hardware/compiler dependent constants. The latter constants for our configuration were measured in section 4.7. With these formulas, expressions for the speedup of both models were derived, together with a lowerbound for the speedup.

Inspection of the formulas for the execution time shows that to minimise the overhead caused by external communications and thus maximise speedup, the shape of a local grid should be chosen in such a way that $C_x\, y + C_y\, x$ is minimised with respect to $x \cdot y$.

## 5.    Parallellisation of the HIRLAM Program

In the HIRLAM program some terms of the primitive equations are solved implicitly. This results in numerically solving a Helmholz equation and the need of global communication. These calculations are performed in procedure SICALL. It is possible to leave out these calculations, but then the time step has to be chosen smaller (approximately a factor 3) to avoid instability. This makes the program more expensive, but we will demonstrate that this increase in computation time can be gained back by parallellising the HIRLAM program.

## 5.1 Communication in the HIRLAM Program

The HIRLAM program uses a 3-dimensional space of gridpoints to model the behaviour of the atmosphere above a rectangular area on earth. These gridpoints (except the ones at the boundary) have to communicate with their neighbour gridpoints in all six directions. As was explained in section 4.1, we have to project one of these dimensions onto the other two.

In the HIRLAM program, communication takes place in all three dimensions, but the amount of vertical communication exceeds the amount of horizontal communication. Also, in typical applications of the HIRLAM program, the number of gridpoints in the vertical direction is much lower than in both horizontal directions. Furthermore, a great number of constants and variables used do not depend on the height. Obviously, the best solution is to project the vertical dimension onto the horizontal plane, so the gridpoints modelling a vertical column of air are always calculated on the same transputer.

The remaining point is how to distribute these "columns" over the transputer grid in the case the number of columns exceeds the number of transputers. In section 4.5 we have seen that when each transputer takes care of a local grid of size x . y, the external communication time is minimised if $(C_x . y + C_y . x)$ is minimal with respect to x . y.

Because the behaviour of the atmosphere is inherently symmetric in the two horizontal directions, the number of communications in both directions $C_x$ and $C_y$ will be approximately the same in the HIRLAM program. Performance therefore will be maximal when x and y are chosen approximately equal. Of course, the sizes of the local grids of the different transputers should all be roughly the same, because the speed of the total system is dictated by the slowest transputer.

At a few spots in the HIRLAM program, diagonal communications take place. These communications have to be replaced by two-step communications. To avoid unnecessary waiting as a result of diagonal communication, the programmer should take care that the particular information is transferred in the previous communication phase to a real neighbour.

In Fig. 1 the subroutines in which communications take place, are marked. The subroutines DYN, VDIFFX and HDIFF contain local communication, whereas the subroutine STATIS contains global communication. This subroutine contains the calculations of some statistical values, like the average pressure tendency. Because these calculations are not essential, STATIS can be omitted in a transputer implementation. Consequently, only local communications are needed in a transputer implementation of the HIRLAM program in explicit mode. The subroutine BDMAST is a replacement for the subroutine SICALL in case the implicit calculations are turned off.

## 5.2 Choosing the Implementation Model

In the previous section, we saw that the HIRLAM program contains communication at four stages during each time step. The decision which implementation model is most appropriate for the HIRLAM program depends on the value of $Q$, introduced in the previous section. We therefore have to take a closer look at the number of array references and the number of communications in the HIRLAM program. The subroutine with the most communication is DYN, so we will look at DYN first.

### .2.1 Calculation of Q for Subroutine DYN

)YN mainly consists of two loops over the (vertical) levels, the first of which is used for iitialising. Inside these two loops there are various loops over the horizontal grid. As a esult of optimisations, all 2-dimensional variables are "flattened" and stored in -dimensional arrays, so these loops are also 1-dimensional.

ig. 7 shows the communication pattern of DYN. Listed are the neighbour values needed in  particular loop. A mark 'O' after a variable name indicates that the neighbour value is old", which means that this value was used before. It is possible to avoid the O-communications" by saving the received neighbour values for subsequent use. A mark N' means "new", so a communication is always necessary here. A neighbour value is iarked with 'N' the first time it is needed, or if the variable was recomputed after the last sage.

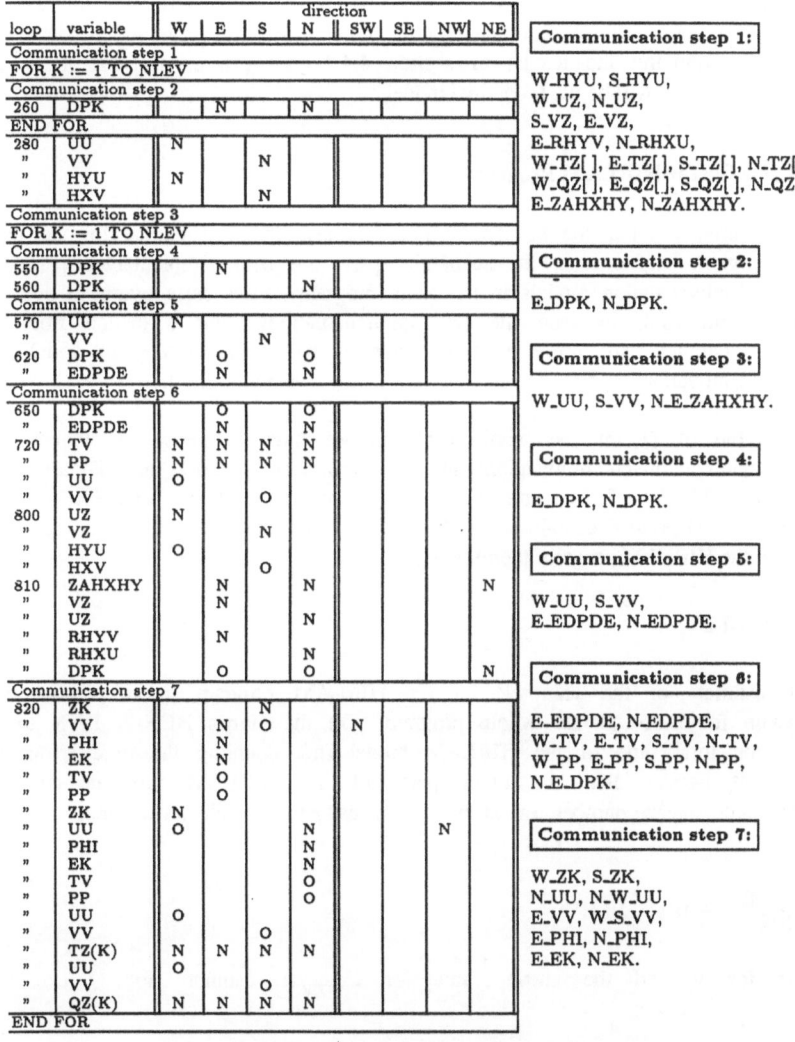

Figure 7  Communication in Subroutine DYN

Four values from diagonal neighbours are needed. These values must be send in two steps, so these communications must be counted twice.

Without the 'O' values, the total number of horizontal communications in the east-west direction, $C_x$, and in the north-south direction, $C_y$, are both 2 + 21.NLEV, where NLEV is the number of vertical levels.

In the Fig. we also see that there are seven communication steps, two of which are outside the vertical loops. The number of communication steps therefore is 2 + 5.NLEV. The values exchanged during each communication step are listed at the right of the Fig. A prefix 'W_' means that the value of the following variable in westward direction is needed; the prefixes 'E_', 'N_' and 'S_' have analogous meaning. For diagonal communications, a double prefix is used. For instance, 'N_E_ZAHXHY' denotes the value of the variable ZAHXHY of the eastward neighbour of the northward neighbour.

The value of $V$, which stands for the number of array references, is for DYN 21 + 187.NLEV. The value of $Q$, computed with these values, is 4.5, independent of the value of NLEV. Applying the results of section 4.4.3, we can expect that the best implementation model for DYN will be the sequential model.

## 5.2.2 Estimation of $Q$ for the HIRLAM Program

The question now arises, whether the sequential model is also the best choice for the rest of the HIRLAM program. Of course, the sequential parts are better implemented parallel, but mixing the two implementation models in the same program causes extra overhead due to the switching back and forth between the two implementations. For a switch from a sequential part to a parallel part, array elements have to be copied into plain variables, while for a switch back, all variables have to be copied back (if they were modified).

To calculate the value of $Q$ for the whole program, we have to know the values of $C_{xy}$ and V. The number of communications in the subroutines VDIFFX and HDIFF are much lower than in DYN. VDIFFX(1) contains 2 . NLEV communications, VDIFFX(2) 12 . NLEV communications and HDIFF contains 6 + 16 . NLEV communications. The total number of communications in the HIRLAM program therefore is:

$$C_{xy} = 10 + 72 . NLEV$$

To get a rough estimate of the value $V$ for the HIRLAM program, it is possible to extrapolate the value for DYN to the whole program. On the Harris HCX-9, DYN took about 18.5 % of the total time of the HIRLAM model (not counting the time spent in STATIS). When we assume that the other parts of the HIRLAM program contain approximately the same mean number of array references per second, an estimate for $V$ will be:

$$V \approx \frac{187 . NLEV}{18.5 \%} \sim 1011 . NLEV$$

Using this value for $V$ and the above value for $C_{xy}$ an estimate for Q can be computed:

$$Q \approx \frac{1011 \cdot \text{NLEV}}{10 + 72 \cdot \text{NLEV}} \approx 14$$

For this value of $Q$, the sequential model is slightly better, although the difference with the parallel model is minimal.

## 5.3 Performance Expectations

We would like to know what running times we have to expect, when running the HIRLAM program on a grid of transputers. In section 4.5, formulas were given for the execution time as a function of a number of variables. One of these variables is $T_{calc}$, the total calculation time for one gridpoint. Because this variable will probably dominate total processing time in the HIRLAM program, it is particularly important to determine $T_{calc}$.

## 5.3.1 Determining $T_{calc}$ for subroutine DYN

First $T_{calc}$ will be determined for subroutine DYN. The values of $C_x$, $C_y$ and V for DYN are known, so $T_{calc}$ can be determined by implementing DYN on a transputer and measuring the running time, for it is the only unknown in the formulas for the execution time (Eqs. 2,4).

We have implemented subroutine DYN in C on one transputer, without external communication and using the sequential model with pointers.

We used the variant with pointers instead of the variant with 1-dimensional arrays because in DYN, many variables are used more than once, so this variant is probably the fastest.

The running time on a T414 was 120 ms per local gridpoint, with NLEV = 9, so $\frac{T^{seq}}{xy}$ = 120 ms.

We recall the formula for the execution time of the sequential model (4):

$$T^{seq} = x\,y\,(T_{calc} + VT_{subs} + CT_{loop} + C_{xy}\,T_{array}) + C_{ext}(T_{chan} + T_{link}) \quad (10)$$

Because the program does not use external communications, the second term of this formula can be dropped. To determine $T_{calc}$ we must subtract $(VT_{subs} + CT_{loop} + C_{xy}\,T_{array})$ from the measured 120 ms. The above formula is actually for the variant with 1-dimensional arrays and not for the pointer variant. However, as can be seen in Fig. 5, the execution times of these two variants do not differ too much for Q = 4.5, so we can use it anyway.

The three terms to subtract can be computed, using the constants determined in section 4.7, as follows (assuming NLEV = 9):

$$VT_{subs} = (21 + 187 \cdot \text{NLEV}) \cdot 0.61\ \mu s = 1039\ \mu s$$

$$CT_{loop} = (3 + 17 \cdot \text{NLEV}) \cdot 0.57\ \mu s = 89\ \mu s$$

$$C_{xy}\,T_{array} = (4 + 42 \cdot \text{NLEV}) \cdot 3.36\ \mu s = 1284\ \mu s$$

The value C generally is the number of calculation steps, because this normally equals the number of loops over the grid. In the current implementation of the HIRLAM program, there are more loops than calculation steps and because we did not change the loop structure during conversion to C, the number of loops is greater than necessary. Therefore, the real number of loops is used here, instead of the number of communication steps (which is $3 + 5 . \text{NLEV}$).

Surprisingly, the three values just computed, are very small compared to the measured running time of 120 ms, only about 2 % of it! The value of $T_{calc}$ for DYN therefore is

$$T_{calc} = 120 \text{ ms} - 1039 \text{ }\mu s - 89 \text{ }\mu s - 1284 \text{ }\mu s = 118 \text{ ms}$$

The communication time for subroutine DYN, in case external communications do take place, is

$$C_{ext}(T_{chan} + T_{link}) = (C_x y + C_y x)(T_{chan} + T_{link})$$
$$= (x + y)(2 + 21 . \text{NLEV})(6.03 \text{ }\mu s + 5.15 \text{ }\mu s)$$
$$= (x + y) \text{ } 2135 \text{ }\mu s$$

which is about 3.5 % of the total processing time, for a 1 x 1 local grid, 1.7 % for a 2 x 2 grid, etc. Clearly, external communications are not a bottleneck in DYN.

## 5. 3.2 Estimating $T_{calc}$ for the HIRLAM Program

To obtain $T_{calc}$ for the whole HIRLAM program, we would have to implement it completely on a transputer. We did not go through this tedious task, involving the translation of several thousands lines of Fortran-77 code into C. Instead, we estimated the total $T_{calc}$ from that of DYN by assuming that DYN takes the same percentage of CP time on a transputer (in C), as it did on the mini-computer (in Fortran-77). We then get for the value of $T_{calc}$ for the total HIRLAM program:

$$T_{calc} \approx \frac{118 \text{ ms}}{18. 5\%} \approx 638 \text{ ms}$$

Using the estimated value of V and counted values C, $C_x$ and $C_y$ for the whole program, we can also compute the values for the other three subterms of the first term of (10). The value for C which is used here, is the minimum number of calculations steps needed for the implementation of the HIRLAM program. The current implementation in Fortran uses more steps (loops). The values for NLEV = 9 are:

$$VT_{subs} = (1011 . \text{NLEV}) .0.61 \text{ }\mu s = 5550 \text{ }\mu s$$
$$CT_{loop} = (4 + 10 . \text{NLEV}) . 0.57 \text{ }\mu s = 54 \text{ }\mu s$$
$$C_{xy}T_{array} = (10 + 72 . \text{NLEV}) . 3.36 \text{ }\mu s = 2211 \text{ }\mu s$$

Also, the second term of (10) can now be computed:

$$C_{ext}(T_{chan} + T_{link}) = (C_x y + C_y x)(T_{chan} + T_{link})$$
$$= (x + y)(5 + 36 \cdot NLEV)(6.03 \ \mu s + 5.15 \ \mu s)$$
$$= (x + y) \ 3678 \ \mu s$$

When we substitute all these values into (10), we get:

$$T^{seq} \approx xy \ (638 \ ms + 5550 \ \mu s + 54 \ \mu s + 2211 \ \mu s) + (x + y) \ 3678 \ \mu s \qquad (11)$$

From this equation, we can conclude that internal communication cost (2211 μs) is very low, compared to calculation time (638 ms). Also, the external communication is not costly. It accounts for only 1.1 % of the total processing time for a local grid of size 1 x 1, 0.6 % for a grid of size 2 x 2, etc. External communication therefore, will not be a bottleneck in the HIRLAM program.

The total running time of a transputer implementation of the HIRLAM program, for a 34 x 34 x 9 grid is plotted in Fig. 8, as a function of the total number of transputers $P_x \cdot P_y$, using (11). The values for $P_x$ and $P_y$ in this Fig. are equal. When $P_x$ and $P_y$ do not divide 34, not all transputers process the same number of gridpoints.

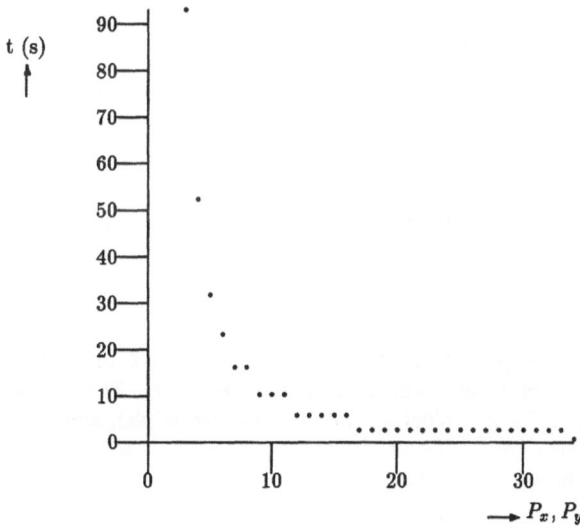

Figure 8 Estimated running time per timestep for HIRLAM (34 x 34 gridpoints, 9 levels) on a T414 network of $P_x \times P_y$ transputers.

## 5.4    Performance with T800 transputers

In the preceding section, we have seen that the calculation time dominates total execution time. Acceleration of the calculations will therefore have a direct effect on the execution time for the HIRLAM program. Most calculations in the HIRLAM program are floating point calculations. Floating point operations are relatively slow on a T414, as compared to a T800 transputer, so performance will be much higher on a network of T800 transputers than on a T414 network.

To get an indication of the running time of the HIRLAM program on a grid of T800 transputers, we have determined a factor by which some typical floating point calculations (taken from DYN) are performed faster on a T800 than on a T414.

Because our transputer system did not contain T800 transputers, and the C compiler we used is not (yet) able to produce code for a T800, we had to use another system for this test. This system is an IBM-AT with a slot-card containing two T800 transputers, which can be programmed in OCCAM-2, using the MEGATOOL programming environment.

The determined speedup for using OCCAM-2 on a T800 instead of C on a T414 was approximately 16. However, the OCCAM-2 compiler seems to produce better code than the C compiler, because a comparison between the two systems, using a version of the program with integers instead of floating point variables, showed that the OCCAM-2 version was approximately 10 % faster.

When we decrease the value of $T_{calc}$ in formula (11) by this factor, we get:

$$T^{seq} \approx xy(40 \text{ ms} + 5550 \mu s + 54 \ \mu s + 2211 \ \mu s) + (x + y) \ 3678 \ \mu s \qquad (12)$$

From this formula, it can be calculated that for a grid of T800 transputers the external communication time will account for at most 13 % (1 x 1 local grids) of the total execution time.

The estimated running time for the HIRLAM program on a grid of T800 transputers is plotted in Fig. 9.

## 5.5    Speedup Calculations for the HIRLAM Program

Instead of plotting the total execution time in Fig. 9, it is also possible to plot the speedup as a function of the total number of transputers, which is done in Fig. 10. The speedup values for all possible values of $P_x$ . $P_y$ are plotted. In the case where the same total number of transputers can be formed by different combinations of $P_x$ and $P_y$, the combination with the smallest difference between $P_x$ and $P_y$ is chosen.

This Fig. clearly shows the almost linear speedup for the HIRLAM program as the number of transputers is increased, up to one transputer for every gridpoint. The point in the upper right corner corresponds to the speedup for $P_x = P_y = 34$, so there is one transputer per gridpoint. For the points with a speedup of about 500, the maximum local grid-size is 1 x 2;  for a speedup of approximately 250, the maximum local grid-size is 2 x 2, etc.

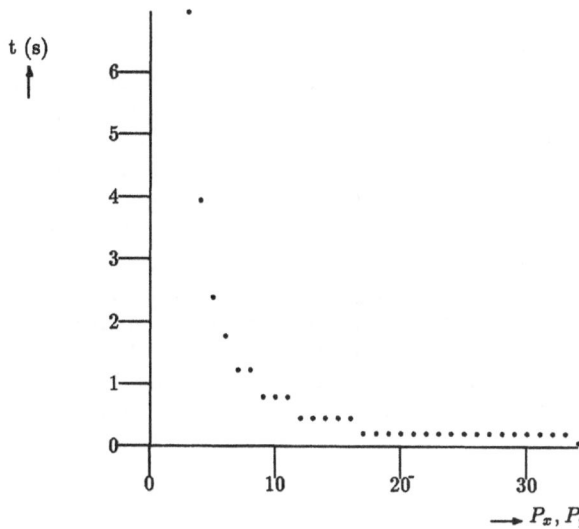

Figure 9    As Fig. 8 but for a T800 network

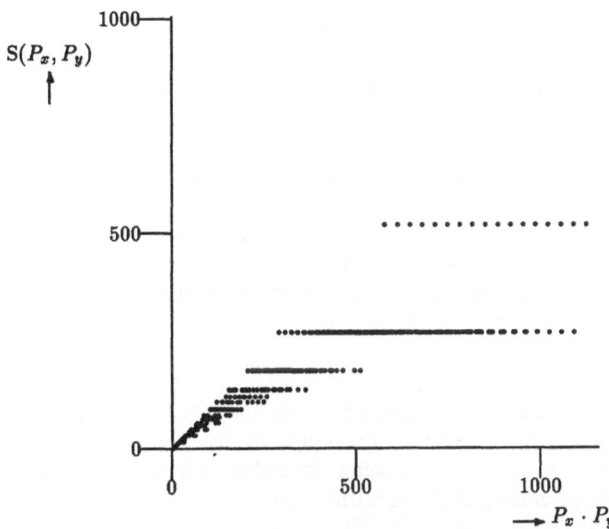

Figure 10  Estimated Speedup for HIRLAM on T800 Transputers

Each of these "lines" in the Fig. corresponds to a particular maximum local grid-size. To maximise efficiency, $P_x$ and $P_y$ should be chosen in such a way that the corresponding speedup value is the very first point of a "line". For these values of $P_x$ and $P_y$, the total number of transputers used for a computation with a particular maximum local grid-size is minimal, and the speedup is almost equal to $P_x \cdot P_y$.

## 5.6    Conclusion

Section 5.2.1 shows that the parts of the HIRLAM program where communication takes place, such as the subroutine DYN, are best implemented using the sequential model. As was argued in section 5.2.2, implementing parts of a program sequentially and other parts parallel, may cause much extra overhead. The estimated value of $Q$ for the HIRLAM program indicates that both implementation models will be approximately equally fast.

Furthermore, one subroutine of the HIRLAM program (DYN), was implemented on a transputer in order to derive an estimate for the running time of the whole program on a grid of transputers. On a network of T414 transputers, both the amount of internal and external communication turned out to be negligable with respect to the calculation time. With T800 transputers, the overhead due to external communication will also be moderate. As a result, speedup for the HIRLAM program will be almost linear in the number of transputers.

## 6    Conclusions and Future Work

### 6.1    Conclusions

The question to be answered by this report is whether it is feasible to implement a weather prediction model like HIRLAM on a network of transputers and if so, whether it can be done in such a way that performance is linear in the number of transputers involved in the calculation.

In section 4, two implementation models for grid-based computations, the parallel and the sequential model, were presented, together with a number of possible optimisations. Which model is best suited for a given application depends on the ratio of the number of variable references and the number of communications. With suitable test programs, a criterion is determined by which the fastest model can be selected for a given application. For the HIRLAM program, it was estimated that both implementation models are approximately equally fast.

From the formulas for the execution time in section 4.5 it follows that for inherently symmetric computation models like atmospheric models, the shape of the subgrid processed by one transputer should be square, in order to minimise inter-processor communication, and therefore maximise speedup. The shape of the global grid is irrelevant here.

With appropriate benchmark programs, a number of hardware/compiler dependent constants were measured, which occur in the formulas for the execution time. With these constants, it was estimated that for the HIRLAM program, running on a grid of T414 transputers, the time used for internal communication is less than 1 % of the total processing time, and the time used for external communication is maximal 1 %, decreasing when the local grid gets bigger. As a consequence, the speedup will be almost linear in the number of transputers. Also, performance will hardly decrease if the local grid is not exactly square.

Because floating point operations on a T800 are approximately 16 times faster than on a T414, the relative overhead due to external communication is somewhat larger for a grid of T800 transputers. However, in the worst case (1 x 1 local grids), it only accounts for approximately 13 % of the total execution time, so even when using T800 transputers, external communication will not be the bottleneck.

Besides the principal questions answered by this report, the feasibility study has initiated several spinoffs like the development of transputer support software, the starting of transputer related projects and extensive contacts with some companies.

## 6.2   Future Work

Several points of further research, have become clear during the feasibility study:

- Further development is needed in implementing HIRLAM completely on transputers, taking the results of this report as a starting point.   The next question to be answered concerns the impact of global communication on a transputer implementation of the HIRLAM model.

- A programming environment for transputers and, more general, parallel systems needs to be developed. The emphasis will be on the development of (specification) languages for writing software and operating software.

- More investigation has to take place to find out if and how various classes of differential equations with a numerical solution algorithm on a grid can automatically be implemented on a grid of transputers.

## Acknowledgement

The authors express their gratitude to M. Veldhorst and M. Kramer for stimulating discussions and to A. Dijkstra and E.H. Kremer for their technical assistance.

## References

HIRLAM, 1988: Forecast Model Documentation Manual, available from KNMI, de Bilt, The Netherlands.

INMOS, 1986: Transputer Reference Manual.

Richardson, L.F., 1922: Weather prediction by Numerical Process, Republished by Dover Publications, 1965.

# Benefits and Challenges from Parallel Processing: The View from NCAR

BILL BUZBEE

Director, Scientific Computing Division, National Center for Atmospheric Research, P.O. Box 3000, Boulder, CO 80307, USA

## ABSTRACT

In the 1990s, supercomputing and parallel processing are expected to be synonymous. Consequently, the National Center for Atmospheric Research (NCAR) is taking steps
to encourage scientists to experiment with and use parallel processing in both shared- and
distributed-memory systems. This exercise has already yielded significant new
capability in global ocean modeling. Probably the area of greatest challenge in parallel
processing is to formulate models and devise algorithms such that they contain a high
degree of concurrency.

This paper also summarizes some current developments and trends in supercomputing at NCAR. A UNIX-specific version of the NCAR Graphics package has just been
completed and it is expected to become the most widely used software package ever
developed at NCAR. An Internet RJE capability that exploits the relatively high bandwidth
of national networks has just been introduced to provide ease of access to supercomputers.
Within the scientific computing community there is widespread speculation that high-
performance, single-user workstations may lower the need for supercomputers. Our
measurements show that the performance of these workstations is one to two orders of
magnitude below that of supercomputers and, thus, these workstations are not capable of
supporting the large simulations for which supercomputers are typically used. On the
other hand, current trends suggest that the workstation will be the center of the
scientist's computational universe in the '90s. So we conclude with a brief discussion of
how workstations, networks, and supercomputers may combine to significantly enhance
scientific productivity.

Topics in Atmospheric and Oceanic Sciences
© Springer-Verlag Berlin Heidelberg 1990

**Parallel Processing At NCAR**

Numerical simulation of atmospheric and related phenomena has long been a major area of research at the National Center for Atmospheric Research (NCAR) in Boulder, Colorado. Because these simulations require an enormous amount of computational power, NCAR has always acquired the most powerful computer systems--supercomputers-- available. But even today's supercomputers do not have the capability to support many of the simulations needed by scientists in this community.

Because of the diminishing rate of growth in the speed of a single processor, there is increasing agreement that in the 1990s supercomputing and parallel processing will be synonymous. Two promising architectures for high-performance parallel processors are: (1) shared-memory systems in which a few (less than 100) high-performance processors share a common memory; and, (2) distributed-memory systems in which thousands, even tens of thousands, of low-cost processors with local memory cooperate via a communications network of some sort (see Figures 1 and 2). Consequently, NCAR is taking steps to encourage scientists to experiment with both types of architectures and, when appropriate, to incorporate parallel processing into their models.

**Parallel Processing on the NCAR CRAY X-MP/48**

In 1986, NCAR installed a CRAY X-MP/48 and this machine can function as a shared-memory system for parallel processing. To encourage use of the X-MP/48 as a

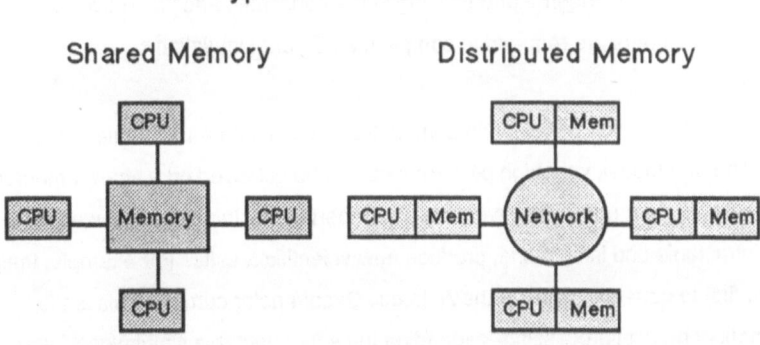

Figure 1

# High-Performance Parallel Computers

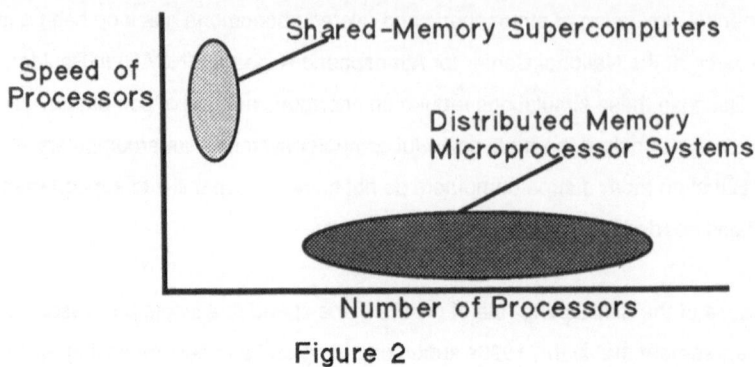

Figure 2

parallel processor, a Mono-Program (MP) job class with relatively favorable charging rates was created. To optimize system performance, MP jobs are given the entire system and certain network functions are disabled to minimize system interruption. Two wall-clock hours per day are allocated to the processing of MP jobs and about four models regularly take advantage of it. Measurements show that the X-MP/48 generally achieves its highest performance when executing MP jobs.

Albert Semtner, Naval Postgraduate School, and Robert Chervin, NCAR, have used the MP job class in their work[1] on eddy resolving, global ocean simulation. To achieve parallel processing, they took an existing model and reconstructed it into independent tasks that process sections in longitude and depth. These independent tasks are parallel processed using the microtasking capabilites of the X-MP/48 system. Peak performance of the X-MP/48 on this model is approximately 450 Mflops. Nevertheless, since a typical simulation uses 0.5 degree grid spacing in the horizontal, and 20 vertical levels, some 200 wallclock hours are required to complete a 25-year simulation.

The work of Semtner and Chervin demonstrates that by careful formulation and implementation of a model, very high performance can be acheived on a shared-memory parallel processor. As a result, they were able to undertake a simulation that was previously intractable and in so doing, produce new scientific results. For example, their model is the first to correctly simulate the Antarctic Circum-polar current. This is the primary benefit of parallel processing--expanding the set of tractable simulations. And that is why parallel processing has such an important role in our future.

**Parallel Processing on the CAPP CM-2**

In the spring of 1988, NCAR joined with the University of Colorado at Boulder and the University of Colorado at Denver to form the Center for Applied Parallel Processing (CAPP). CAPP is administered by the University of Colorado at Boulder with Professor Oliver McBryan as the Director. Whereas, research over the past decade has shown the viability of parallel processing, most applications of it have involved model problems and low-performance computers. The objective of CAPP is to solve "real world" problems using systems with at least one hundred processors and to achieve performance levels that equal or exceed those of supercomputers.

In the fall of 1988, the Defense Advanced Research Projects Agency (DARPA) granted CAPP an 8,000 processor Connection Machine, Model 2 (CM-2) manufactured by Thinking Machines Corporation. As part of its contribution to CAPP, NCAR provides space and operational support for the machine. The CM-2 is a distributed-memory architecture in which each processor has 8,000 bytes of local memory. When configured with 64,000 processors, the machine has achieved from one to five Gflops while executing a variety of applications. More than 40 reserachers from at least 10 organizations are using the CAPP machine to develop and test parallel models and algorithms.

NCAR is particularly interested in the potential of the CM-2 to support climate and ocean simulations. Our research plan is to start with a relatively simple model and determine if it can be formulated for the CM-2 such that high performance is achieved. If so, we will then progress through a sequence of increasingly sophisticated models to see if they can be so formulated. The first step was taken in the spring of 1988 when the shallow water model was implemented [2] on a 16,000 processor CM-2. When the performance is extrapolated to a 64,000 processor system, we estimate that the model will run at over 1.5 Gflops. The CRAY X-MP/48 achieves about 0.5 Gflops when all four of its processors are applied to this problem. So we will now attempt a spectral model on the CM-2.

Distributed-memory machines such as the CM-2 will probably have a relatively limited domain of application as compared to shared-memory supercomputers. But machines like the CM-2 offer both high performance and low cost. Many people in the high-performance computing community believe that by tightly coupling distributed-memory and shared-memory systems, we will reap the benefits of both.

## Some Challenges from Parallel Processing

Achieving high performance on a parallel processor requires careful attention to model formulation and implementation. The first prerequisite is concurrency in the model. That in turn requires good knowledge of parallel algorithms. When suitable parallel algorithms are not known, they will have to be developed and the associated research requires detailed knowledge of computer architecture. Once sufficient concurrency is available, the next challenge is to implement the model such that there are no significant overheads from task management, e.g., task synchronization, intertask communication, contention for memory, etc. Supercomputer users do not normally possess this expertise. Rather, we must provide a critcal mass of experts who will provide training in parallel processing and assist with algorithmic research as needed.

## Some Current Developments and Trends in Computing at NCAR

Under funding from the National Science Foundation, the NCAR Scientific Computing Division provides a state-of-the-art supercomputer facility that is used by some 1,200 atmospheric and oceanic scientists at NCAR and at universities throughout the United States. To properly serve the scientific community, supercomputers must be accompanied by an infrastructure of support including graphics capability and remote access. To prepare for the future, we must measure the performance of a variety of systems to correctly assess their potential. We must also stay abreast of new technology such as distributed computing between workstations and supercomputers for improving scientific productivity.

## A UNIX Version of NCAR Graphics

Graphics has long been a key requirement of the scientific computing community. Thus, development of the NCAR Graphics package began some 25 years ago and today it is one of the most widely used graphics packages in science and engineering. Actually, there are three versions of the package--a portable generic version, a VMS version, and a UNIX version. Because the generic version is portable, it has been implemented on almost every conceivable type of computer. Portable software is often difficult to install on a specific system because the package has to be mapped onto the unique hardware and software attributes of that system. In order to achieve portability, software engineers have to scarifice efficiency. Due to these requirements, a VMS version of the package was

NCAR Graphics

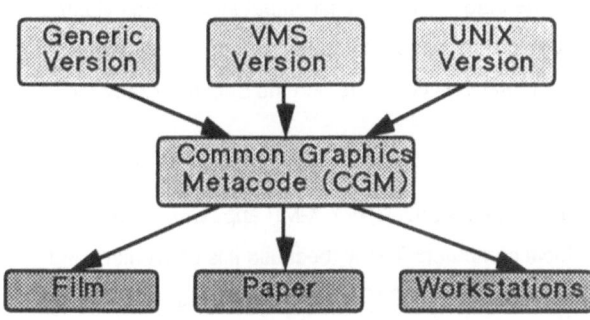

Figure 3

developed in 1987, and a UNIX version was completed in 1988. The UNIX version is designed to auto-install on many systems and, on many functions, its performance is an order of magnitude higher than that of the generic version. Given the growing usage of UNIX in science and engineering, we expect the UNIX version of NCAR Graphics to be one of the most widely used software packages ever developed at NCAR.

All the latest versions of the NCAR Graphics package use the Common Graphics Metacode (CGM) to produce device-independent representation of images. The advantage of CGM is that one computer can be used to produce the image, e.g., a CRAY X-MP, it can then be shipped via the network to another computer, e.g., the scientist's local computer, for processing into an image on a printer, on film, or on a display screen (see Figure 3).

**The NCAR Internet RJE**

The collection of TCP/IP networks in the United States, known as the Internet, provides connectivity to virtually every university in the country. The Internet also offers higher bandwidth than other comparably priced communication options. Since more than half the scientists using our supercomputers in 1988 were situated at universities located far from NCAR, SCD developed an Internet Remote Job Entry (RJE) capability. The Internet RJE uses standard TCP/IP functions in combination with a unique gateway computer at NCAR, thus no NCAR-developed software has to be installed on the remote computer. To use the Interent RJE, a scientist works on their local computer to develop a job and/or data for an NCAR supercomputer, then with a single command ships the job to the Internet RJE gateway at NCAR. Via electronic mail, the gateway acknowledges to the scientist the receipt of the job and then submits it to a CRAY X-MP job management system. Once the job has executed, the gateway collects all output into a single file and automatically ships the file back to the scientist's local computer. A second electronic mail message notifies the scientist that the job is complete and the output is ready for examination.

The Internet RJE essentially makes an NCAR CRAY X-MP appear as an attached processor to the remote scientist's local computer. Today, because it is convenient and easy to use, well over 90% of all remote usage of NCAR's supercomputers is via the Internet RJE. We routinely use this system to deliver output files of ten megabytes which may contain either text or graphics metacode. The Interent RJE provides a nice example of how national networks are changing the practice of science and engineering.

**Supercomputer Performance**

Like many other comparable organizations, NCAR finds it valuable to continually measure the performance of state-of-the-art supercomputers. However, whereas other organizations tend to use the metric of Mflops, we prefer to measure wallclock time to complete a job since this is the metric of primary interest to users. NCAR has also developed a widely used Community Climate Model (CCM) and a simplified version of it is used as a convenient benchmark. The CCM is formulated to execute from either disk, or fast, secondary memory such as the Solid State Disk on a CRAY X-MP. Figure 4 displays performance comparisons for several state-of-the-art supercomputers. (Note the impact that disk performance can have on wallclock time.)

## Performance Comparisons
### Community Climate Model (CCM)
### Five-Day Simulation

| System | I/O | CPU Sec. | CPU Ratio | Wall Clock Sec. | Wall Clock Ratio |
|--------|-----|----------|-----------|-----------------|------------------|
| CRAY Y-MP | SSD | 187 | .6 | 193 | .6 |
| CRAY X-MP | SSD | 289 | 1.0 | 320 | 1.0 |
| CRAY X-MP | Disk | 288 | 1.0 | 387 | 1.2 |
| CRAY 2 | Disk | 414 | 1.4 | 419 | 1.3 |
| NEC SX-2 | XMU | 241 | .8 | 318 | 1.0 |
| NEC SX-2 | Disk | 242 | .8 | 1069 | 3.3 |
| ETA 10E | Disk | 625 | 2.2 | 793 | 2.5 |
| IBM 3090/600E | Disk | 855 | 3.0 | n/a | n/a |

Figure 4

## Mini- and Super-mini Class Systems
### Community Climate Model [1]
### Five-day Simulation

| Computer System [2] | I/O Device | CPU Time (secs) | CPU Time Ratio w X-MP | Wall Clock (secs) | Wall Clock Ratio w X-MP |
|---|---|---|---|---|---|
| CRAY X-MP | SSD | 289 | 1.0 | 320 | 1.0 |
| CRAY X-MP | Disk | 288 | 1.0 | 387 [3] | 1.2 |
| Alliant FX/80 (4 CEs) | Disk | 3080 | 10.7 | 3220 [3] | 10.1 |
| SUN 4/280 | Disk | 10300 | 35.6 | 10920 [3] | 34.1 |
| IBM 4381/P14 | Disk | 11000 | 38.1 | 14420 [3] | 45.1 |
| VAX 8550 | Disk | 12900 | 44.6 | 16734 [3] | 52.3 |

1 Version CCM0B of the NCAR Community Climate Model.
2 Tests are on one processor, unless otherwise indicated.
3 Wall clock time from run on a quiet, but not dedicated system.

### Figure 5

Over the past year, several manufacturers have introduced powerful, single-user systems that are often referred to as "super-workstations" or "desktop supercomputers." The advertised performance of these machines is often comparable to that of super-computers, so many people are speculating about their long-term impact on supercomputers. In an effort to provide some factual information in this area, we have measured the performance of the CCM on several relatively powerful minicomputers. We chose this set of mini-computers because they have the requisite processor speed, memory capacity, and disk capacity to execute the CCM. Their capabilities are comparable to that of the next generation of super workstations. (Figure 5 displays the results.) Simply put, supercomputers surpass the performance of super workstations by one to two orders of magnitude. Undoubtedly, super workstations will be tremendously valuable in supporting visualization and other applications, but they do not yet have the overall system capability--processor speed, memory capacity, and secondary storage--to support large, long-running simulations that typify supercomputer applications. In fact, at NCAR we define supercomputers to be those systems that solve problems that cannot be solved by any other class of equipment.

### The Scientist of the '90s

There is growing agreement that the desktop computer will be the center of the scientist's computational universe during the 1990s. This will be a dramatic change because for decades the supercomputer, or mainframe system, has been the center of that universe--that is, all resources, including human time, were subservient to the requirements of the "central system."  To be more precise, in the 1990s the desktop system will consititute and define the scientist's computational environment and resources such as supercomputers, mass storage, etc., will be "subservient" via high-speed networks.

The combination of powerful desktops and supercomputers linked by high-speed networks will greatly increase computational productivity.  Desktops in the '90s will have large color screens, large-memory capacity, iconic interfaces, audio I/O, and a high-speed port. They will continue to be rich in software and low in cost relative to supercomputers. Both desktop systems and supercomputers will have some version of POSIX as the operating system. A common operating system will foster ease of use and portability of software. POSIX will also foster online documentation. "Windows" on the desktop will encompass both local and supercomputer processes. Supercomputers will have wonderfully large memories. Hopefully, these memories will eliminate the need to develop out-of-memory models and that, in turn, will greatly increase software productivity.

## Summary

In the 1990s the combination of parallel processing and large memory on super-computers will make many problems solvable that are intractable today.  And the combination of easy-to-use desktops and supercomputers via high-speed networks will substantially improve the productivity of computational scientists. In a nutshell, the '90s promise to be the most exciting decade yet in the era of supercomputing!

## References

1. Semtner, A. and R. Chervin, "A simulation of the global ocean circulation with resolved eddies," to appear in *J. Geophys. Res.*

2. Sato, R. and P. Swarztrauber, "Benchmarking the Connection Machine 2," *Proceedings of the Supercomputing Conference*, Orlando, Florida, Nov. 14-18, 1988, pp. 304-309.

# Multitasking on the ETA$^{10}$

GEORGE MOZDZYNSKI

Control Data Limited, 3 Roundwood Avenue, Stockley Park, Uxbridge, UX11 1AG, U.K.

## ABSTRACT

Numerical weather forecasting has always had a voracious appetitite for the fastest computers of the day. Today, these supercomputers typically achieve their peak performance by employing multiple processors which work together on a single program to meet time critical schedules.

This paper introduces the ETA10 Multitasking Library, the software which allows programmers to run their programs in parallel on an ETA10 computer system. Examples are presented on how this library has been used on some weather code kernels.

## 1. INTRODUCTION

The ETA10 is a multiprocessor computer system with up to eight processors, where each processor is capable of performing over 1,000 million floating point operations per second. A brief overview of the ETA10 architecture and model characteristics is contained in section 2.

For a large number of users this level of performance may be sufficient to avoid the need to multitask. The same can be said for computer installations where a workload consists of a large number of independent jobs. In this case, as long as each CPU has one or more user tasks to execute, there can be little benefit in multitasking. On the other hand, for workloads that are characterised by small numbers of dependent jobs, multitasking may be the only solution to meet time critical schedules and provide better system resource utilisation. Weather forecasting is such a workload. In either case, the availability of high performance CPUs is a significant factor in running these workloads effectively. Section 3 presents ($R_\infty$, $N_{\frac{1}{2}}$) performance measures for some programs that were recently (Nov 1989) run on an ETA10-G series CPU (7 ns cycle) at ETA Systems, St. Paul, Minnesota. As the ETA10 has a similar CPU architecture and same basic instruction set as the CYBER 205 (20 ns cycle) it is interesting to compare the performance of these systems.

Topics in Atmospheric and Oceanic Sciences
© Springer-Verlag Berlin Heidelberg 1990

On the ETA10 each CPU has it's own Local Memory, which is connected to a much larger Shared Memory and a much smaller Communications Buffer Memory. To share data between CPUs it is therefore necessary to move data between Shared Memory or Communications Buffer Memory and the Local Memory. This is accomplished today by making explicit Fortran level calls to the ETA10 Multitasking Library, which is described in section 4.

Section 5 describes some initial experience in multitasking weather codes on an ETA10 system, in particular the approaches used for grid-point and spectral models.

Finally, section 6 reviews some performance issues, in particular an assessment of the overhead associated in sharing data between processors on an ETA10 system.

## 2. ETA10 ARCHITECTURE

The ETA10 is a multiprocessor system with up to 8 processors, the models and major characteristics being summarised in Table I.

| MODEL | CYCLE TIME n s | COOLING | PEAK MFLOPS per cpu 64 Bit | PEAK MFLOPS per cpu 32 Bit | CPUs | SHARED MEMORY m words |
|---|---|---|---|---|---|---|
| ETA10-G | 7.0 | LN2 | 571 | 1142 | 2-8 | 128-256 |
| ETA10-E | 10.5 | LN2 | 380 | 760 | 1-8 | 64-256 |
| ETA10-Q | 19.0 | AIR | 210 | 420 | 1-2 | 8-64 |
| ETA10-P | 24.0 | AIR | 166 | 332 | 1-2 | 8-64 |

Table I

Summary of ETA10 Models and main characteristics

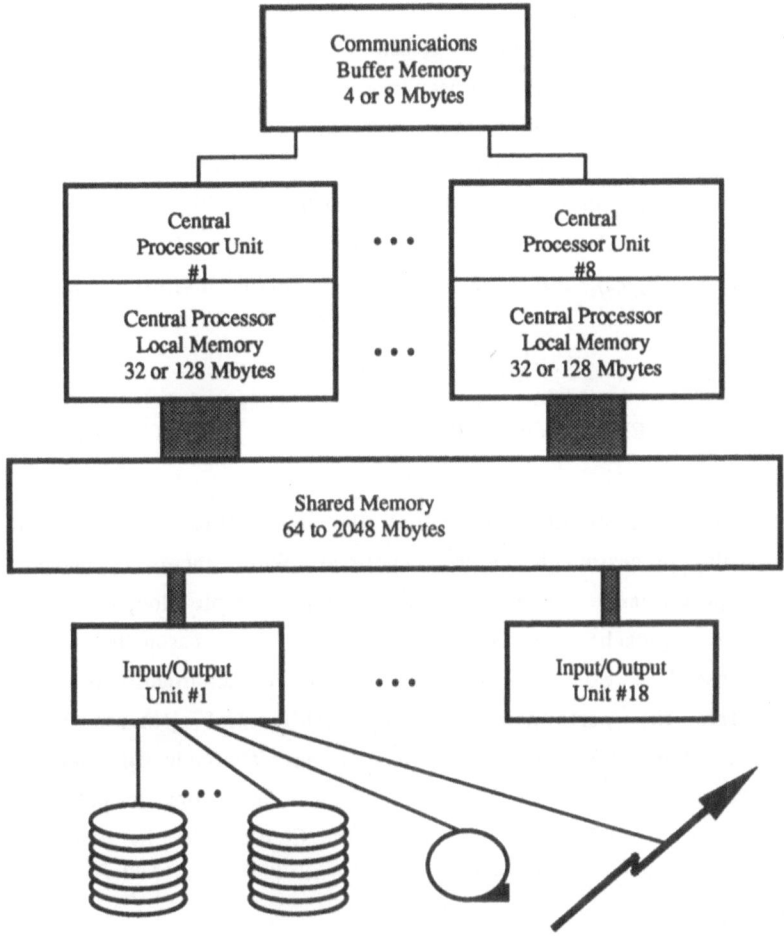

Figure 1    ETA10 Function Diagram

Each CPU has it's own Local Memory of 4 million words (32 Mbytes) or optionally 16 milion words (128 Mbytes), and is connected to a Shared Memory of up to 256 million words (2048 Mbytes), as shown in Figure 1.

Shared Memory is used for a variety of purposes, the main ones being, acting as a buffer device between Local Memories and disks, tapes, and networks, providing a very fast paging device supporting the virtual addressing mechanism in each CPU, and finally, as a means of sharing data between CPUs.

Data can be simultaneously transfered between each Local Memory and Shared Memory at a rate of one word per cycle,giving an aggregate bandwidth of up to 9 Gbytes per second. Furthermore, as Shared Memory transfers are performed by a

Shared Memory Interface and not the CPUs themselves, CPUs are free to continue computing while transfers are in progress.

Input/Output Units (1 to max 18) are connected to the Shared Memory, and provide the mechanism for interfacing disk, tapes and networks to an ETA10 system. IOUs can transfer data to and from Shared Memory at rate of up to 55 Mbytes per second, giving an aggregate IOU/SM bandwidth of up to 990 Mbytes per second.

All processor types (CPUs and IOUs) are connected to the Communications Buffer Memory (4 or 8 Mbytes), and have test and set mutual exclusion capabilities for data located in CBM.

## 3. MEASURESUREMENT OF $( R_{\infty}, N_{\frac{1}{2}})$ ON AN ETA10-G SERIES CPU

Whether programs are multitasked or not the availability of high performance vector and scalar processors is clearly beneficial. While there are many ways to measure the performance of CPUs, from timing simple loops through large applications, the approach taken here is the former. The reason for this approach is partly because less work was involved, but more importantly because it was easier to relate the performance to the architectural features being used. Furthermore, as the ETA10-G series only became available internally at ETA Systems in the past two months (Oct 1988), it was interesting to compare it with the other models of the ETA10 series and also the CYBER 205, by running programs to measure the $( R_{\infty}, N_{\frac{1}{2}} )$ performance of these systems. $R_{\infty}$ and $N_{\frac{1}{2}}$ are defined by [Hockney] as

$R_{\infty}$    The asymptotic (i.e maximum) performance in millions of floating point operations per second MFLOPS

$N_{\frac{1}{2}}$    The half performance length, is the vector length necessary to achieve half the asymptotic performance

These performance measures are calculated for some simple operations using program MULT which is listed in Appendix A, and results tabulated in Table II for single precision (64 Bit) arithmetic and in Table III for half precision (32 Bit) arithmetic.

| Operation | CPU | Cycle Time nanosecs | $R_\infty$ Mflops | $N_{\frac{1}{2}}$ Flop | t0 $\mu$secs |
|---|---|---|---|---|---|
| Dyadic A(I)=B(I)*C(I) | ETA10-G | 7.0 | 283 | 38 | .136 |
| | ETA10-E | 10.5 | 189 | 38 | .204 |
| | ETA10-P | 24.0 | 83 | 39 | .471 |
| | CYBER 205 | 20.0 | 100 | 99 | .993 |
| All vector triad A(I)=D(I)*B(I)+C(I) | ETA10-G | 7.0 | 284 | 40 | .142 |
| | ETA10-E | 10.5 | 189 | 41 | .218 |
| | ETA10-P | 24.0 | 83 | 40 | .486 |
| | CYBER 205 | 20.0 | 100 | 104 | 1.042 |
| Cyber 205 triad A(I)=S*B(I)+C(I) | ETA10-G | 7.0 | 570 | 53 | .094 |
| | ETA10-E | 10.5 | 380 | 53 | .140 |
| | ETA10-P | 24.0 | 166 | 53 | .320 |
| | CYBER 205 | 20.0 | 200 | 159 | .797 |
| Scalar code A(I)=B(I)*C(I) | ETA10-G | 7.0 | 7.0 | 3.5 | .497 |
| | ETA10-E | 10.5 | 4.6 | 3.5 | .747 |
| | ETA10-P | 24.0 | 2.0 | 3.5 | 1.705 |
| | CYBER 205 | 20.0 | 2.9 | 2.7 | .908 |

Table II

Measured values of $R_\infty$ and $N_{\frac{1}{2}}$ on ETA10 Series and Cyber 205 CPUs,

using 64 Bit arithmetic (Single precision)

As the SECOND() function on the CYBER 205 and the ETA10 are only accurate to a microsecond, the operation being timed was repeated 10,000 times in DO loop 111. Descriptor notation was used in the program over standard Fortran as this resulted in slightly better object code. The decision to use descriptor syntax seemed reasonable as this provided a more accurate measure for t0 (the time for vector startup) and thus $N_{\frac{1}{2}}$. The scalar code version for the dyadic operation reverted to standard Fortran with full scalar optimisation selected, i.e. OPT=DPRS. The 111 DO loop overhead was found to be insignificant due to overlap between vector and scalar (branch) operations and for this reason was not subtracted from the timing loop. Results from program MULT were plotted to determine if there were any non-linear effects, and none were evident. A least squares method was then used to provide a fit to the linear equation (1).

$$t = a * n + t0 \qquad\qquad (1)$$

where t and n are the CPU time and vector length respectively as recorded by program MULT, and a and t0 the unknowns. $R_\infty$ and $N_\frac{1}{2}$ are then determined from (2) and (3).

$$R_\infty = \frac{1}{a} \qquad\qquad (2)$$

$$N_\frac{1}{2} = t0 * R_\infty \qquad\qquad (3)$$

From the results in Table II it is clear that $R_\infty$ for both ETA10 and CYBER 205 vector operations exhibit the same performance relative to the clock cycle. This is to be expected as the performance of simple ( *, +, - ) vector operations for both machines is determined simply by the number of vector pipelines (both have 2) and the cycle time. That is, each of the two vector pipelines can produce 1 result per clock using 64 Bit arithmetic or 2 results per clock using 32 Bit arithmetic for 'unlinked' operations. For 'linked' operations such as the 'Cyber 205 Triad' (one example of a linked triad supported on the ETA10 and CYBER 205) the performance is doubled, giving 2 and 4 results per clock for 64 and 32 Bit arithmetic respectively.

In terms of $N_\frac{1}{2}$ and t0 the CYBER 205 and ETA10 are not the same. The CYBER 205 is a machine that required relatively long vectors to achieve good performance. For this reason performance programming on the CYBER 205 was the art of finding long vectors, which typically meant code restructuring or even a rewrite. For weather codes, finding long vectors typically meant performing vector operations across two or in some cases three dimensions of grid point space.

The vector startup on the ETA10 has clearly been reduced over the CYBER 205 without sacrificing peak performance. This improvement has been achieved by employing two new architectural features on the ETA10 known as *vector shortstop* and *back to back vectors*. With vector shortstop a small buffer is contained in the vector pipelines which is used to pre-fetch or to save the first few elements of a vector being generated, to reduce the startup of subsequent vector operations. Back to back vector operations are two or more vector hardware instructions that have no intervening scalar instructions. As DO loop 111 in program MULT has a scalar branch for each iteration, no benefit is being made of back to back vectors. The timing specification for ETA10 vector instructions are not surprisingly more complicated than the 205's, and include other variables beside vector shortstop and back to back. This was evident by simple inspection of the t/n plots in Figure 2. Whereas the CYBER 205 had no visible spread, the ETA10 plots had a significant spread, for example, in the case of the dyadic vector operation for 64 Bit arithmetic, the $N_{\frac{1}{2}}$ varied from 27 to 70

FLOPS, as compared to the least squares fit of 38 FLOPS.

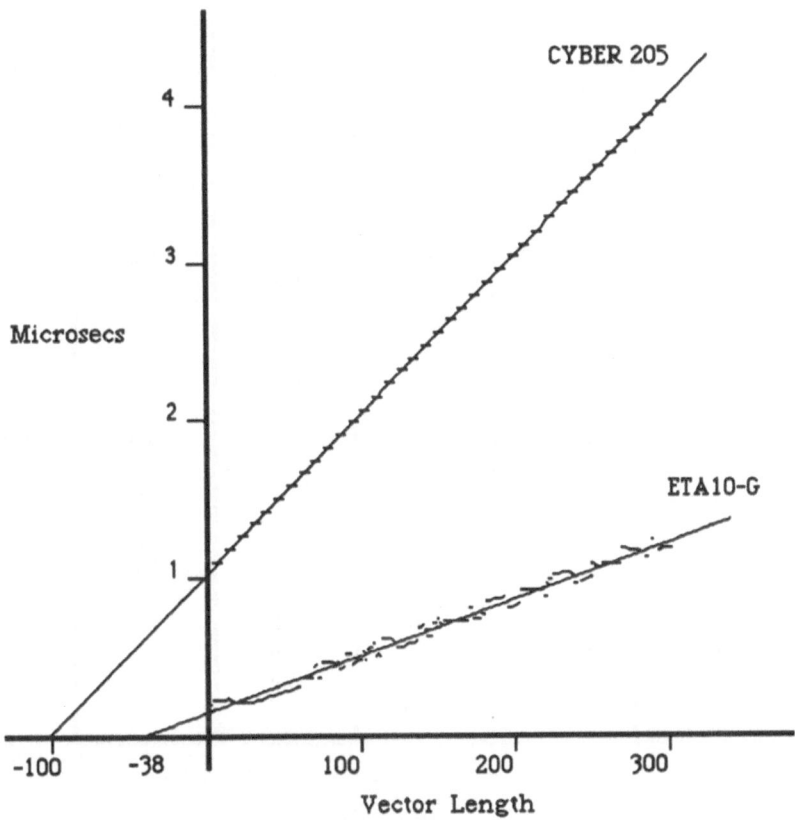

Figure 2

Vector Dyadic timing for ETA10-G (7 ns cycle) and CYBER 205 (20ns clock)

For Scalar operations, the ETA10 is about 20 percent slower than the CYBER 205 relative to the clock cycle. This is mainly due to the ETA10 requiring more cycles for load/store operations, as with other scalar operations.

Table III confirms that the ETA10 performs twice as fast on 32 Bit arithmetic over 64 Bit arithmetic, while scalar operations proceed at the same rate.

| Operation | CPU | Cycle Time nanosecs | $R_\infty$ Mflops | $\frac{N\frac{1}{2}}{Flop}$ | t0 $\mu$secs |
|---|---|---|---|---|---|
| Dyadic A(I)=B(I)*C(I) | ETA10-G | 7.0 | 563 | 82 | .145 |
| | ETA10-E | 10.5 | 375 | 82 | .219 |
| | ETA10-P | 24.0 | 164 | 81 | .493 |
| | CYBER 205 | 20.0 | 200 | 200 | 1.001 |
| All vector triad A(I)=D(I)*B(I)+C(I) | ETA10-G | 7.0 | 563 | 84 | .149 |
| | ETA10-E | 10.5 | 375 | 85 | .226 |
| | ETA10-P | 24.0 | 164 | 84 | .509 |
| | CYBER 205 | 20.0 | 200 | 201 | 1.003 |
| Cyber 205 triad A(I)=S*B(I)+C(I) | ETA10-G | 7.0 | 1130 | 108 | .096 |
| | ETA10-E | 10.5 | 752 | 107 | .142 |
| | ETA10-P | 24.0 | 329 | 107 | .324 |
| | CYBER 205 | 20.0 | 400 | 320 | .799 |
| Scalar code A(I)=B(I)*C(I) | ETA10-G | 7.0 | 7.0 | 3.2 | .462 |
| | ETA10-E | 10.5 | 4.6 | 3.2 | .694 |
| | ETA10-P | 24.0 | 2.0 | 3.2 | 1.585 |
| | CYBER 205 | 20.0 | 2.9 | 2.8 | .956 |

Table III

Measured values of $R_\infty$ and $N\frac{1}{2}$ on ETA10 Series and Cyber 205 CPUs,

using 32 Bit arithmetic (Half precision)

# 4. ETA10 MULTITASKING LIBRARY

The ETA10 Multitasking Library provides two complementary user interfaces for multitasking, known as *Fortran directives* and *Library Calls*.

The main differences between the two interfaces are that Library Calls use a subroutine call interface and are available today, while Fortran directives use a Fortran comment interface and are due to be available in the next release of the EOS operating system. Other differences are summarised in table IV.

|  | Fortran directives | Library Calls |
|---|---|---|
| Syntax | C#MTL dir ( args... ) | CALL MTxxxx ( args... ) |
| Advantages | Easier to use Fortran Comments | Language independent i.e. can use with Fortran, Pascal,C |
| Disadvantages | Fortran only | Low level |
| Availability | EOS 1.3 (July 1989) | Now |

Table IV Differences between Fortran directives and Library calls

These two user interfaces are essentially a high-level and low-level method of multitasking on the ETA10. The approaches used in both are identical, in fact, the underlying software that support the directives end up by issuing multitasking library calls.

The multitasking library allows users to multitask their programs by providing library calls or directives that operate on library objects. There are four types of library object, namely, *tasks, counters, datalinks* and *data buffers*, the system locations of these objects being shown in figure 3.

A *task* is a set of subroutines in a program that can execute in parallel with other tasks. Tasks execute in the Local Memory associated with each CPU.

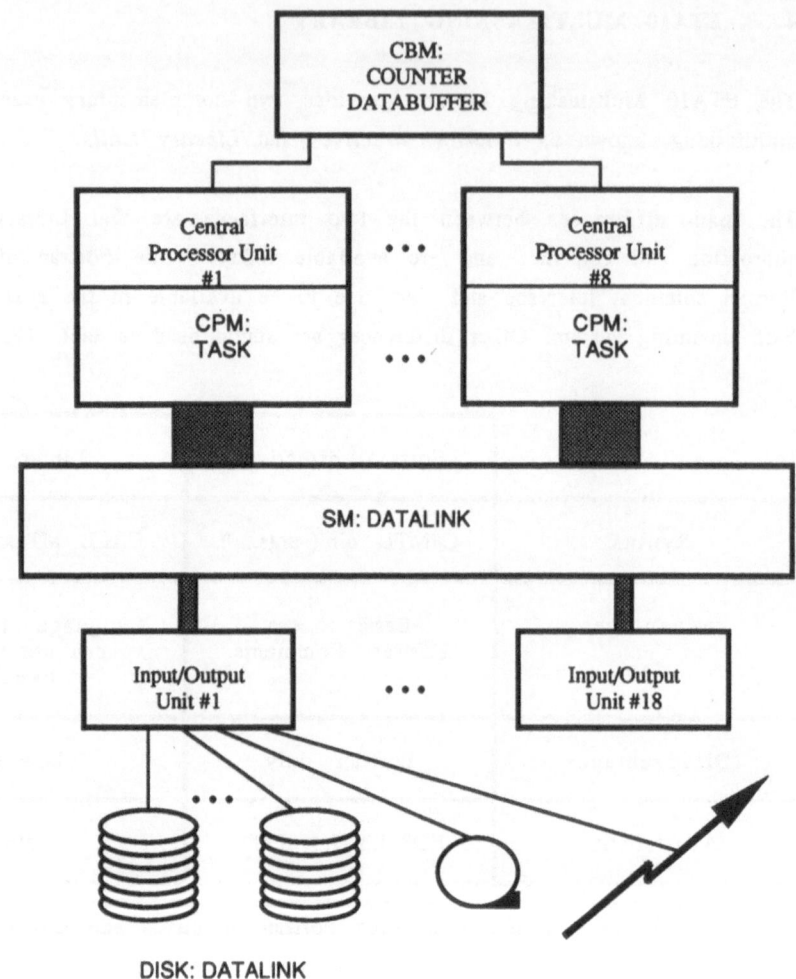

Figure 3

Location of ETA10 Multitasking Library Objects
( TASKs, COUNTERs, DATALINKs and DATABUFFERs )

A *counter* is a mechanism for synchronising task activities. Counters are located in the Communications Buffer Memory, and are updated by use of test and set mutual exclusion CPU hardware instructions. A base/limit hardware protection scheme is used to guarantee the separation between tasks belonging to different jobs.

A *datalink* is a file based mechanism for sharing data in Shared Memory. Datalinks are simply files that are located in Shared Memory that contain no buffer space in any CPU Local Memory. Standard Fortran files, on the other hand, have a small amount of buffer space in Local Memory and for that reason cannot be used for data sharing. Note that it is possible to create a file by using BUFFER OUT and then use it as a datalink in a multitasked program. As datalinks are files, it is not necessary to preallocate space as they can be extended by simply 'writing' to them, exactly like Fortran files. In terms of resource management, datalinks can be migrated to disk by the operating system depending on space contention in Shared Memory and priority of the associated tasks.

And finally a *databuffer* is a mechanism for sharing small amounts of data. Databuffers are small blocks of data contained in the Communications Buffer Memory that can be accessed directly from the *application domain* via the multitasking library. A domain is a hardware mechanism that determines the address space and priveleges available to a process.

For each of the above objects types, there are library calls to create an object, perform operations on it, and finally calls to terminate it. A summary of library calls is contained in Table V. The parameter options for these calls are excluded for brevity, however, the test programs in the appendices give a good feel for the level of detail required.

|  | TASK CALLS | COUNTER CALLS | DATALINK CALLS | DATABUFFER CALLS |
|---|---|---|---|---|
| DEFINE (CREATE) OBJECT | MTDEF | CTDEF | DLDEF | DBDEF |
| TERMINATE (DELETE) OBJECT | MTTERM | CTTERM | DLTERM | DBTERM |
| OPERATE ON OBJECT | MTRUN MTWAIT | CTPOST CTWAIT CTBARR | DLWRIT DLREAD DLSTAT | DBPUT DBGET |

Table V  Summary of Multitasking Library Calls

## 5. WEATHER CODES

In this section 1 would like to review how the ETA10 Multitasking Library is being applied to the multitasking of some weather codes. An issue of particular interest on the ETA10 is how the computational data is partitioned and managed across the CPU Local Memories and the Shared Memory. I will do this by describing some methods used on both a Grid Point and a Spectral weather code, although in practice these would not necessarily be mutually exclusive.

### 5.1 Grid Point Codes (Semi-Implicit)

From a simple point of view, the multitasking of a semi-implicit grid point code is the problem of partitioning the grid among the available processors and the problem of passing edge data between tasks at each timestep. If an ETA10 Local Memory is large enough to contain it's part of the grid, a *static* aproach (Figure 4) can be used, whereby the bulk of the grid in each CPU remains in the same place and only data at the edges need be moved to and from Shared Memory by use of datalink library calls. The edge data is managed in a separate small datalink for efficiency reasons. The time step loop for a static memory management approach would have something like the following structure :-

```
DO  STEP = 1, NSTEPS
    read neighbour task's edge data from Shared Memory      (DLREAD)
    compute
    write this task's edge data to Shared Memory            (DLWRIT)
    barrier                                                 (CTBARR)
ENDDO
```

The multitasking overheads for this approach are effectively a few milleseconds per timestep, however, some extra computation would be required to handle conditions at the edges, to take into account the *interaction range*. As the ETA10 now supports 16MWord Local Memories, grids of up to 120 MWords can be used with the static approach. For problem sizes that are larger than the sum of the Local Memory space, the same code and approach could still be used, however, it is unlikely to be as efficient as a dynamic memory management approach (Figure 5) due to paging system costs. The dynamic scheme works by partitioning the grid into smaller parts (segments), which are then pipelined through the processors.

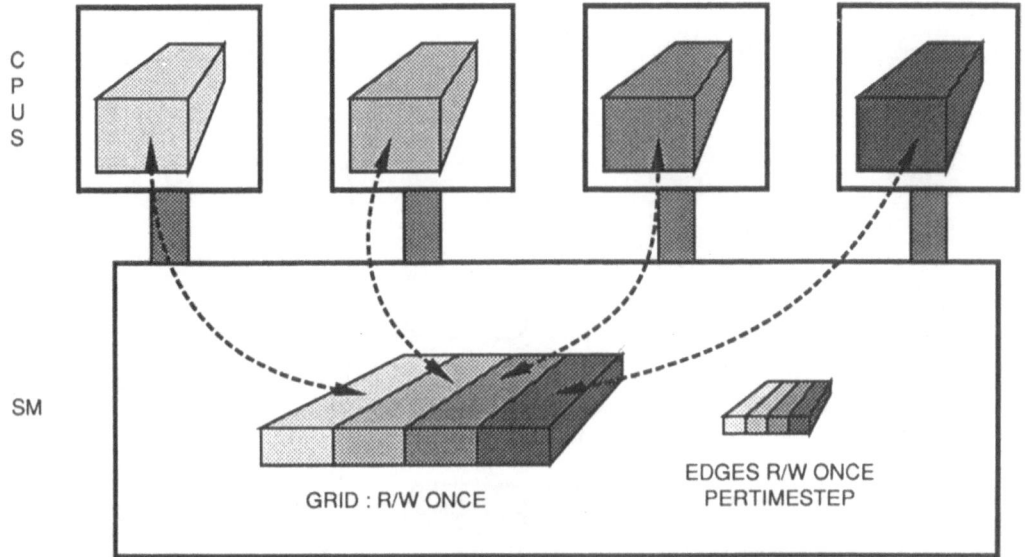

FIGURE 4  STATIC MEMORY MANAGEMENT

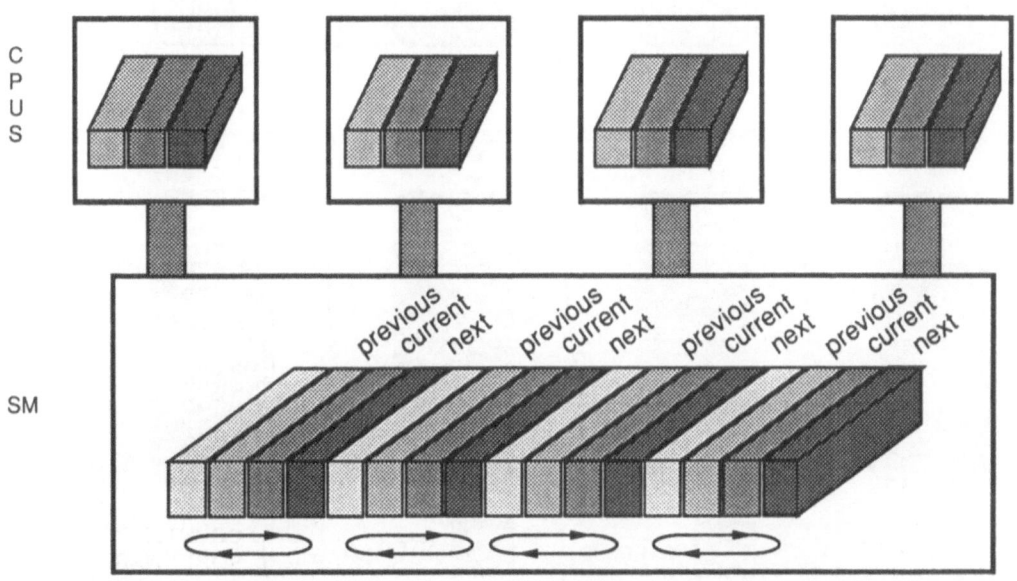

FIGURE 5   DYNAMIC MEMORY MANAGEMENT
SEGMENTS PIPELINED THROUGH CPU LOCAL MEMORIES

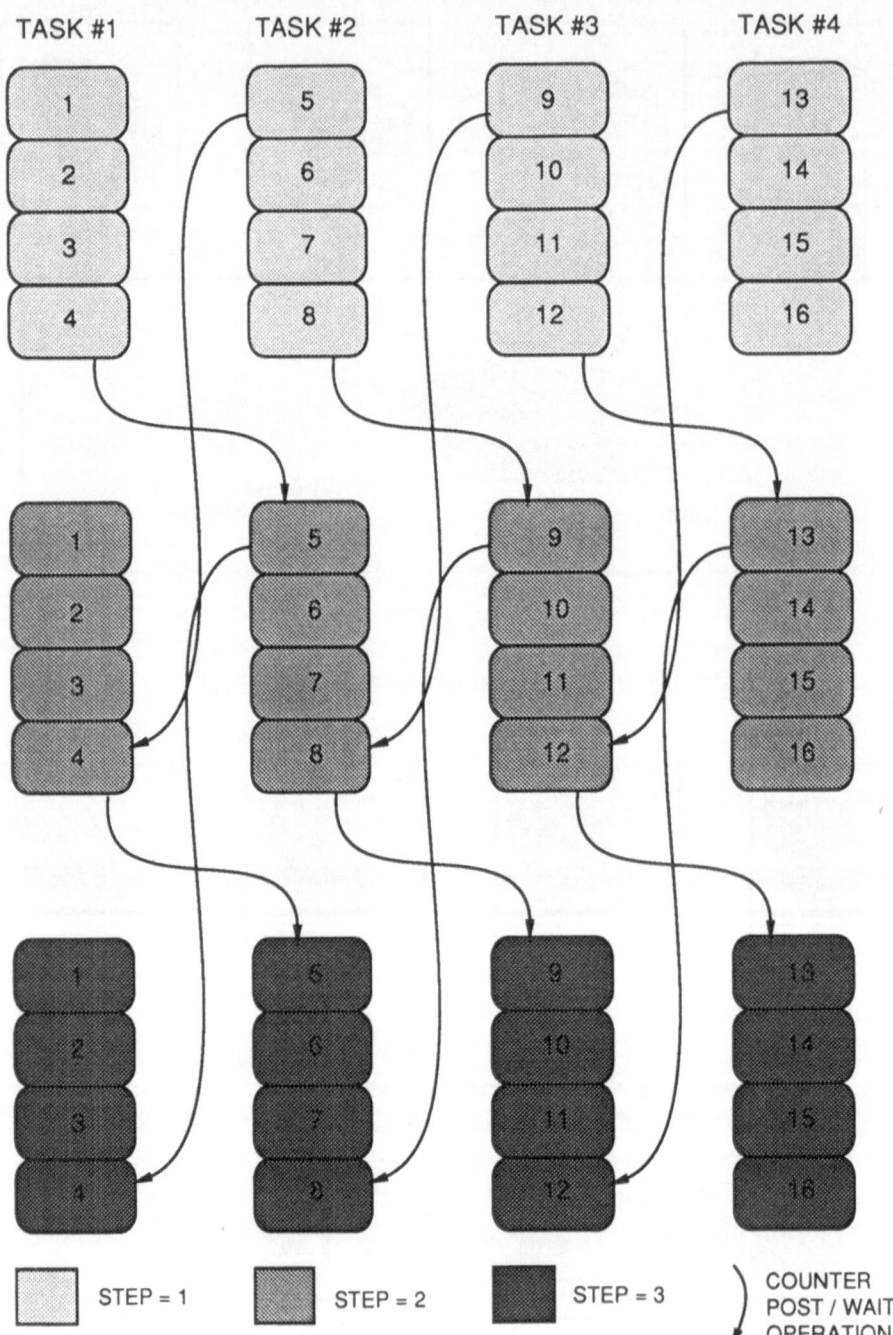

TASK #1    TASK #2    TASK #3    TASK #4

| STEP = 1 | STEP = 2 | STEP = 3 | COUNTER POST / WAIT OPERATION |

FIGURE 6  Dynamic Memory Management synchronisation
example for 16 segments, 4 tasks/CPUs, and 3 timesteps

In the scheme we used, each processor was given a fixed number of segments to process, although a worklist approach providing better load leveling could also be used. By using asynchronous datalink transfers for the input of the next segment and the output of the previous segment, datalink I/O can be totally overlapped by computation of the current segment. Further, no edges datalink is required as it's purpose is accomplished by the segment transfers to and from Shared Memory. The time step loop for a dynamic memory management scheme would have the following structure (synchronisation excluded) :-

```
DO  STEP = 1, NSTEPS
        DO  SEGMENT = 1, NSEGS
        wait  for  current  segment  read  to  complete        (DLSTAT)
        initiate  read  of  next  segment                      (DLREAD)
        compute  current  segment
        wait  for  previous  segment  write  to  complete      (DLSTAT)
        initiate  write  of  current  segment                  (DLWRIT)
        ENDDO
    ENDDO
```

Synchronisation in the dynamic scheme is only performed on the first and last segments, and is very efficient as in the steady state (no other workload in system) all counter posts are performed ahead of their respective waits, as can be seen in Figure 6.

## 5.2    SPECTRAL CODES

The partitioning scheme used in grid point models work on the principle that the calculations performed in a single timestep are localised, that is, they are dependent on the state of relatively near neighbours, and as a result readily map to local/shared or even networked memory architectures. Unfortunately, this near neighbour relationship is no longer true for spectral models, where in a single timestep a point in the grid is affected by all other points in the grid to a greater or lesser extent. Does this present a major problem for all architectures other than those with a single common memory (e.g. Cray X-MP). Not so. [SNELLING] describes a scheme whereby a copy of the spectral data is kept in each processor's Local Memory, and the complexity of the multitasking strategy is contained in the method used to perform the associative reduction of this data at each timestep. (Figure 7)

This associative reduction is nothing more than the following code :-

```
IF ( STRM .EQ. 1 ) THEN
        CALL DLWRIT ( IDSPEC, SPEC, 1, SPECLEN, TOKEN, 1, STATUS )
        CALL CTPOST ( IDSEQ( STRM+1 ), 1, 0, STATUS )
ELSE
        CALL CTWAIT ( IDSEQ( STRM ), 1, 0, STATUS )
        CALL DLREAD ( IDSPEC, 1, SPECBUF, SPECLEN, TOKEN, 1, STATUS )
        DO  JBUF = 1, SPECLEN
                SPECBUF(JBUF) = SPECBUF(JBUF) + SPEC(JBUF)
        ENDDO
        CALL DLWRIT ( IDSPEC, SPECBUF, 1, SPECLEN, TOKEN, 1, STATUS )
        IF ( STRM .LT. NSTRMS ) THEN
                CALL CTPOST ( IDSEQ( STRM+1 ), 1, 0, STATUS )
        ENDIF
ENDIF
```

A parallel method (Figure 7) of performing the same associative reduction is then a simple extention of the above code :-

```
IF ( STRM .EQ. 1 ) THEN
        DO  ISEG = 1, NSEG
        ISPEC = (ISEG-1)*LENSEG+1
        LSPEC = LENSEG
        IF ( ISEG .EQ. NSEG ) THEN
                LSPEC = LENSPEC-(LENSEG*(NSEG-1))
        ENDIF
        CALL DLWRIT ( IDSPEC, SPEC(ISPEC), ISPEC, LSPEC, TOKEN, 1, STATUS )
        CALL CTPOST ( IDSEQ( STRM+1, ISEG ), 1, 0, STATUS )
        ENDDO
ELSE
        DO  ISEG = 1, NSEG
        ISPEC = (ISEG-1)*LENSEG+1
        LSPEC = LENSEG
        IF ( ISEG .EQ. NSEG ) THEN
                LSPEC = LENSPEC-(LENSEG*(NSEG-1))
        ENDIF
        CALL CTWAIT ( IDSEQ( STRM, ISEG ), 1, 0, STATUS )
        CALL DLREAD ( IDSPEC, ISPEC, SPECBUF, LSPEC, TOKEN, 1, STATUS )
        DO  JBUF = 1, LSPEC
        SPECBUF(JBUF) = SPECBUF(JBUF) + SPEC(ISPEC-1+JBUF)
        ENDDO
        CALL DLWRIT ( IDSPEC, SPECBUF, ISPEC, LSPEC, TOKEN, 1, STATUS )
        IF ( STRM .LT. NSTRMS ) THEN
                CALL CTPOST ( IDSEQ( STRM+1, ISEG ), 1, 0, STATUS )
        ENDIF
        ENDDO
ENDIF
```

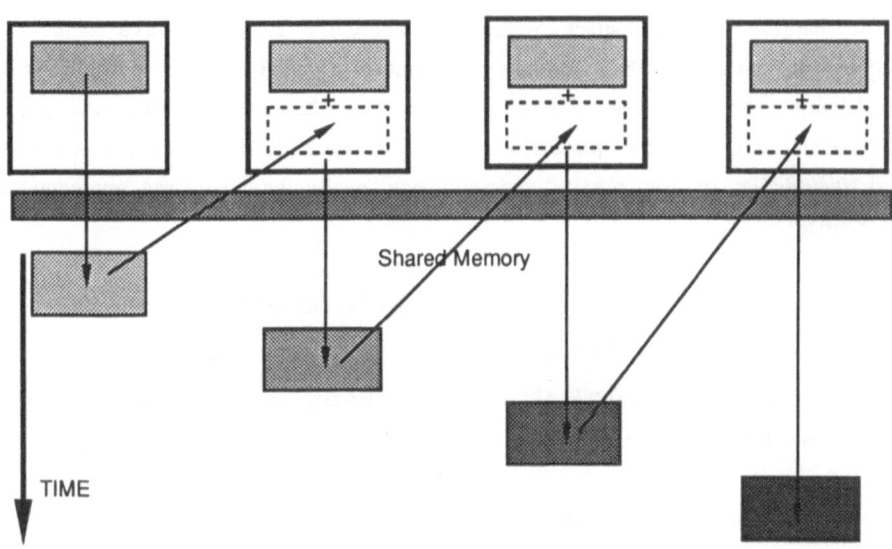

**Serial Associative Reduction**

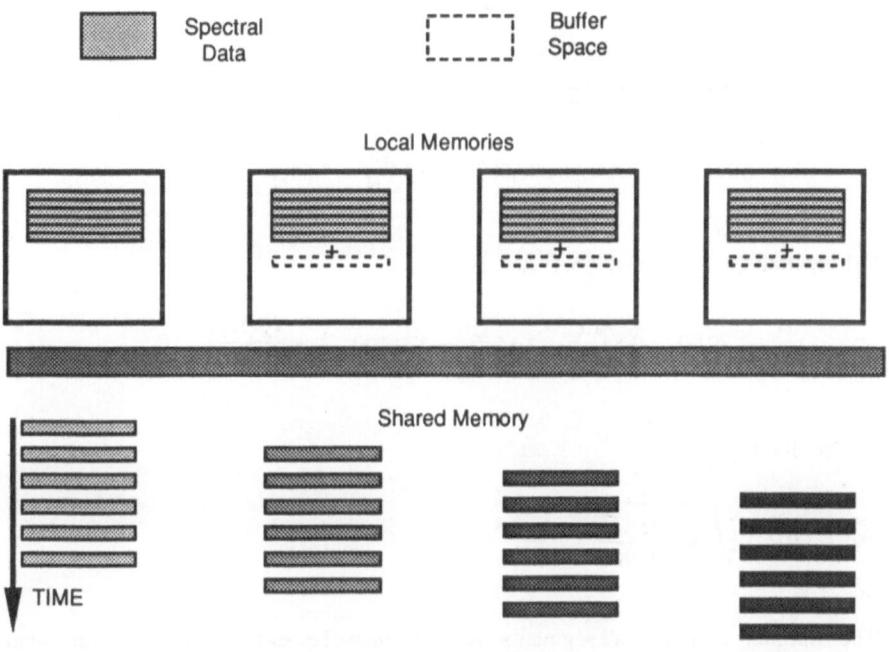

**Parallel Associative Reduction**

**Figure 7   Serial and Parallel Methods for Associative
Reduction of Spectral Data**

The parallel spectral space reduction is clearly more effective than the serial approach, however, this prompts the question what is the optimal number of segments and segment length. On the one hand, increasing the number of segments improves the parallelism, on the other it increases the overhead. The characteristics of the parallel aproach can easily be expressed mathematically as follows.

Let L be the length of the spectral data and N the number of segments. The segment length is then $\frac{L}{N}$.

The time t to wait for (CTWAIT), read (DLREAD), update (i.e. add), write (DLWRIT) and post (CTPOST) a single segment based on today's overheads is then

$$t = 297,000 + 3.54 \frac{L}{N} \text{ cycles} \tag{4}$$

The time T to perform the whole associative reduction for C CPUs is then

$$T = t (N + C - 1) \tag{5}$$

Substituting from (4) we get

$$T = 297,000N + 3.54L + 297,000C + 3.54\frac{LC}{N} - 3.54\frac{L}{N} - 297,000 \tag{6}$$

Differentiating T with respect to N

$$\frac{dt}{dn} = 279,000 - 3.54\frac{LC}{N^2} + 3.54\frac{L}{N^2} = 279,000 - 3.54\frac{L(C-1)}{N^2} \tag{7}$$

In the limit $\frac{dt}{dn} --> 0$

$$N' = \sqrt{3.54\frac{L(C-1)}{297000}} \tag{8}$$

The optimal number of segments N' can then be easily tabulated for some known problem sizes and numbers of CPUs (Table VI). The above logic was inserted into a multitasking model simulator for an ECMWF T106 sized problem that only contained the required multitasking calls in a 2 Scan [see Dent] strategy.

| Spectral Data Length L 64 Bit words | Number of CPUs C | | | | | | |
|---|---|---|---|---|---|---|---|
| | 2 | 3 | 4 | 5 | 6 | 7 | 8 |
| 900,000 (T106) | 3 | 5 | 6 | 7 | 8 | 8 | 9 |
| 3,500,000 (T213) | 7 | 9 | 12 | 13 | 15 | 16 | 18 |

Table VI    Optimal number of segments N' for parallel associative reduction, tabulated for known spectral data lengths and numbers of CPUs.

In this strategy the whole spectral array is read twice and written once in addition to the parallel associative reduction. The multitasking overheads could then be easily determined by running for a large number of timesteps (1000). The overhead was found to be 131 seconds at 2 CPUs to 185 seconds at 4 CPUs. It is evident from this that the multitasking overheads for this class of problem are acceptable especially when one considers that for the rest of the time the processors cannot interfere with one another, one of the benefits of having independent local memories. It should also be noted that the effective data transfer rate between Local Memory and Shared Memory will improve with the availability of 255K (64 bit words) large page software support. This will further reduce the multitasking overhead, as today, only 65K large page support is provided by the operating system.

# 6. MEASUREMENT OF ( $R_\infty$ , $S_\frac{1}{2}$ ) ON A MULTIPROCESSOR ETA10-E

The availability of Local Memory in a multiprocessing environment has its advantages, when compared with the 'shared' memory model of multitasking. Firstly, if data is to be shared this can only be done by explicit use of datalink or databuffer library calls. While this may be considered a lot of effort from the programming point of view, it does minimise the possibility of inadvertent data sharing and the associated problem of reproducability of results. Further protection of shared data can also be achieved by use of the access permission ('R', 'W', 'RW') attributes of datalinks, and by the use of separate datalinks for key variables.

Another more obvious advantage of independent local memories is the absence of bank conflicts which occur in a common memory architecture when there is insufficient memory bandwidth to support all processors hitting memory hard, such as for vector load/store operations. This overhead can be as much as 10% of the total time and in some cases greater.

However, the above advantages must be weighed against the cost required to share data in a local/shared memory architecture, and also the synchronisation cost. As there are a number of ways to both synchronise and share data on an ETA10, it was decided to measure each of these separately, although in real programs, synchronisation would normally be used together with data sharing. In fact, on an ETA10 1 can see no reason to synchronise other than to share data.

$S_\frac{1}{2}$ is defined by [Hockney and Snelling] as a measure of synchronisation overhead in terms of how many floating-point operations could have been performed in the time of the overhead. Table VII summarises the ($R_\infty$ , $S_\frac{1}{2}$ ) performance of a two processor ETA10-E system using the 3 methods of synchronisation as measured by programs MULT1, MULT2 and MULT3 contained in Appendix A. Two observations can be made on the contents of Table VII. Firstly, that the $R_\infty$ performance of 2 ETA10 CPUs is exactly 2 times the single $R_\infty$ performance, as you would expect with independent local memories. The second observation is that the synchronisation overhead today on an ETA10 is very high, making the ETA10 more suitable to problems where the multitasking granularity can be measured in tens of milleseconds or larger.

| | | $R_\infty$ MFLOPS | $S_{\frac{1}{2}}$ FLOP | t0 $\mu$secs |
|---|---|---|---|---|
| TASKS | Standard | 379 | 662,000 | 1,745 |
| | Busy | 380 | 456,000 | 1,198 |
| COUNTERS | Standard | 379 | 290,000 | 763 |
| | Busy | 377 | 172,000 | 456 |
| BARRIERS | Standard | 380 | 273,000 | 720 |
| | Busy | 378 | 60,000 | 159 |

Table VII  Measured values of $R_\infty$ and $S_{\frac{1}{2}}$ when dyadic vector operations are split between two processors of an ETA10-E using 64 bit arithmetic. The cost of synchronisation is tabulated using three methods, namely *tasks* (MTRUN, MTWAIT), *counters* (CTPOST, CTWAIT), and *barriers* (CTBARR). For each of these synchronisation methods the option to perform *standard* (OS blocking) or *busy* (spin) waits is measured separately.

It should be noted that the overheads in Table VII are for an initial release of the ETA10 Multitasking Library and EOS operating system, and are expected to reduce with some simple optimisation.

Table VIII summarises datalink and databuffer performance as measured by programs MULTDL and MULTDB contained in Appendix A. While datalinks achieve exceptionally good transfer rates to/from Shared Memory, the startup time is quite large. This startup time is understandable when one considers that a datalink is a file system based object, requiring a long path through the operating system before the first word is transferred. The exact opposite is true for databuffers, which have a relatively slow transfer rate but good startup. Given this situation, it is a simple matter of selecting either datalinks or databuffers depending on the data transfer length and overall data object length. The selection of a datalink or databuffer for a data object is done automatically by the Fortran directives implementation of the multitasking library, but is not part of the library calls interface.

| | | Transfer Rate (MBytes/Sec) | Startup Time ($\mu$secs) |
|---|---|---|---|
| DATALINKS | DLWRIT | 502 | 1,344 |
| | DLREAD | 502 | 1,323 |
| DATABUFFERS | DBGET | 14.7 | 107 |
| | DBGET | 7.8 | 93 |

Table VIII   Performance of Datalink and Databuffer transfers on an ETA10-E CPU

## 7.   SUMMARY

The ETA10 is a multiprocessor system where $R_\infty$ performance for each CPU as been measured in excess of 1 GFLOPS.

Lessons have been learnt from the CYBER 205 predecessor, specifically, a major improvement in short vector performance.

Multitasking on the ETA10 is relatively new, and while a multitasking library subroutine call interface is provided today, a more user friendly directive interface is expected in the near future.

The cost of synchronisation and data sharing is very high mainly due to an initial software implementation. Nevertheless this indicates that the ETA10 architecture is better suited to problems where the multitasking granularity can be measured in tens of milleseconds.

And finally, our initial experience in multitasking weather codes suggests that the ETA10 is well suited to these class of problems, both in terms of computational performance and memory hierarchy.

# REFERENCES

Dent, D., 1986, The ECMWF Model: Past, Present and Future, Multiprocessing in Meteorological Models, Springer-Verlag

ETA10 Computer System, Multitasking Library User's Guide & Reference, PUB-1120

Hockney, R.W., 1986 : ( $R_\infty$ , $N_{\frac{1}{2}}$ , $S_{\frac{1}{2}}$ ) Measurements on the 2 CPU Cray X-MP, Multiprocessing in Meteorological Models, Springer-Verlag

Snelling, D.F., 1987 : A High resolution Parallel Legendre Transform Algorithm, E.C.M.W.F, submitted to the International Conference on Supercomputing

# APPENDIX A

Program used to measure $( R_\infty , N\frac{1}{2} )$ on an ETA10-G series CPU

```
        PROGRAM MULT
        PARAMETER ( NMAX = 2000 )
        PARAMETER ( NOPS = 1 )
        PARAMETER ( NREPEAT = 10000 )
        DIMENSION A( NMAX ), B( NMAX ), C( NMAX ), D( NMAX)
        DESCRIPTOR AD, BD, CD, DD
        DATA B/ NMAX * 1.0 /, C/ NMAX*1.0 /, D/ NMAX * 1.0 /, S/ 1.0/

        DO 20 N = 2, NMAX, 2
                ASSIGN AD, A( 1;N )
                ASSIGN BD, B( 1;N )
                ASSIGN CD, C( 1;N )
                ASSIGN DD, D( 1;N )
                T1 = SECOND ( )

                DO 111 I = 1, NREPEAT
                        AD = BD * CD
C                       AD = DD * BD + CD
C                       AD = S * BD + CD
111             CONTINUE

                T2 = SECOND ()
                T = ( T2 - T1 ) / NOPS
                PRINT *, N, T/NREPEAT
20      CONTINUE

        STOP
        END
```

Program used to measure $( R_\infty , S_{\frac{1}{2}} )$ when a job is split between two CPUs of an ETA10-E series system, using the task (MTRUN, MTWAIT) method of synchronisation.

```
PROGRAM MULT1
PARAMETER ( NMAX = 10000 )
COMMON /LOCAL/ A( NMAX ), B( NMAX ), C( NMAX)
DATA B/ NMAX*1.0 /, C/ NMAX*1.0 /
CALL MTDEF ( IDT, 'MTASK', 2, 1, 0, ISTAT )
DO 20 N = 2, NMAX, 100
        T1 = RTC ()
        NHALF = N / 2
        NH1 = NHALF + 1
        CALL MTRUN ( IDT, 'DOALL', NH1, N, 0, ISTAT )
        CALL DOALL ( 1, NHALF )
        CALL MTWAIT ( IDT, 0, ISTAT )
        T2 = RTC()
        T = T2 - T1
        PRINT *, N, T
20   CONTINUE
     CALL MTTERM ( IDT, 0, ISTAT )
     STOP
     END


     SUBROUTINE DOALL ( N1, N2 )
     PARAMETER ( NMAX = 100000 )
     COMMON /LOCAL/ A( NMAX ), B( NMAX ), C( NMAX)
     DO 10 I = N1, N2
10   A(I) = B(I) * C(I)
     RETURN
     END


     FUNCTION RTC
     CALL Q8CLOCK( , , I )
     RTC = I * 142 * 10.5 * 1.0E-9
     RETURN
     END
```

110

Program used to measure $( R_{\infty} , S\frac{1}{2} )$ when a job is split between two CPUs of an ETA10-E series system, using the counter (CTPOST, CTWAIT) method of synchronisation.

```
      PROGRAM MULT2
      PARAMETER ( NMAX = 10000 )
      COMMON /LOCAL/ A( NMAX ), B( NMAX ), C( NMAX), IDC1, IDC2
      DATA B/ NMAX*1.0 /, C/ NMAX*1.0 /, IDC1/ 1 /, IDC2/ 2 /
      CALL MTDEF ( IDT, 'MTASK', 0, 1, 0, ISTAT )
      CALL CTDEF (IDC1, 0, 0, ISTAT )
      CALL CTDEF (IDC2, 0, 0, ISTAT )
      CALL MTRUN ( IDT, 'DOALL', 0, ISTAT )
      DO 20 N = 2, NMAX, 100
            T1 = RTC ()
            CALL CTPOST ( IDC1, 1, 0, ISTAT )
            NHALF = N / 2
            DO 10 I = 1, NHALF
10          A(I) = B(I) * C(I)
            CALL CTWAIT ( IDC2, 1, 0, ISTAT )
            T2 = RTC()
            T = T2 - T1
            PRINT *, N, T
20    CONTINUE
      CALL MTWAIT ( IDT, 0, ISTAT )
      CALL MTTERM ( IDT, 0, ISTAT )
      STOP
      END
      SUBROUTINE DOALL
      PARAMETER ( NMAX = 100000 )
      COMMON /LOCAL/ A( NMAX ), B( NMAX ), C( NMAX), IDC1, IDC2
      DO 20 N = 2, NMAX, 100
            CALL CTWAIT ( IDC1, 1, 0, ISTAT )
            NH1 = N/2 + 1
            DO 10 I = NH1, N
10          A(I) = B(I) * C(I)
            CALL CTPOST ( IDC2, 1, 0, ISTAT )
20    CONTINUE
      RETURN
      END
```

Program used to measure ( $R_{\infty}$ , $S\frac{1}{2}$ ) when a job is split between two CPUs of an ETA10-E series system, using the barrier (CTBARR) method of synchronisation.

```
PROGRAM MULT3
PARAMETER ( NMAX = 10000 )
COMMON /LOCAL/ A( NMAX ), B( NMAX ), C( NMAX), IDC
DATA B/ NMAX*1.0 /, C/ NMAX*1.0 /, IDC/ 1 /
CALL MTDEF ( IDT, 'MTASK', 0, 1, 0, ISTAT )
CALL CTDEF (IDC, 2, 1, ISTAT )
CALL MTRUN ( IDT, 'DOALL', 0, ISTAT )
DO 20  N = 2, NMAX, 100
          T1 = RTC ()
          NHALF = N / 2
          DO 10  I = 1, NHALF
10        A(I) = B(I) * C(I)
          CALL CTBARR ( IDC, 0, ISTAT )
          T2 = RTC()
          T = T2 - T1
          PRINT *, N, T
20   CONTINUE
     CALL MTWAIT ( IDT, 0, ISTAT )
     CALL MTTERM ( IDT, 0, ISTAT )
     STOP
     END


     SUBROUTINE DOALL
     PARAMETER ( NMAX = 100000 )
     COMMON /LOCAL/ A( NMAX ), B( NMAX ), C( NMAX), IDC
     DO 20  N = 2, NMAX, 100
          NH1 = N/2 + 1
          DO 10  I = NH1, N
10        A(I) = B(I) * C(I)
          CALL CTBARR ( IDC, 0, ISTAT )
20   CONTINUE
     RETURN
     END
```

Programs used to measure datalink and databuffer performance on an ETA10-E system

```
      PROGRAM MULTDL
      PARAMETER ( NMAX = 100000 )
      COMMON /LOCAL/ A(NMAX ), IDD
      CALL DLDEF ( IDD, 'DLFILE', 'RW', 8, 0, ISTAT )
      CALL DLWRIT ( IDD, A(1), 1, NMAX, TOKEN, 1, ISTAT )
      DO 20  N = 2, NMAX, 100
              T1 = RTC()
              CALL DLWRIT ( IDD, A(1), 1, N, TOKEN, 1, ISTAT )
C             CALL DLREAD ( IDD, 1, A(1), N, TOKEN, 1, ISTAT )
              T2 = RTC ()
              T  = T2 - T1
              PRINT *, N, T
   20 CONTINUE
      CALL DLTERM ( IDD, 0, ISTAT )
      STOP
      END




      PROGRAM MULTDB
      PARAMETER ( NMAX = 300 )
      COMMON /LOCAL/ A(NMAX ), IDD
      CALL DBDEF ( IDD, NMAX, 0, ISTAT )
      CALL DBPUT ( IDD, A(1), 1, NMAX, 0, ISTAT )
      DO 20  N = 2, NMAX, 100
              T1 = RTC()
              CALL DBPUT ( IDD, A(1), 1, N, 0, ISTAT )
C             CALL DBGET ( IDD, 1, A(1), N, 0, ISTAT )
              T2 = RTC ()
              T  = T2 - T1
              PRINT *, N, T
   20 CONTINUE
      CALL DBTERM ( IDD, 0, ISTAT )
      STOP
      END
```

# Multitasking the Meteorological Office Forecast Model on an ETA[10]

A. DICKINSON

Meteorological Office, London Road, Bracknell, Berks. RG12 2TE, U.K.

## 1. INTRODUCTION

The United Kingdom Meteorological Office (UKMO) is currently in the process of replacing its Cyber 205 computer by a compatible but significantly more powerful ETA[10] computer system. The ETA[10] configuration is based on four processors each of which is twice as powerful as the Cyber 205 and has twice the local memory. The processors have access to 64 million words of shared memory and the overall system is expected to have a peak performance of about 3000 MFLOPs. A comparison of the major features of the Cyber 205 and ETA[10] computers is given in Table 1.

|  | Cyber 205 | ETA[10] |
|---|---|---|
| Clock speed (ns) | 20 | 10.5 |
| Number of processors | 1 | 4 |
| Local memory (Mw) | 2 | 4 |
| Shared memory (Mw) | – | 64 |
| Disk space (Gbytes) | 5 | 20 |
| Peak performance (MFLOPs) | 400 | 3000 |

Table 1. The main performance features of the UKMO Cyber 205 and ETA[10] computers.

In common with most other supercomputers available today, the full power of the ETA[10] may only be harnessed to a single problem through the use of multi-tasking. The software which allows programs to create and control the running of parallel tasks on the ETA[10] is known as the Multi-tasking Library (MTL) and is described by Mozdzynski (1988). It is expected that all of the time-critical components of the UKMO operational suite will eventually be multi-tasked so that they each use all of the four processors

Topics in Atmospheric and Oceanic Sciences
© Springer-Verlag Berlin Heidelberg 1990

available on the ETA[10]. These programs include the global, regional and mesoscale models, along with data assimilation and much of the post-processing.

As yet, only preliminary multi-tasked versions of some of these codes have been developed. In the remainder of this paper we therefore concentrate on just one of these programs - the global forecast model. This is used to illustrate the multi-tasking techniques being applied to our operational suite, and to give an indication of the levels of performance that might eventually be achieved.

## 2.    THE FORECAST MODEL

### 2.1    Formulation

The current version of the UKMO operational global model as run on the Cyber 205 uses 15 σ-levels on a 1.5°x1.875° latitude-longitude grid. The governing equations are integrated using the split-explicit scheme of Gadd (1978). At each timestep the gravity-inertia terms are integrated in three short 'adjustment' steps followed by a single 'advection' step which includes the horizontal advection and diffusion terms. The effects of the physical parametrizations are also added in during the advection stage. These include convection, large-scale precipitation, climatological radiation, boundary-layer effects and gravity wave drag. At high latitudes stability filtering is applied to prevent the small east-west grid length from imposing unacceptable restrictions on the length of a timestep. A full description of the model is given by Bell and Dickinson (1987).

|                       | Cyber 205 | ETA[10]    |
|-----------------------|-----------|------------|
| Number of levels      | 15        | 20         |
| Number of points E-W  | 192       | 288        |
| Number of points N-S  | 121       | 217        |
| Timestep (seconds)    | 900       | 600        |
| Grid box (degrees)    | 1.5x1.875 | 0.833x1.25 |

Table 2. Global model dimensions used on the Cyber 205 and as planned for the ETA[10].

A higher resolution 0.833°x1.25° version of this model is planned for the ETA[10]. Its dimensions are shown in Table 2. In addition to this increase in vertical and horizontal resolution, revisions to the model's physical parametrizations are also planned. These include the introduction of a fully interactive radiation scheme.

## 2.2    Vectorization

The vector and scalar instruction sets of the Cyber 205 and ETA[10] computers are identical. Both machines are pipelined vector processors and so it is desirable to maximise the length of vectors used in calculations so as to minimise the overheads due to vector start-up, which reflects the time taken for the first pair of data elements to pass through the vector pipes. The forecast model uses 32-bit arithmetic throughout in order to take advantage of the factor of two speed-up over 64-bit arithmetic on the Cyber 205. For 32-bit adds or multiplies on a Cyber 205 a vector of length 200 is only 50% efficient whilst a vector of length 10000 is 95% efficient.

On the Cyber 205 the forecast model code has been designed to maximise vector lengths and to be totally contained in memory so as to minimise I/O. Model data is organised so that the grid point values of each variable are stored as horizontal fields. A vector can therefore span one level of the model giving vector lengths of around 23000 for most calculations in the dynamics. This approach is also well suited to efficient programming of the physical parametrizations, since these codes generally have no horizontal dependencies. Their conditional nature means that results are often required only at subsets of the integration area. The control store instruction allows unwanted results to be masked out, although it is often more efficient to use the compress and expand instructions to form contiguous sub-vectors containing only those points at which calculations are required (Dickinson, 1982).

Advances in the design of the ETA[10] have reduced the cost of vector start-up in certain circumstances; for example when vector instructions occur back to back some of the cost of the start up phase is masked by the execution of the previous vector instruction. If all vectors are back to back, then a 32-bit vector add or multiply of length 60 is 50% efficient and a vector of length 2000 is 95% efficient. The ETA[10] also has a vector

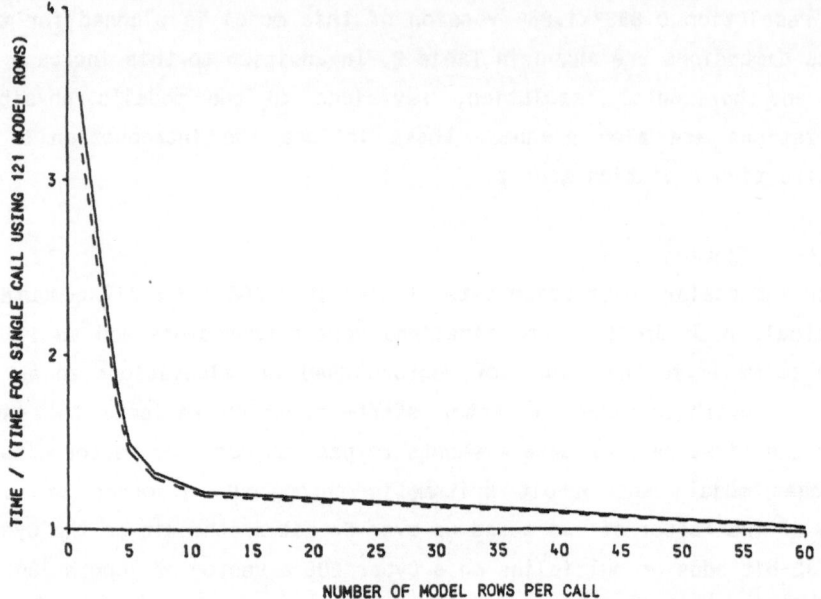

Figure 1. Timings of one timestep of the global model physics using
a range of vector lengths. This is achieved by using multiple calls
to each physics routine using a fixed number of model rows. The timings
are scaled against the cost when using one call for the full integration
area of 121 rows. Plots are (———) Cyber 205 and (– – –) ETA[10].

short stop feature which, when circumstances permit, allows the results of
one vector operation to feed directly into the next without the cost of
start-up.

Any gain in short vector performance in the forecast model resulting from
these design changes will obviously depend on the instruction sequences
used in the model code. Timings from the model's physical parametrizations
using a range of vector lengths are shown in Figure 1. It appears that
speeds in excess of those implied by the ratio of the clock times of the
two machines can be obtained on the ETA[10] at short vector lengths, but that
the basic speed advantage from using long vectors in the forecast model
code still remains.

2.3    Multi-tasking
The strategy chosen to multi-task the forecast model has the advantage
that only minor changes are needed to the original program, retaining the
'in-memory', horizontal fields approach adopted for the Cyber 205. As shown
in Figure 2, the forecast domain is split into N latitude sections, where N

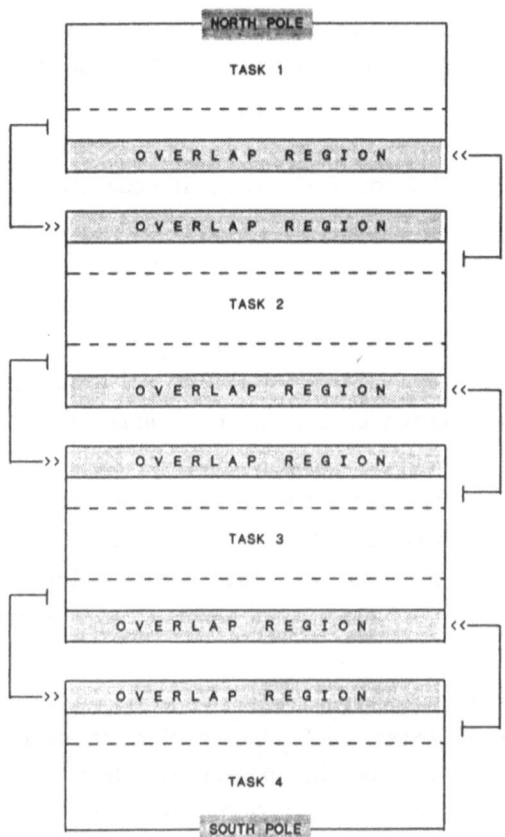

Figure 2. Task sectioning by dividing a model level across 4 processors. The shaded regions are additional rows used to provide boundary conditions. These are overwritten once per timestep with data from the adjacent task.

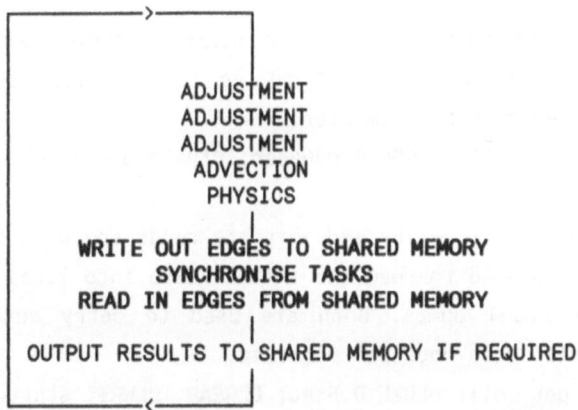

Figure 3. The timestep structure within a task. Note that there is only one synchronisation point per timestep.

is the number of processors, with each processor being allocated its own geographic area. Single tasks are initiated in each processor at the beginning of an integration and each task then works on its own region of the globe exchanging boundary conditions with adjacent tasks via shared memory. The data required by each task is contained totally in local memory, with the forecast fields being overwritten at each timestep. Model fields are only written to shared memory at forecast write-up times.

Flow through a model timestep is shown in Figure 3. Synchronisation of the tasks is carried out only once per timestep in order to facilitate load balancing. An overlap region consisting of 5 extra rows is required along shared boundaries. The precise number of rows in the overlap region depends on the integration scheme being used. In this case, one row is required for each of the adjustment steps and two for the advection step which is fourth order. Some saving in the cost of these extra calculations is obtained by dropping off rows as they become redundant at each sub-timestep.

The following MTL calls are used in the model to control the initialisation and synchronisation of tasks, and the sharing of data between parallel tasks. The timings quoted below and in Section 3 have been obtained using a pre-release version of ETA OS 1.2. A fuller description of these calls and more detailed timings are given in Mozdzynski (1988).

MTDEF, MTRUN define and initiate the running of parallel tasks. These routines are called only once for each task at the beginning of a forecast. Typical cost per call: MTDEF 3sec; MTRUN 0.002sec.

CTDEF, CTBARR are used to define and perform barrier operations, allowing the synchronisation of tasks. CTDEF is called once per integration and CTBARR is called once per timestep.
Typical cost per call: CTDEF 0.0002sec; CTBARR 0.001sec.

DLDEF, DLREAD, DLWRIT define and initiate reads and writes to shared memory files. DLREAD is used to read the initial data into local memory and DLWRIT to write out model dumps. Both are used to carry out the exchange of boundary overlap data once per timestep.
Typical cost per call: DLDEF 0.5sec; DLREAD, DLWRIT start-up 0.001sec, transfer rate 500 Mbytes/sec.

CBDEF, CBGET CBPUT define and initiate reads and writes to the communications buffer (CB). CBDEF is called once per CB array. CBGET and CBPUT are used to pass small amounts of miscellaneous data between tasks. Typical cost per call: CBDEF 0.0003sec; CBGET, CBPUT start-up 0.0001sec, transfer rate 10 Mbytes/sec.

## 3. TIMINGS

At the present time the UKMO ETA[10] is configured as two separate systems, the first allowing access to three processors and the second using the remaining single processor. All the ETA[10] timings quoted below have been obtained on the three processor system.

Timings for 24 timesteps of the model are shown in Table 3, the data time of the integrations being 00z 15/08/88. In making these measurements the time taken to initiate the tasks and read in data from disk has been ignored, since this is a once-and-for-all cost, which at the present time is dominated by the expense of calling MTDEF (3 secs per call) at the beginning of each task. The portion of code timed is essentially 24 passes through the loop shown in Figure 3.

|  | Time in seconds | Speed-up over C205 |
|---|---|---|
| Cyber 205 | 66.22 | |
| ETA[10] single task | 37.54 | 1.76 |
| ETA[10] multi-tasked on 3 CPUs | 13.75 | 4.82 |

Table 3. Wall clock timings for 24 timesteps of the forecast model.

## 3.1 Single processor timings

In single task mode the ETA[10] run achieves a 1.76 times speed-up over the Cyber 205. In fact, the ratio of clock times (Table 1) implies that a speed-up of at least 1.9 times should be obtained. An examination of the times taken by each routine shows that most do indeed run at between 1.9 and 2 times the Cyber 205. The common feature of the remaining slower routines is that they all contain gather or scatter instructions.

There are two reasons why these instructions lead to less efficient code on the ETA[10]. Firstly, the current version of the compiler recodes the scatter instruction to use short vector lengths. This 'work around' was designed to circumvent hardware problems with this instruction in development versions of the ETA[10]. These have long since been corrected and it is expected that the next release of the compiler will remove this fix, allowing this instruction to be executed directly.

Secondly, the speed of gather or scatter instructions is also affected by the number of bank conflicts encountered. For 32-bit arithmetic there are 512 banks on the Cyber 205 but only 128 on the ETA[10]. The choice of 192 points around a latitude circle can therefore lead to bank conflicts when using constant stride arithmetic, since $2*192 = 3*128$. This effect will be alleviated once the enhanced resolution of 288 points around a latitude circle is used. As well as having fewer banks, the bank busy time on the ETA[10] is 6 cycles as compared to 4 cycles on the Cyber 205. This will add to the relative cost of successive gathers or scatters from the same bank, such as might occur when reading consecutive values from the same element of a look-up table. It is expected that coding changes will remove many of these bank conflicts, for example by introducing multiple copies of look-up tables.

## 3.2   Multi-processor timings and load balancing

The latitudinal distributions of the computations done in the physics and dynamics are quite different. The physics, on the one hand, are weighted towards the equatorial regions, where the cost of convection is large (see Figure 10 of Dent, 1988). The stability filtering, on the other hand, ensures that the polar regions are the most expensive part of the dynamics.

| Task number | Number of latitude rows | Number of vectors |
|---|---|---|
| 1 | 37 | 3.7 million |
| 2 | 48 | 2.0 million |
| 3 | 36 | 3.7 million |

Table 4. Distribution of work among the 3 tasks used in the 24 timestep multi-tasked integration.

These effects are to some extent self cancelling over a full timestep, and so it is natural to synchronise at this frequency. Even so, the cost of filtering is such as to require smaller task sizes in the polar regions.

The task sizes used in the timed multi-tasked integration are shown in Table 4. These were held fixed throughout the integration. The 3 CPU multi-tasked run shows a speed-up of 4.82 times the Cyber 205, and a multi-tasking efficiency of 90% relative to the single tasked ETA[10] run. A breakdown of the contributions to the total cost of the multi-tasked integration is given in Figure 4. In this context efficiency is defined by

$$\frac{100*(\text{single-tasked time})}{(\text{multi-tasked time})*(\text{number of processors})}.$$

It can be seen that the main multi-tasking overhead is due to the tasks being out of balance (5%). This effect is compounded by paging over the first few timesteps as the model determines its working set. Other small fluctuations in the timestep-by-timestep cost of a task are also apparent. These appear to be caused by intermittent polling from system routines which, as yet, are not part of the operating system shell. The other major overhead is from the extra computations due to the overlapping edges (3%) and the expense of exchanging these boundary values via shared memory (1%).

The integrations carried out so far indicate that once an acceptable level of task balancing has been achieved, there may be no need to change the loading in each CPU throughout the remainder of an integration, since the amount of work done by each task is mostly influenced by seasonal rather than diurnal effects. The appropriate weighting for a forecast could therefore be calculated from timing information gathered during the previous operational run. Nevertheless, it is possible to rebalance tasks as the integration proceeds, for example by adapting the technique used to exchange the boundary overlap rows, though the cost of this still needs to be fully assessed.

In all probability, it will be necessary to rebalance the tasks once the interactive radiation scheme is included in the model. This is because

# MULTI—TASKING EFFICIENCY ON 3 PROCESSORS

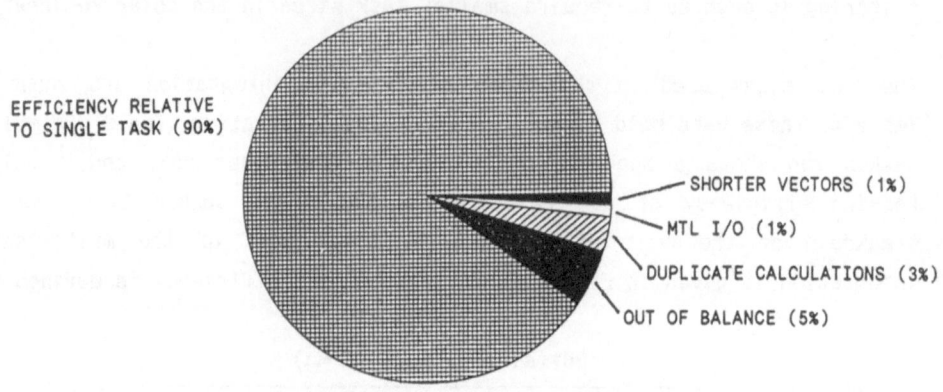

Figure 4. Breakdown of the cost of the 3 CPU multi-tasked integration.

interactive radiation is called only once every 3 hours (18 timesteps), but it accounts for 25% of the integration time. However, it is known that the computational loading of interactive radiation is more evenly distributed than in the multi-tasked timing run described above (see Table 4). Use of interactive radiation will therefore lead to an increase in the relative cost of the equatorial task during a radiation timestep, unless the distribution of work among the tasks is adjusted accordingly.

## 4.    DISCUSSION

Having achieved multi-tasking with the current model on a 3 CPU ETA[10], it is of interest to speculate on the level of performance that might be realised with the higher resolution version of the global model (see Table 2) on a 4 CPU ETA[10].

Initially the use of 4 processors will lead to a reduction in efficiency, partly because the use of an additional processor will make it more difficult to achieve load balancing, and partly because the cost of the boundary overlaps will become proportionally larger. However, these effects will be mitigated once the resolution is increased. The ratio of overlap rows to model rows is 20:121 in the 3 CPU version of the current model.

This ratio will reduce to 30:217 in the 4 CPU high resolution model. In addition, the high resolution model will include more detailed physical parametrizations, and so the relative cost of the dynamics will be reduced. This will lead to an equivalent reduction in the cost of the boundary overlap calculations since these are only done within the dynamics. The increase in problem size will also make it more easy to achieve load balancing, since more work will be done between synchronisation points and the extra rows will allow finer control over the amount of work given to each processor.

Thus, taking these points into consideration, it seems reasonable to expect that the 4 CPU enhanced resolution model will achieve a level of efficiency of at least 90%, although the cost of rebalancing during an interactive radiation timestep has yet to be assessed. A speed-up in excess of 7 times the Cyber 205 should also be possible once the problems associated with the gather and scatter instructions, discussed in section 3.1, have been overcome.

## 5.    REFERENCES

Bell, R. S. and A. Dickinson, 1987: The Meteorological Office numerical weather prediction system. Met. Office Scientific Paper No 41.

Dent, D., 1988: The ECMWF model: past, present and future. Multiprocessing in meteorological models, Springer-Verlag.

Dickinson, A., 1982: Optimizing numerical weather forecast models for the Cray-1 and Cyber 205 computers. Comput. Phys. Commun., 26, pp 459-468.

Gadd, A. J.,1978: A split explicit integration scheme for numerical weather prediction. Quart. J. Roy. Met. Soc., 104, pp 567-582.

Mozdzynski, G., 1988: Multi-tasking on the ETA[10]. The proceedings of the ECMWF workshop on the use of parallel processors in meteorology: 5-9 Dec 1988. ECMWF, Reading, U.K.

# Multiprocessing of a Mesoscale Model

SADAMU SAITOH and KEN HAYAMI

C&C Information Technology Research Laboratories, NEC Corporation, Miyamae-ku, Kawasaki 213, Japan

## 1. INTRODUCTION

Computers have been used in meteorology as efficient tools for numerical research as well as weather prediction. Recent progress of supercomputers enable us to perform numerical experiments which require huge computations. In many cases such numerical experiments were only a dream for meteorologists.

Mesoscale processes are one of the most exciting research topics in meteorology. Saitoh and Tanaka (1987, 1988) developed a mesoscale numerical model to research into the mechanism of meso-$\beta$ scale frontal rainband formation, where the importance of the role played by conditional symmetric instability (CSBI) proposed by Bennets and Hoskins (1978) and Emanuel (1979) was shown. The model, which is based on non-hydrostatic dynamical equations assuming zonal symmetry and parameterized warm precipitation processes, succeeded to show how CSBI circulation leads to the meso-$\beta$ scale precipitation area. One of the most interesting problems to be solved is to investigate the interaction process between synoptic, meso-$\beta$ and meso-$\gamma$ scale circulations in datail. From a numerical modeling point of view, both the fine space and wide integral domain should be incorporated in order to simulate explicitly synoptic, meso-$\beta$ and meso-$\gamma$ scale circulations at the same time. Such a large scale numerical experiment requires large computer resources. Similar situations are often seen in numerical modeling of the atmospheric processes. The most striking feature of

Topics in Atmospheric and Oceanic Sciences
© Springer-Verlag Berlin Heidelberg 1990

the current development of computers is characterized by parallel processing. Vector processing and multiprocessing are the typical way to implement the concurrency of problems, and they can be taken advantages at the same time. Then, assuming a MIMD machine composed of several number of vector processors, the ability of the system is derived most effectively when both the vector and multiprocessor parallelisms are fully explored.

The main purpose of the present paper is to study the parallelism of the mesoscale numerical model that we have constructed. Although, our model is designed for the frontal rainband problem, variety of mesoscale problems can be treated by the model with minor modifications. The following disscussions are applicable even when other kinds of mesoscale problems are solved.

The mesoscale numerical model is discribed in Section 2. Vector processing of the model using NEC SX-2 is presented in Section 3. Possibility of multiprocessing of the model is discussed in Section 4. Finally, conclusions are given in Section 5.

## 2.  THE MESOSCALE MODEL

The present mesoscale computational model is designed to investigate the mechanism of the formation of frontal rainbands. Water processes are introduced by using Kessler's (1969) parameterization scheme. Thorough description of the numerical experiments are given in Saitoh and Tanaka (1987; 1988). Here we briefly present only the gist of the model.

### 2.1  Basic equations

We use a non-hydrostatic and elastic system of equations which is similar to the equations used in numerical studies of the three-dimensional sea breeze in Florida (Tapp and White, 1976), three dimensional convective cloud (Klemp and Wilhelmson, 1978), and mountain lee waves (Durran and Klemp, 1983).

The equations of the conservation of momentum in x-, y-, and z-directions are given by

$$\frac{\partial u}{\partial t} + C_p \frac{\partial \pi}{\partial x} = f_u , \qquad (2.1)$$

$$\frac{\partial v}{\partial t} + C_p \frac{\partial \pi}{\partial y} = f_v , \qquad (2.2)$$

$$\frac{\partial w}{\partial t} + C_p \frac{\partial \pi}{\partial z} = f_w , \qquad (2.3)$$

$$\frac{\partial \pi}{\partial t} + \frac{\bar{C}^2}{C_p \, \bar{p} \, \bar{\theta}_v^2} \left[ \bar{p} \, \bar{\theta}_v \left( \frac{\partial u}{\partial x} + \frac{\partial v}{\partial y} \right) + \frac{\partial (\bar{p} \, \bar{\theta}_v)}{\partial z} \right]$$

$$= f_\pi , \qquad (2.4)$$

respectively. The equations of potential temperature, water vapor, cloud water, and rain water contents are given by the same form by

$$\left( \frac{\partial}{\partial t} + u \frac{\partial}{\partial x} + v \frac{\partial}{\partial y} + w \frac{\partial}{\partial z} \right) \phi = M_\phi + D_\phi . \qquad (2.5)$$

$$(\phi = \theta , \; q_v, \; q_c, \; q_r)$$

The right-hand side terms of Eq.(2.1), (2.2), (2.3) and (2.4) are given by

$$f_u = -(u\frac{\partial}{\partial x} + v\frac{\partial}{\partial y} + w\frac{\partial}{\partial z})u + f_v + D_u , \quad (2.6)$$

$$f_v = -(u\frac{\partial}{\partial x} + v\frac{\partial}{\partial y} + w\frac{\partial}{\partial z})v - f_u + D_v , \quad (2.7)$$

$$f_w = -(u\frac{\partial}{\partial x} + v\frac{\partial}{\partial y} + w\frac{\partial}{\partial z})w$$

$$+ \left[ \frac{\theta}{\overline{\theta}} - 1 + 0.61(q_v - \overline{q}_v) - q_c - q_r \right] g + D_w , \quad (2.8)$$

$$f_\pi = -(u\frac{\partial}{\partial x} + v\frac{\partial}{\partial y} + w\frac{\partial}{\partial z})\pi$$

$$- \frac{R}{C_v}(\frac{\partial u}{\partial x} + \frac{\partial v}{\partial y} + \frac{\partial w}{\partial z})\pi + \frac{C^2}{C_p\overline{\theta}_v^2}\frac{d\theta_v}{dt} , \quad (2.9)$$

respectively. Moist processes represented by M in Eq.(2.5) are defined as

$$M_\theta = -\frac{L}{C_p\overline{\pi}}(\Delta + E_r) , \quad (2.10)$$

$$M_{q_v} = \Delta + E_r , \quad (2.11)$$

$$M_{q_c} = -\Delta - A_r - C_r , \quad (2.12)$$

$$M_{q_r} = \frac{1}{\overline{\rho}}\frac{\partial}{\partial z}(\overline{\rho}V_T q_r) - E_r + A_r + C_r . \quad (2.13)$$

$D_\xi$ , ( $\xi$ = u , v , w , $\theta$ , $q_v$, $q_c$, $q_r$ ), represents the diffusion terms. In the above equations, u , v and w are the velocity component in the x-, y- and z- directions, respectively, $\pi$ the non-dimensional pressure perturbation, $\theta$ the potential temperature, $q_v$, $q_c$ and $q_r$ are the mixing ratios of water vapor, cloud water and rain water, respectively, R the gas constant of dry air, $C_p$ and $C_v$ the specific heat of dry air at constant pressure and constant volume, respectively, $\rho$ the air density, f the Coriolis parameter, g the gravity acceleration, C the speed of sound, L the latent heat of condensation, $\Delta$ the rate of condensation or evaporation of cloud water, $E_r$ the rate of evaporation of rain, $A_r$ the rate of autoconversion, $C_r$ the rate of collection, $V_T$ the terminal speed of rain, $\theta_v$ is the virtual potential temperature, and upper-bars denote the basic field which is usually taken to be in a certain balanced state.


## 2.2 Numerical framework

Time integration of the governing equations described in Section 2.1 is performed by using the finite difference technique. Time discritization is made by the split semi-implicit scheme, which is composed of the small time step and large time step calculations (see Fig. 2.1). Terms which are important to ensure the numerical stability of the acoustic waves are named 'acoustic terms' (left-hand side terms in Eqs.(2.1), (2.2), (2.3) and (2.4), and are updated on small time steps. While, right-hand side terms in Eqs.(2.1), (2.2), (2.3), (2.4), and the whole contents of Eq.(2.4) are updated on large time steps, and are kept constant while several small time step calculations are made. The leapfrog scheme is used for large time step integrations. Small time step is treated semi-implicitly, where only vertical differentiations are treated implicitly using the Crank-Nicolson scheme, while other terms are integrated by the forward explicit time step scheme. Microphysical calculations are made on the large time steps and composed of tentative evaluation and its adjustment. Space discretization is made on a staggered grid.

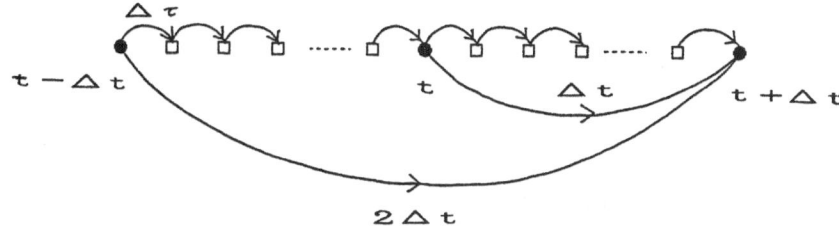

Fig. 2.1  Time stepping method.  Time integration is composed of
    small and large time stepping.  Circles represent the large time
    step points and the squares the small time step points.  Several
    small time step calculations with time increment of $\Delta \tau$ are made
    while the single large time step calculation with time increment
    $\Delta t$ is performed.

## 2.3  Structure of the program

A flow diagram of the numerical model is shown in Fig. 2.2.  The
initial field of the experiment is arranged by the subroutine INIT.  In
order to increment the time  from t to  t+$\Delta t$ ( $\Delta t$ represent  the
large time step interval), a series of subroutines SSTEP, LSTEP, FUVW,
BOUNDARY, TFILT and RECORD are called successively.  Among these subroutines, most of the computation time is consumed by LSTEP and SSTEP.

Figure 2.3 shows the structure of subroutine LSTEP, which is composed  of two steps of calculations.  In the first step,  equations of
$\theta$, $q_v$, $q_c$ and $q_r$ are stepped forward with time increment $\Delta t$ taking only the dynamical terms into account.  In the second step, the ad-
justments due to the moist processes are made.

The structure of  SSTEP is  shown  in Fig. 2.4,  in which ISTMAX
times of small time step  calculations are  required to  increment the
time by $\Delta t$ (ISTMAX=$\Delta t / \Delta \tau$).  At first,  u  and  v  are stepped forward by $\Delta \tau$,  then tridiagonal equations with respect to the vertical
velocity are solved on each horizontal grid point.  Finally, the evaluation of the non-dimensional pressure is performed.

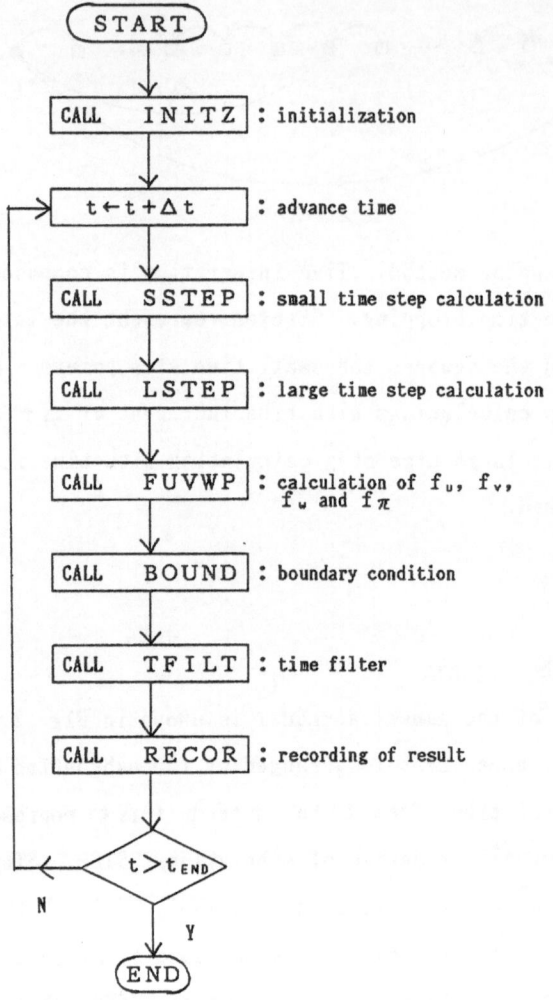

Fig. 2.2    Block diagram of the mesoscale model.    In order to step
forward by $\Delta t$, a series of subroutines are called from the main
program.

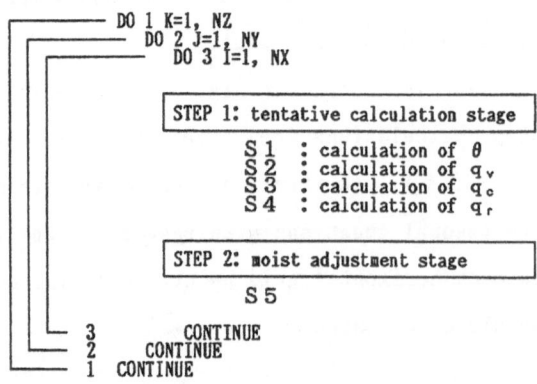

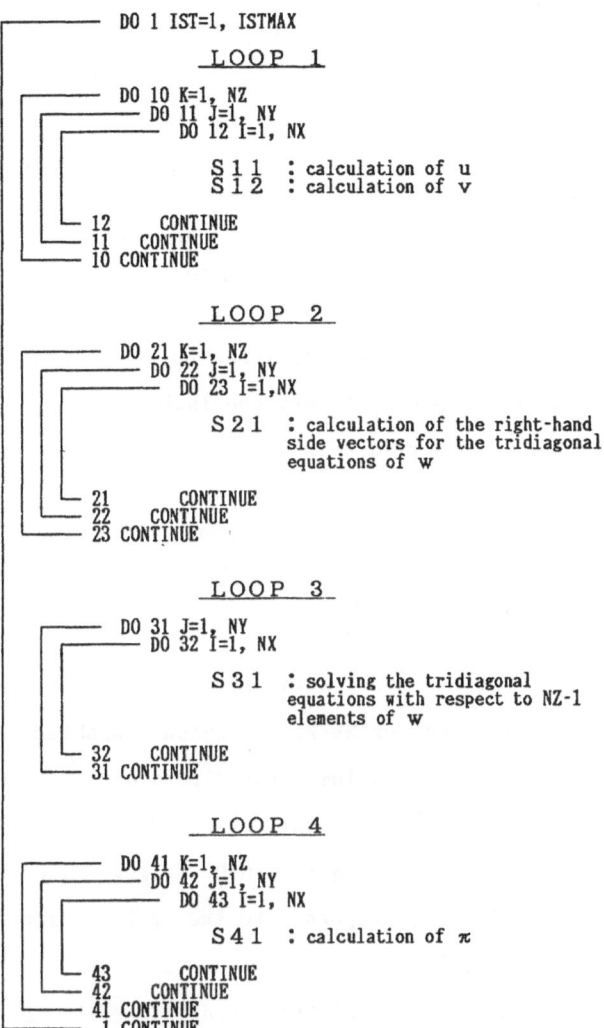

```
          DO 1 IST=1, ISTMAX
                LOOP  1
          DO 10 K=1, NZ
            DO 11 J=1, NY
              DO 12 I=1, NX

                  S 1 1  : calculation of u
                  S 1 2  : calculation of v

          12      CONTINUE
          11    CONTINUE
          10 CONTINUE


                LOOP  2
          DO 21 K=1, NZ
            DO 22 J=1, NY
              DO 23 I=1,NX

                  S 2 1  : calculation of the right-hand
                           side vectors for the tridiagonal
                           equations of w

          21      CONTINUE
          22    CONTINUE
          23 CONTINUE


                LOOP  3
          DO 31 J=1, NY
            DO 32 I=1, NX

                  S 3 1  : solving the tridiagonal
                           equations with respect to NZ-1
                           elements of w

          32    CONTINUE
          31 CONTINUE


                LOOP  4
          DO 41 K=1, NZ
            DO 42 J=1, NY
              DO 43 I=1, NX

                  S 4 1  : calculation of π

          43      CONTINUE
          42    CONTINUE
          41 CONTINUE
           1 CONTINUE
```

Fig. 2.4     Structure of subroutine SSTEP.  See text for detail.

Fig. 2.3     Structure of subroutine LSTEP.  LSTEP is composed of two
step calculations.  S1, S2, S3, S4 and S5 represents the tasks
(a series of operations).  Among the tasks, S1, S2, S3 and S4
can be executed, concurrently,  while execution of S5 should be
started after   the calculations of STEP 1 have finished.

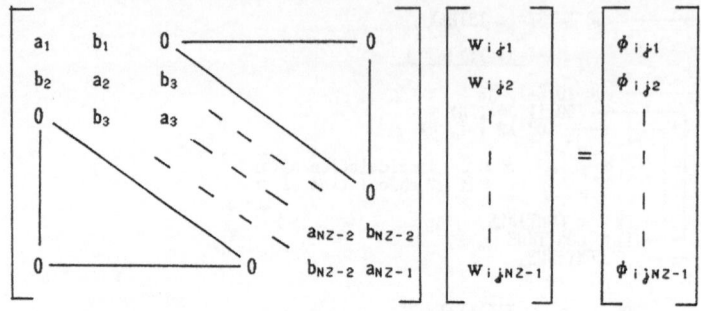

Fig. 2.5    Tridiagonal equations for vertical velocity w.  The
coefficient matrix appears for any combination of i and j, where
i and j represents the grid point in x- and y- directions,
respectively.


3.    VECTOR PROCESSING ON NEC SX-2

In order to demonstrate the vector processing efficiency of the
model on the NEC supercomputer SX-2, we chose a problem of frontal-
rainband formation.    In the problem, two-dimensional version of the
model is used in which 125 ( horizontally )  x  20 ( vertically ) grid
points are incorporated and typically 8640 of large time step calculations are made, which corresponds to the 250km x 10km integral
domain and 2day integration. As an example of the result of the numerical experiment, the development of the cloud induced by the CSBI circulation is shown in Fig. 3.1. It is demonstrated, from the dynamical
point of view, that the formation of warm-frontal rainbands is most
likely explained by the slantwise meridional circulation induced by
CSBI.  Thorough description of the problem and the physical interpre-
tations are given in Saitoh and Tanaka (1987, 1988).   Here, only some
important computational features are presented.

Vector processing of the model took 204s CPU time, while 1024s
was required when processed in scalar mode.   Then, the speed up ratio
by the vector processing is approximately 5.   It may seem that the
speed up ratio by vector operation is not so large in spite of the

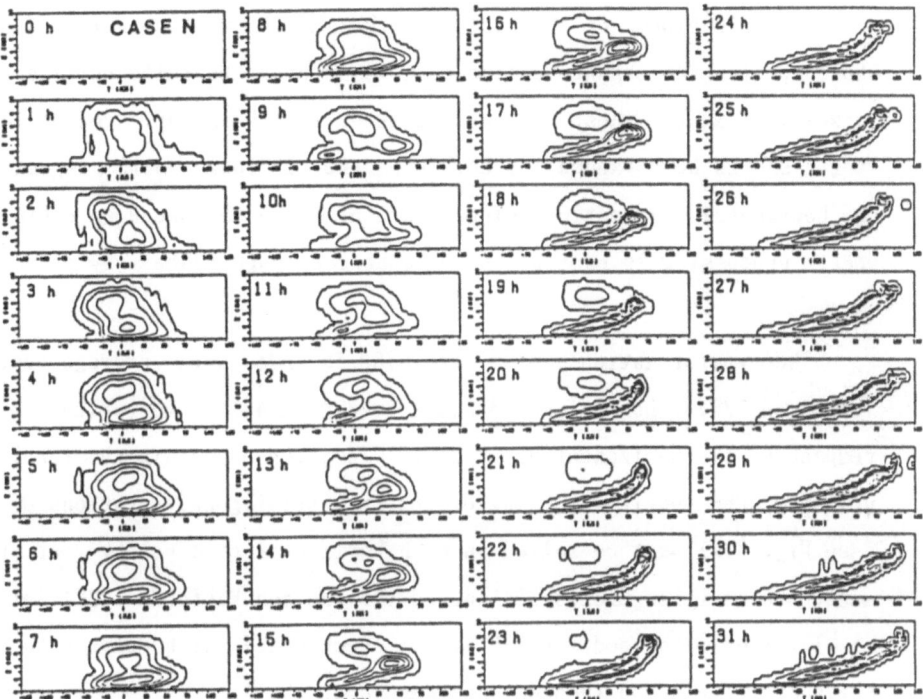

Fig. 3.1   Vertical cross sections of cloud water content.  The contour interval is $0.2g \ kg^{-1}$ and the outermost curve corresponds to $0g \ kg^{-1}$.  A slantwise cloud develops along the slantwise updraft of CSBI circulation.

vectorization ratio.  One of the reasons for the low speed up ratio by vector processing is that very high speed arithmetic operatins are performed on SX-2 even in scalar mode. When vector processing is made, 51% and 26% of the total CPU time is consumed by the subroutines LSTEP and SSTEP, respectively.   Analysis of the program by using the ANALY-ZER/SX shows that the vectorization ratio of the code is 95% as a whole.

In subroutine LSTEP,  89% of the code is vectorized in the hori-zontal direction with vector length 125.  In principle, the whole por-tion within the DO loop  in Fig. 2.3 is vectorizable.   However, actu-ally,  a portion for the saturation   adjustment remains unvectorized.

Minor modifications  of the code will realize the vectorization of the whole portion and enhance the speed up.  In order to fully utilize the capacity of SX-2, whose  vector  register length is 256,  vector loop length should be expanded  by combining the  DO loops  with respect to the horizontal and vertical directions, by which the speed up ratio by vector processing is expected to be enhanced approximately by a factor of 2.

The  vectorization  ratio of  SSTEP is 100% with  average vector length of 125.   Loops 1, 2, 3 and 4 in Fig. 2.4  are vectorized  with respect to the horizontal direction.   In order to obtain the vertical velocity, the  positive definite symmetric tridiagonal equations of w (see Fig. 2.5) must be solved on each horizontal grid point.   At present, the modified Choleski's method (Melosh and Bamford, 1969; Melosh et al., 1967)  is used,  which can be  vectorized  with respect to the horizontal direction (Hockney and Jesshope, 1981).

The mesoscale numerical model was developed,  originally, on  the scalar processor.  When, the original scalar code was executed on SX-2 without any artificial modifications for  vectorization,  the vectorization ratio was approximately 80%. After the tuning for vectorization was made, vectorization ratio increased upto 95%.   From  our experience,  the utilization of work variables was one of the most effective way to enhance the vectorization ratio.


## 4.   MULTIPROCESSING

In the present study, we assume a  general  computer model  which consits of p equal vector  processors  and a central  memory shared by the processors.   Inter-processor communication and synchronization is assumed to be done by an interconnection network. A program for a MIMD machine generally consits tasks of certain sizes.   Some of the tasks can be executed concurrently by allocating the tasks to different processors. At some stage of the program, one or more  processors may become idle until other processors finish their tasks.  This kind of CPU time loss is called synchronization overhead. In addition, a certain amountof CPU time may elapse when parallel tasks are allocated to processors.

## 4.1  General strategies for multiprocessing

The following should be taken into account when a program is implemented on the multiprocessing environment. First of all first, the size of the tasks allocated to different processors should be of the same order in order to decrease synchronization overhead. Secondly, the granularity of the task should be large enough in order that the time loss by multitask generation is sufficiently compensated by the speed up due to multiprocessing.

When the two kinds of parallel processing, vector pipelining and multiprocessing, are utilized at the same time, the capacity of the computer system is fully exploited. Consider a situation where there is a double DO loop (see Fig. 4.1a), the processes within the loop can be executed concurrently with respect to both DO-variables I and J. The inner loop may be execueted by vector processing, while the outer loop may be assigned to the multiprocessors. If p processors are available the task shown in Fig. 4.1 is devided into p independent tasks and are executed concurrently on p processor resources. Therefore, when the number of the processor p is not so large, say lesser than 10, the priority should be given to vector processing rather than the multiprocessing. The speed up by multiprocessing is at most p, while the speed up by vector processing may be as high as 50. As the inner loop length increases, the speed up by vector processing is enhanced (this is true until the loop length exceeds the vector register length of the computer). Hence, in some cases, the exchange of inner and outer loop is recommended.

136

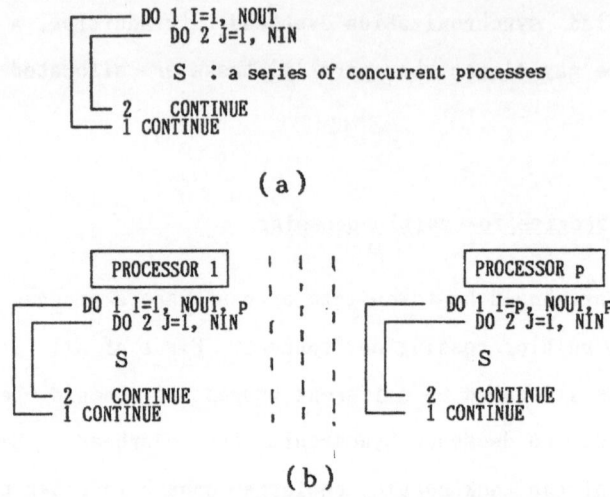

(a)

(b)

Fig. 4.1a  A double DO loop where a series of processes S can be
executed concurrently with respect to both I and J.

4.1b  Allocation of tasks to p processors. Tasks of the same
size are assigned to each processor.


## 4.2 The pararell processing bottleneck

After parallelism of the tasks are fully implemented, a problem
illustrated in Fig. 4.2 arises.   Suppose $\alpha$% of the execution time is
consumed by  the intrinsically scalar operations and  the remainder by
the cocurrently  executable operations.  When this code is executed on
a parallel computer  whose  speed up ratio  due to parallel processing
(by both the vector and multi processing,  or either of which) is  $\xi$,
the ratio of the execution time consumed by the  scalar operation be-
comes  $\alpha / (\alpha + (1-\alpha)/\xi)$ .  When, for example,  $\alpha$ =10% and  $\xi$ =50,
the relative weight of the scalar processing increses due to the  par-
allel processing as much as 85%.  So that careful coding and selection
of the algorithms for intrinsicaly scalar operations becomes even more
important when the code is executed on parallel computers.   From the
computer architecture  point of view, speed up ofthe scalar operations
is as important as the parallelization of the computer resources.

Execution time

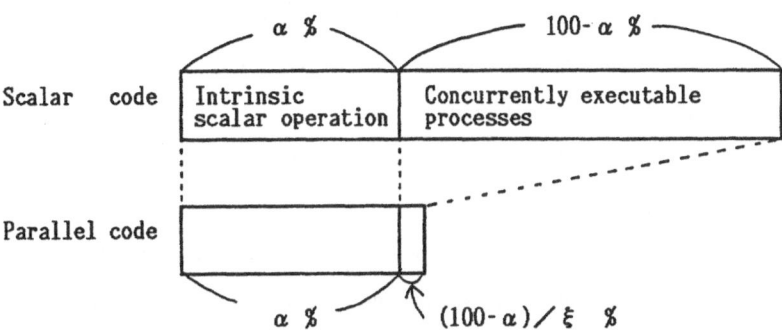

Fig. 4.2   A schematic diagram which shows a bottle neck due to
intrinsic scalar operations.  $\alpha$ represents the ratio (%) of CPU
time required for the execution of intrinsic scalar operations
when the code is   executed on a scalar processor.  $\xi$ is the
speed up due to parallel  processing (due to both vector and
multiprocessing).


## 4.3  Multiprocessing of the mesoscale model

Among  the  alternative ways of mapping  subroutine LSTEP on different processors for the purpose of multiprocessing,  the most prob-
able  one is the domain decomposition, in which the outer-most DO loop
is decomposed  and  mapped to p processors as shown in Figs. 4.1a  and
4.1b.  At the same time, vector operation is incorporated with respect
to the inner-most loop.   For the typical mesoscale problems, a large
number of grid points are required in  the horizontal  direction  than
in  the vertical  direction.  Hence, vectorization should be made with
respect to the horizontal direction in order to ensure sufficient vector length. This way of processor allocation has some advantages.   In
the first place, the granularity of the task is large  enough to overcome  multitasking  overhead. Secondly, it seldom  requires  any  modifications of the code even when the number of processors is  changed.
This aspect is especialy important  under the  multi-user  environment
in  which  the number  of  the  processors available varies  depending
on the  other jobs waiting to be executed and/or being executed.

The multiprocessing strategy of SSTEP is the same as that of LSTEP. Loops 1, 2, 3 and 4 in Fig. 2.4 should be vectorized with respect to the inner horizontal direction, and multiprocessing is applied with respect to the outer loops.

## 5. CONCLUSIONS

Parallel processing of the mesoscale numerical model was discussed. Variety of mesoscale phenomena can be treated by the model with minor modifications. Main conclusions of the present study are;

(1) The mesoscale numerical model was executed on NEC supercomputer SX-2. The vectorization ratio of 95% was attained, and the speed up ratio by vector processing was approximately 5 in comparison with the scalar mode operation. The vectorization ratio can be enhanced still more since, in principle, the whole portion of the model is concurrently executable.

(2) Vector processing and multiprocessing can be applied almost the whole portion of the model, where the inner DO loop is vectorized and outer DO loop is mapped to different processors.

In the present paper, we have discussed on the strategies for parallel processing assuming a MIMD machine composed of a rather small number of vector processors. When the number of processors becomes significant larger, other strategies may become more effective, which we shall leave for future consideration.

### ACKNOWLEDGEMENT
The numerical model discussed here was constructed by the first author (S.S.) in collaboration with Prof. Hiroshi Tanaka of the Nagoya University. The first author would like to thank Prof. Hiroshi Tanaka for his heartiful encouragements, Mr. Yoshiki Seo of the C&C System Labs., NEC Corporation for many useful suggestions on multiprocessing. We would like to acknowledge Drs. Katsuhiro Nakamura and Yoshihiro Nagai of C&C Information Technology Research Labs., NEC Corporation for their continuous encouragements. The numerical calculations were done on NEC supercomputer SX-2.

### REFERENCES

Bennets, D.A. and B.S. Hoskins, 1979: Conditional symmetric instability - a possible explanation for frontal rainbands. Quart. J. Roy. Meteor. Soc., 105, 945-962.

Emanuel, K.A., 1979: Inertial instability and mesoscale convective system. Part I. Linear theory of inertial instability. J. Atmos. Sci.,36, 2425-2449.

Durran, D.L. and J.B. Klemp, 1983: A compressible model for the simulation of moist mountain waves. Mon. Wea. Rev., 111, 2341-2361.

Hockney, R.W. and C.R. Jesshope, 1981: Parallel computers, Bristol, Adam Hilger Ltd, 423pp.

Kessler, E., 1969: On the distribution and continuity of water substance in atmospheric circulation. Meteor. Monogr., 10, No.32, 84pp.

Klemp, J.B. and R.B. Wilhemson, 1978: The simulation of three-dimensional convective storm dynamics. J. Atmos. Sci., 35, 1070-1096.

Melosh, R.J. and R.M. Bamford, 1969: Efficient solution of load-deflection equations. J. Structural Div., Proc. ASCE, 661-676.

Melosh, R.J., T. Lang, L. Schmele and R. Bamford, 1967: Computer analysis of large structural systems. AIAA Paper, No.67, 955.

Melosh, R.J., 1969: Manipulation errors in finite element analysis. Japan-U.S. Seminar on Matrix Methods of Structural Analysis and Design, Tokyo.

Saitoh, S. and H. Tanaka, 1987: Numerical experiments of conditional symmetric baroclinic instability as a possible cause for frontal rainband formation. Part I. A basic experiment. J. Meteor. Soc. Japan, 65, 675-708.

Saitoh, S. and H. Tanaka, 1988: Numerical experiments of conditional symmetric baroclinic instability as a possible cause for frontal rainband formation. Part II. Effects of watrer vapor supply. J. Meteor. Soc. Japan, 66, 39-53.

Tapp, M.C. and P.W. White, 1976: A nonhydrostatic mesoscale model. Quart. J. Roy. Meteor. Soc., 102, 277-296.

# Parallel Processing at Cray Research, Inc.

MARK FURTNEY

Cray Research, Inc., 1440 Northland Drive, Mendota Heights, MN 55120, USA

*ABSTRACT*

Multiprocessor computer systems are an integral part of Cray's future hardware directions. Increasing numbers of high-performance CPUs that can timeshare a large workload and/or cooperate on a single user program form the foundation for current and future systems. The software to support the use of multiple processors for a single user program has evolved through three generations: macrotasking, microtasking, and Autotasking. This paper briefly describes these three generations of software, with an emphasis on the just-released Autotasking. Included in the presentation of Autotasking will be design criteria and goals, tradeoffs in the design, and results from Autotasking on a variety of codes and kernels.

## 1. Introduction

Cray Research, Inc., introduced its first multiprocessor supercomputer in 1982, the dual-CPU CRAY X-MP/2. To support individual programs using both CPUs, a library of synchronization primitives was introduced (macrotasking), and it became the de-facto industry standard. In 1984, the UNICOS operating system, based on AT&T's UNIX System V, was introduced. This system provided new, easy-to-use tools (background processes, pipes) for utilizing multiple CPUs to accomplish complex tasks. As users and developers gained experience with macrotasking, several unexplored avenues for parallel exploitation were opened. The collection of these techniques (microtasking) addressed some of the weaknesses of macrotasking, was generally easy to use, and provided good performance in both batch and dedicated computing environments. Microtasking evolved from macrotasking, and provides substantial advantages over macrotasking; it still required that programmers occasionally do some detailed data dependence analyses to use it safely. Unlike vectorization with CRAY Fortran compilers, microtasking is not automatic, and that is the next evolutionary step. Autotasking has grown from the excellent performance experiences with microtasking, and provides an automatic mechanism for exploiting parallelism without programmer intervention. Since there will probably never be a replacement for the experienced Fortran programmer, Autotasking provides a variety of techniques by which the knowledgeable

Topics in Atmospheric and Oceanic Sciences
© Springer-Verlag Berlin Heidelberg 1990

user can fine-tune a program and exploit levels of parallelism not visible to the Auto-tasking system.

Section 2 describes briefly the macrotasking libraries, and experiences with macrotasking. Section 3 describes microtasking and how it has addressed some of the shortcomings of macrotasking. Section 4 discusses Autotasking: how it evolved from microtasking, how it fits in the compiling system, and how it can be used. Some of the benefits of the Autotasking system are discussed, and the results of Autotasking several large Fortran programs are reported in this section also. A short Summary section concludes this paper.

## 2. Macrotasking

Macrotasking is a technique whereby programs are modified with explicit calls to a special Fortran-callable library of synchronization routines (these routines are also callable from C and other languages). This collection of routines, termed the *macrotasking library*, provides the primitives necessary to allow a single program to execute correctly on multiple CPUs. These library routines interact directly with the operating system for the creation of extra tasks, and with the hardware to provide the necessary synchronization between concurrently-executing tasks. The macrotasking library contains four sets of routines: one (tasks) for task creation and manipulation, and three for synchronization (locks, events, barriers).

Macrotasking works best when the amount of work to be partitioned over multiple processors is large. When the work to be partitioned is not large compared to the synchronization time, excess synchronization time may become noticeable. When applying macrotasking to an existing code that did not consider multiprocessing as one of its design considerations, it may be necessary to do a significant amount of code restructuring. This can lead to an opportunity to introduce new errors. When the work is not easy to partition into equal-sized tasks, load imbalance may occur, producing lower speedups than anticipated. Further, macrotasked programs may display markedly different performance characteristics when comparing batch versus dedicated executions. Because of these and other reasons, macrotasking of existing large engineering and scientific applications codes did not appeal to many programmers. Microtasking is the direct result of efforts to address some of the characteristics of macrotasking that made it unpopular.

## 3. Microtasking

Microtasking is a technique for multiprocessing programs which is based on exploiting parallelism in DO-Loops. The primary design goals of microtasking have been to provide good performance over a wide range of problem sizes, to make it easy to use, and to work well in both batch and dedicated environments. Microtasking employs a master/slave relationship between CPUs. When the master processor enters a region that can benefit from parallel execution, it alerts the slaves, which may then enter the computation.

Programmers who employ microtasking determine where parallelism exists, then place comment directives in the text of their programs. See the bottom half of Table 2 for a description of the microtasking directives. A preprocessor reads these directives, and translates them and their associated DO-Loops into a form acceptable to the compiler. Code is generated that allows the program to use extra CPUs *if they are available*. If idle CPU cycles are available on the system, microtasked codes can use them as accelerators for completing a particular loop. If idle cycles are not available, the micro-tasked code (executing in the master processor) does not slow down to summon them or to wait for them. Loop iterations are handed out to the next CPU ready for work (self-scheduling), resulting in excellent load-balancing. The code that performs this protected "handing out" of iterations is extremely efficient, around 40 clock periods on a CRAY X-MP system. This permits the profitable exploitation of very fine-grained parallelism. Microtasking also makes good use of other CRAY X-MP features (like hardware deadlock detect) to provide very fast mechanisms for getting CPUs to join a "fray" (a code segment that may utilize multiple CPUs) and to remove CPUs from a potential busy-wait situation (for example when a CPU is withdrawn from a fray by the operating system and the other CPUs must wait for a DO-Loop to be completed before continuing).

When microtasking on the CRAY-2 system (which does not have such a rich set of parallel synchronization hardware), alternate methods were developed to handle these situations. These methods have proven to be so powerful that they are being moved to the CRAY X-MP microtasking design for comparison.

Although microtasking generates its computational efficiency by spreading DO-Loop iterations over multiple CPUs, it is based on subroutine boundaries, not loop boundaries. This has been done because subroutines provide a natural break in a code where the scope of variables (which subroutines and tasks can "see" particular instances of particular variables) can be easily manipulated into a form that permits correct parallel execution. This implies that some portions of microtasked subroutines are executed redundantly by whichever CPUs show up. In many instances, this redundant computation merely sets up a local context in which the parallel loop may execute (without slowing the master processor down by forcing it to broadcast local context).

On other occasions, there may be a fair amount of redundant code, or the code not inside parallelizable DO-Loops may not be safe to execute by multiple CPUs. This is the first of the drawbacks to microtasking. The second drawback is that programmers must still find the parallelism, and sometimes this can be difficult. Just as microtask-ing evolved from macrotasking and grew on its strengths while addressing its weaknesses, so Autotasking has evolved from microtasking. Autotasking retains the many advantages of microtasking, and relieves some of the shortcomings.

## 4. Autotasking

Autotasking is a technique whereby the compiling system detects opportunities for parallel exploitation, and generates code to execute these parallel regions on multiple

CPUs. Autotasking retains the many advantages of microtasking (self-scheduling for good load-balancing, very low synchronization overhead, uses idle CPU cycles when available, excellent performance in both batch and dedicated environments, original source code unmodified, etc.), while adding several new advantages. Autotasking can be completely automatic, where the compiling system does all the analysis and generates parallel code, or it can work with the programmer to provide support for potentially higher levels of parallelism exploitation. Autotasking works on DO-Loop boundaries, but is easily expandable to "parallel regions" and to subroutine boundaries.

Autotasking inside the compiling system can be thought of as a three-phase operation: *dependence analysis, translation,* and *code generation* (as illustrated by Diagram 1). Programmers may optionally pass directives to any of the three phases (see Tables 1 and 2 for an outline of these directives). For example, a programmer may know that a certain large program segment accounts for only a percent or two of total run-time, and so may choose to disable the extensive dependence analysis done in this region of the code because there will be no payoff (and to save compilation time). Or, a programmer may know that a DO-Loop that contains an external call may be safe to execute in parallel, and wish to alert the compiling system to that fact (the dependence analyzer does not look beyond subroutine boundaries). Sections 4.2 and 4.4 describe these directives.

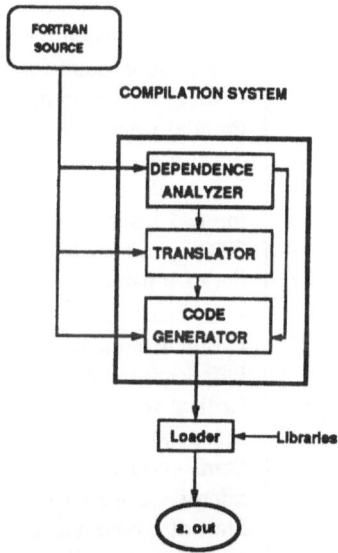

Diagram 1. 3-Phase Compiling System

| CFPP$ Directive | Description |
|---|---|
| CONCUR | Enable concurrency analysis |
| NOCONCUR | Disable concurrency analysis |
| INNER | Enable concurrency analysis for vectorizable loops |
| NOINNER | Disable concurrency analysis for vectorizable loops |
| VECTOR | Enable vectorization analysis for innermost loops |
| NOVECTOR | Disable vectorization analysis |
| SKIP | Disable dependence analysis |

Table 1. Directives to the Dependence Analyzer phase

| CMIC$ Directive | Description |
|---|---|
| CASE | Independent code block separator |
| CONTINUE | Parallelism extension to external |
| GUARD | Start a critical section |
| DO ALL | Independent DO-Loop iterations where the DO-Loop is the entire Parallel Region |
| END CASE | Independent code block terminator |
| END DO | Termination point for Reduction |
| END GUARD | Critical section terminator |
| END PARALLEL | Parallel Region terminator |
| PARALLEL | Start of a Parallel Region |
| DO PARALLEL | Independent DO-Loop iterations where the DO-Loop is inside a Parallel Region |
| SOFT EXIT | GOTO on next line cause a branch out of a parallel region |
| ALSO PROCESS[+] | Independent code block separator |
| CONTINUE[+] | Parallelism extension to external |
| DO GLOBAL[+] | Independent DO-Loop iterations |
| END GUARD[+] | Critical section terminator |
| END PROCESS[+] | Independent code block terminator |
| GUARD[+] | Start a critical section |
| MICRO[+] | microtasking subroutine follows |
| PROCESS[+] | Start of a code block |
| STOP ALL PROCESS[+] | GOTO on next line cause a branch out of a parallel region |

Table 2. Directives to the Translator phase

[+] Microtasking directive

## 4.1. Functions of the Dependence Analyzer

The dependence analyzer performs a wide variety of program optimizations. It looks for parallel constructs and may perform some source code transformations to produce faster-executing code. Payoffs from dependence analysis come in four major areas:
- Enhanced vectorization
- Recognition and generation of parallel
    constructs (concurrentization)
- Automatic in-line expansion
- Special code sequence recognition

### 4.1.1. *Enhanced Vectorization*

The dependence analysis phase recognizes vectorization opportunities and uses a host of techniques to try to produce vectorizable code. These techniques include statement reordering, ambiguous subscript resolution, reference reordering, splitting calls out of loops, loop nest restructuring, and loop exchanges (to get stride of one and/or longest vector length on the innermost loops). In a recent vectorization study [1] of 100 loops, the vectorized loop count went from 51 to 72 when using the new compiling system, showing a substantial improvement.

### 4.1.2. *Concurrentization*

The dependence analysis phase recognizes parallelization opportunities, and inserts directives to the next phase (translation), which tell it how to exploit this inherent parallelism. Again, the dependence analyzer may do some code transformations to produce parallel opportunities. In general, it will attempt to concurrentize on the outermost loop possible.

### 4.1.3. *Automatic In-line Expansion*

Automatic inlining of subroutines by the dependence analyzer also has a big payoff. Not only does in-line expansion remove the overhead of the call, it also can then possibly concurrentize on a more outer loop (from the original code before in-line expansion). In many cases this is very profitable.

### 4.1.4. *Specific Code Sequence Recognition*

The dependence analyzer also recognizes some special code sequences, including matrix multiply, first and second order linear recurrences, dot product, and search for maximum or minimum. It then generates calls directly to optimized library routines, which can perform these functions in parallel.

### 4.2. Parallel Processing Directives (to the Dependence Analyzer)

Directives to the dependence analyzer begin with "**CFPP$** " in columns 1-6 and run from columns 7-72. Uppercase and lowercase may be used freely in the directive text. Optional parameters to individual directives are delimited by brackets []. The following is a full list of the directives and their optional parameters with a discussion of each.

**CFPP$ CONCUR** *[(n)] [s]*

**CFPP$ NOCONCUR** *[s]*

**CFPP$ INNER** *[(n)] [s]*

**CFPP$ NOINNER** *[s]*

**CFPP$ VECTOR** *[s]*

**CFPP$ NOVECTOR** *[s]*

**CFPP$ SKIP** *[s]*

Throughout this section, the optional parameter *s* refers to the "scope" of the directive, and *n* refers to the "concurrency threshold". As directives are encountered in the Fortran source text, these parameters are treated as if they were in a push-down stack. The allowable values for *s* are:

| s | Description |
|---|---|
| L[OOP] | For the next DO-Loop (but not its inner Loops) |
| R[OUTINE] | For the rest of the current subprogram |
| F[ILE] | For the rest of the file |

When *s* is not specified, the default scope for a directive is LOOP. The "concurrency threshold" parameter, *n*, must be a positive integer constant. When *n* is specified, and the dependence analyzer can determine the DO-Loop tripcount, concurrency analysis is enabled only when (tripcount >= *n*). When *n* is specified and tripcount cannot be determined, concurrency analysis is enabled, but the dependence analyzer must generate the (tripcount >= *n*) test in the *if* clause of the **DO ALL** or **DO PARALLEL** directive (see next section).

**CFPP$ CONCUR** *[(n)] [s]*

**CFPP$ NOCONCUR** *[s]*

The **CONCUR** directive enables concurrency analysis, telling the dependence analyzer phase to look for parallelization opportunities over the scope (*s* parameter) specified. When *n* is not specified, concurrency analysis is enabled unconditionally. When *n* is specified, the rules outlined above are followed. At the beginning of a file sent to the compiling system, the default is **CONCUR FILE**. When concurrency analysis is enabled, the **INNER** directives are activated. The **NOCONCUR** directive disables concurrency analysis over the scope (*s* parameter) specified. Note that both **CONCUR** and **NOCONCUR** imply **VECTOR** (see below). That is, vectorization analysis is enabled whenever concurrency analysis changes.

**CFPP$ INNER** *[(n)] [s]*

**CFPP$ NOINNER** *[s]*

The **INNER** directive enables concurrency analysis for innermost vectorizable DO-Loops over the scope (*s* parameter) specified. The **INNER** directive is only recognized when concurrency analysis is enabled. That is, when the dependence analyzer phase cannot find exploitable parallelism on an outer DO-Loop, and **INNER** is enabled, and it detects an innermost vectorizable DO-Loop, the dependence analyzer will issue a **DO ALL** or **DO PARALLEL** directive for that loop. (See the next section for a discussion of the **DO ALL** and **DO PARALLEL** directives.) The **NOINNER** directive disables concurrency analysis for innermost vectorizable DO-Loops over the scope (*s* parameter) specified. At the beginning of a file sent to compiling system, the default is **NOINNER FILE**. Note that the **INNER - NOINNER** directives refer only to vectorizable innermost DO-Loops. Nonvectorizable innermost DO-Loops are subject to concurrency analysis under the control of the **CONCUR** directive.

**CFPP$ VECTOR** *[s]*

**CFPP$ NOVECTOR** *[s]*

The **VECTOR** directive enables vectorization analysis for innermost DO-Loops over the scope (*s* parameter) specified. At the beginning of a file sent to compiling system, the default is **VECTOR FILE**. The **NOVECTOR** directive disables vectorization analysis over the scope (*s* parameter) specified. The **NOVECTOR** directive is activated only when **NOCONCUR** is in control, and is ignored otherwise. Note that the **NOVECTOR** directive is employed only by the dependence analyzer phase, and therefore blocks only phase 1 dependence analysis. That is, **CFPP$ NOVECTOR** does not imply **CDIR$ NOVECTOR**. **CDIR$ NOVECTOR**, however, implies both **CFPP$ NOCONCUR LOOP** and **CFPP$ NOVECTOR LOOP**.

**CFPP$ SKIP** *[s]*

The **SKIP** directive disables both concurrency and vectorization analysis. **SKIP** is a shorthand for consecutive **NOCONCUR** and **NOVECTOR** directives. The principal use of this directive is turn off phase 1 (dependence analyzer) analysis in portions of program that do not contribute to any significant run-time, thereby saving some compile time.

## 4.3. Functions of the Translator

The primary function of the translator phase is to rewrite the Fortran-code-with-directives into pure Fortran for use by the code generation phase. Directives (whether written by the dependence analyzer phase or the human programmer) are expanded into a series of special function calls and compiler intrinsics that together implement the requested parallel processing functionality. A primary consideration in this rewriting exercise is to enforce the scoping requirements detailed in the directives. (See the next section for a discussion of individual directives and data scoping parameters.) Every variable in each parallel segment of code has a "scope" of either *private* or *shared*, as declared on the directive. There is only one copy of each *shared* variable, and it is available to all processors that contribute to the computation. There are potentially many copies of *private* variables, one per contributing processor.

## 4.4. Parallel Processing Directives (to the Translator)

Translator directives begin with "CMIC$ " in columns 1-6 and run from columns 7-72. Directives can be continued by using **CMIC$\*** in columns 1-6 on continuation statement(s), where "\*" can be any non-blank, non-zero character. That is, directives follow the Fortran continuation rules. Parameters on directives (for example, *private*) may be repeated as needed, and need not be ordered. Uppercase and lowercase may be used freely in the directive text. Optional parameters to individual directives are delimited by brackets []. A full list of the directives and their optional parameters follows with a discussion of each.

**CMIC$ PARALLEL** *[if(expr)] [shared(var[,...])]*
    *[private(var[,...])]*

**CMIC$ END PARALLEL**

**CMIC$ DO ALL** *[if(expr)] [shared(var[,...])]*
    *[private(var[,...])] [savelast] [single]*
    *[chunksize(n)] [numchunks(m)]*
    *[guided] [vector]*

**CMIC\$  DO PARALLEL** *[single]* *[chunksize(n)]*
    *[numchunks(m)]* *[guided]* *[vector]*

**CMIC\$  END DO**

**CMIC\$  GUARD** *[n]*

**CMIC\$  END GUARD** *[n]*

**CMIC\$  CASE**

**CMIC\$  END CASE**

**CMIC\$  SOFT EXIT**

**CMIC\$  CONTINUE**


**CMIC\$  PARALLEL** *[if(expr)]* *[shared(var[,...])]*
    *[private(var[,...])]*
**CMIC\$  END PARALLEL**

The **PARALLEL - END PARALLEL** directive pair delimits a *parallel region*, which provides a technique for modifying some variables' scope to allow correct multiprocessing to occur. Parallel regions provide a powerful mechanism with which the knowledgeable programmer can increase the efficiency of parallel computations by reducing the cost of parallel startup and spreading this cost over multiple parallel exploitation opportunities. Parallel regions are combinations of redundant code blocks and partitioned code blocks (for example, **DO PARALLEL**, or **CASE**, described below). The **PARALLEL** directive indicates where multiple processors enter execution, which may be different from where they demonstrate a direct benefit (partitioned code block). The following paragraphs explain the optional parameters:

*if* - When specified, a run-time test is performed to choose between uniprocessing and multiprocessing. When not specified, multiprocessing is chosen.

*expr* - The logical expression that determines (at run-time) whether multiprocessing will occur. When *expr* is True, multiprocessing is enabled.

*shared* - The variable(s) listed here will have GLOBAL scope. That is, they are accessible to the master and the slave tasks. The *shared* clause identifies those variables that by default are not GLOBAL, but for the purposes of parallel exploitation, need to be. By default, GLOBAL variables are those that appear in a COMMON block, the argument list, or in a DATA or SAVE statement, and all others are LOCAL.

*private* - The variable(s) listed here will have LOCAL scope. That is, each task (master and slaves) will have its own private copy of these variables. The *private* clause identifies those variables that by default are GLOBAL, but for the purposes of parallel exploitation, need to be LOCAL. By default, GLOBAL variables are those that appear in a COMMON block, the argument list, or in a DATA or SAVE statement, and all others are LOCAL.


CMIC$ DO ALL *[if(expr)]* *[shared(var[,...])]*
    *[private(var[,...])]* *[savelast]* *[single]*
    *[chunksize(n)]* *[numchunks(m)]*
    *[guided]* *[vector]*
CMIC$ DO PARALLEL *[single]*
    *[chunksize(n)]* *[numchunks(m)]*
    *[guided]* *[vector]*

The DO ALL and DO PARALLEL directives indicate that the DO-Loop that begins on the next line will be executed in parallel by multiple processors. No directive is used to end a DO ALL or DO PARALLEL loop. The "DO ALL *[parameters]*" directive is a special shorthand for the following three directives (note that no END PARALLEL directive is needed when DO ALL is used):

    CMIC$ PARALLEL *[parameters]*
    CMIC$ DO PARALLEL *[parameter]*
    :
    CMIC$ END PARALLEL

That is, the DO ALL initiates a parallel region whose only code is a DO-Loop with independent iterations. Note that when using DO ALL, the loop index variable is *private*. The following paragraphs explain the optional parameters:

*savelast* - This directive specifies that *private* variables' values (from the final iteration of a DO ALL) will persist in the master task after execution of the iterations of the DO ALL. By default, *private* variables are not guaranteed to retain the last iteration values. Note that *savelast* can only be used with DO ALL, and that if the full iteration set is not completed (for example, due to a SOFT EXIT), the values of the the *savelast* variables are indeterminate.

The rest of the parameters (*single, chunksize, numchunks, guided, vector*) specify the work distribution policy for the iterations of the parallel DO-Loop. By default, the iterations are handed out one at a time (that is, *single* is the default). Only one of the following five work distribution algorithms can be chosen for a given DO-Loop:

*single* - Parcel out iterations to available processors one at a time.

*chunksize(n)* - Break the iteration space into chunks of size "*n*", where *n* is an expression (for best performance, *n* should be an integer constant). *Chunksize(64)* is an analog of microtasking's LONGVECTOR directive.

*numchunks(m)* - Break the iteration space into "*m*" chunks of equal size (with a possible smaller residual chunk).

*guided* - Use Guided Self Scheduling [2] to partition the iteration space. This mechanism does a good job at minimizing synchronization overhead while providing decent dynamic load balancing.

*vector* - This scheduling algorithm is used only in the case of stripmining an innermost vectorized loop. It implies *guided* chunks down to a minimum strip of 64.

The following examples illustrate some typical uses of parallel regions and **DO PARALLEL** directive:

```
      CMIC$ PARALLEL
            X = 3.14159265
            Y = ZZ/3.1333345
            Z = SQRT(A+Y)
      CMIC$ DO PARALLEL
            DO 400   I = 1,IMAX
               :   Code using X,Y,Z
         400 CONTINUE
      CMIC$ END PARALLEL
```

Example 1 - Redundant initialization

In Example 1, each processor redundantly calculates X, Y and Z for its own use. In general this is much faster than having one processor (the master) calculate the values and then broadcasting them to other processors.

```
      CMIC$ PARALLEL
      CMIC$ DO PARALLEL
            DO 200   I = 1,IMAX
               :
         200 CONTINUE
      CMIC$ DO PARALLEL
            DO 400   J = 1,JMAX
               :
         400 CONTINUE
      CMIC$ END PARALLEL
```

Example 2 - Multiple Partitioned blocks

In Example 2, the parallel startup time is amortized over two DO-Loops. Note that the startup cost may be comparable to the cost of a subroutine call.

```
CMIC$ PARALLEL
        DO 600   I = 1,IMAX
             :
CMIC$ DO PARALLEL
        DO 200   J = 1,JMAX
             :
  200    CONTINUE
CMIC$ DO PARALLEL
        DO 400   K = 1,KMAX
             :
  400    CONTINUE
  600 CONTINUE
CMIC$ END PARALLEL
```

Example 3 - Multiple Partitioned blocks

In Example 3, no parallelism in the DO 600 loop could be found, but by using a parallel region, the startup time is amortized over IMAX*2 parallel loops.

```
        SUM = 0.0
CMIC$ PARALLEL  PRIVATE(XSUM)
        XSUM = 0.0
             :
CMIC$ DO PARALLEL
        DO 200   J = 1,JMAX
             :
        XSUM = XSUM+(A(J)*B(J))
             :
  200    CONTINUE
CMIC$ GUARD
        SUM = SUM+XSUM
CMIC$ END GUARD
CMIC$ END DO
CMIC$ END PARALLEL
```

Example 4 - Reduction Computation

In Example 4, the sum reduction computation on SUM is performed at full concurrent/vector speed. Each arriving processor has its own private copy of XSUM, which is added to the global SUM under protection of the GUARD (discussed below). The END DO directive forces late-arriving processors (those which do not get any iterations of the DO 200 Loop) to jump around the summation into SUM and wait for it to be complete before continuing.


CMIC$  GUARD [n]

**CMIC$  END GUARD** *[n]*

The **GUARD - END GUARD**  directive pair delimits a critical region, and provides the necessary synchronization to protect (or guard) the code inside the critical region. A *critical region* is a code block which is to be executed by only one processor at a time, although all processors in the parallel region execute it.  The optional parameter *n* is an expression that serves as a mutual exclusion flag (using the low-order 6-bits of the value).  That is, "GUARD 1" and "GUARD 2" can be concurrently active, but two "GUARD 7"s cannot.  For optimal performance, *n* should be an integer constant, and the general expression capability is provided only for the unusual case that the critical region number must be passed to a lower level routine.  When no *n* is provided, the critical region blocks only other instances of itself, but no other critical regions.  Critical regions may appear anywhere in a program.  That is, they are not limited only to parallel regions.

**CMIC$  CASE**
**CMIC$  END CASE**

The **CASE** directive serves as a separator between adjacent code blocks that are concurrently executable.  The **CASE** directive may appear only in a parallel region.  The **END CASE** directive serves as the terminator for a group of one or more parallel **CASE**s.  In the following example, subroutines ABC, DEF and XYZ are concurrently executable.

```
CMIC$ CASE
      CALL ABC
CMIC$ CASE
      CALL DEF
CMIC$ CASE
      CALL XYZ
CMIC$ END CASE
```

The work in all of ABC, DEF and XYZ completes before execution continues with the code below the **END CASE**.  A special form of the **CASE - END CASE** directive pair is to use it to force only a single processor to execute a code block in a parallel region, as in the following example.

```
CMIC$ PARALLEL
        :
CMIC$ CASE
      CALL XYZ
CMIC$ END CASE
        :
CMIC$ DO PARALLEL
      DO 200  I = 1,IMAX
```

```
    200  CONTINUE
    CMIC$ END PARALLEL
```

In the above example, only one processor calls **XYZ**.


## CMIC$  SOFT EXIT

The **SOFT EXIT** directive indicates that the GOTO statement on the next line branches outside the currently executing partitioned code block or parallel region.


## CMIC$  CONTINUE

The **CONTINUE** directive indicates that the external called on the next line has been specially prepared by the programmer for execution in parallel. The dependence analyzer will not generate this directive, nor can it prepare the called subprogram for this special form of processing. This is an important optimization technique for some programs.


### 4.5. Unique Features of Autotasking

Autotasking offers several features that allow more parallelism to be found and exploited. The three-phase compiling system gives programmers a great deal of freedom in selecting the forms of parallel processing most efficient for individual types of computation. Users also find great range in the forms of directives that can direct the dependence analyzer and the translator, and they are encouraged to combine their own knowledge of the problem domain with the dependence analyzer's output to create faster-running programs. The concept of a parallel region allows the computational overhead of processor startup to be minimized and amortized over multiple exploitable sections of code. Parallel regions also allow the efficient parallel exploitation of many forms of reduction computations. The *savelast* parameter on the **DO ALL** construct allows parallel loops that must carry scalar (or array) elements from the last iteration out of the loop body to be executed correctly and efficiently in parallel. The in-line expansion feature of the dependence analyzer is performed before parallelism analysis is performed, leading to many cases of outer parallel loops surrounding inner (in-lined) loops. The **SOFT EXIT** construct allows loops that contain a jump to outside their range to be safely processed in parallel. The **CONTINUE** construct permits an important cross-subroutine optimization to occur under user direction. The introduction of the five forms of parallel loop iteration partitioning (*single, chunksize, numchunks, guided* and *vector*) add a new dimension to tuning particular DO-Loops. *Guided* and *vector* are particularly useful for a variety of programs.

## 4.6. Some Results Using Autotasking

This section shows a series of results of using Autotasking on a variety of production codes and benchmarks. As expected, some codes contain very little or no parallelism, some contain a modest amount, and some are almost entirely parallel. It is hard to determine what proportion of codes in any particular problem domain fall into each of the three categories, but it is clear that there are many codes of each type. Independent of the amount of parallelism in a given program, users can almost always benefit from the messages generated by the dependence analyzer that describe why vectorization and/or concurrentization have been inhibited.

| CPUs | Time (secs) |
|------|-------------|
| 1 | 424.8 |
| 2 | 239.3 |
| 3 | 176.7 |
| 4 | 146.0 |

Case 1: Magnetohydrodynamics Code (CRAY X-MP/48)

The program in Case 1 is a large Magnetohydrodynamics code. When working out the maximum theoretical speedups possible for a code with this level of parallelism (just under 90%), we see that the implementation of Autotasking produces speedups very close to the maximum speedups possible. This is an example of the very low overhead cost associated with synchronization in Autotasking.

| CPUs | Microtasking Time (secs) | Autotasking Time (secs) |
|------|--------------------------|-------------------------|
| 1 | 258.311 | 250.219 |
| 2 | 131.582 | 157.210 |
| 3 | 92.751 | 124.939 |
| 4 | 73.638 | 109.824 |
| 5 | 55.516 | 97.847 |
| 6 | 53.712 | 93.981 |
| 7 | 51.927 | 89.466 |
| 8 | 50.861 | 87.003 |

Case 2a: Microtasking vs. Autotasking, CRAY Y-MP/832

Case 2a is representative of many programs that have been microtasked, then Autotasked (the microtasking directives were removed for the Autotasked run). The programmer took about two weeks to do the microtasking; the Autotasking system took about 210 milliseconds. It is very difficult for an automatic system to compete with human programmers. As in many cases like this, the human programmer microtasked a series of DO-Loops that contained external calls. The programmer checked the called routines and found it was safe to execute them in parallel. The Autotasking system does not look beyond subroutine boundaries, so was forced to make the "safe" judgment. It judged that the external call was not safe to execute in parallel, so it marked the loops as non-parallelizable because of the external calls. Programmers confronted with this situation can utilize the **CFPP$ CNCALL** directive to the dependence analyzer to indicate that an external call in the next loop should be considered safe for parallel execution.

| CPUs | Microtasking Time (secs) | Autotasking Time (secs) |
|---|---|---|
| 1 | 166.122 | 171.724 |
| 2 | 86.113 | 87.651 |
| 3 | 59.753 | 59.601 |
| 4 | 46.819 | 45.603 |
| 5 | 40.364 | 37.440 |
| 6 | 34.145 | 32.026 |
| 7 | 33.999 | 28.268 |
| 8 | 28.863 | 25.628 |

Case 2b:  Microtasking vs. Autotasking, CRAY Y-MP/832

Case 2b, which compares microtasking and Autotasking, occurs much less frequently. As in Case 2a, the time required was about two weeks for microtasking of Case 2b versus 190 milliseconds for Autotasking. In this case however, Autotasking found several areas for parallel execution not found by the programmer. In particular, some concurrentizable reductions were found (there is no mechanism in microtasking for this construct) and exploited.

| CPUs | CRAY Y-MP | | CRAY-2 | |
|---|---|---|---|---|
| | Actual | Theoretical | Actual | Theoretical |
| 2 | 1.891 | 1.893 | 1.868 | 1.890 |
| 3 | 2.674 | 2.695 | 2.609 | 2.687 |
| 4 | 3.355 | 3.419 | 3.278 | 3.434 |
| 8 | 5.459 | 5.727 | - | - |

Case 3a:  Speedups: LU Decomposition (500x500)

| CPUs | CRAY Y-MP | | CRAY-2 | |
|---|---|---|---|---|
| | Actual | Theoretical | Actual | Theoretical |
| 2 | 1.930 | 1.941 | 1.907 | 1.933 |
| 3 | 2.788 | 2.828 | 2.717 | 2.805 |
| 4 | 3.611 | 3.665 | 3.429 | 3.622 |
| 8 | 6.262 | 6.595 | - | - |

Case 3b:  Speedups: LU Decomposition (1000x1000)

Cases 3a and 3b illustrate speedups over single-CPU versions of large LU-Decomposition computations for an 8-CPU CRAY Y-MP system and a 4-CPU CRAY-2 system. Although these machines are similar in many respects, their different memory speeds and vector start-up times combine to produce slightly different Maximum Theoretical Speedups. This is a good example of why it is not a good idea to compare speedups between machines. This is especially important when considering machines with very different characteristics.

| Kernel Number | Megaflops: Unitasking | Megaflops: Autotasking | Megaflops: Autotasking Plus Mods |
|---|---|---|---|
| 1 | 275.4 | 1556 | 1685 |
| 2 | 74.2 | 70.2 | 446.0 |
| 3 | 89.9 | 122.5 | 285.8 |
| 4 | 149.5 | 236.7 | 236.7 |
| 5 | 118.4 | 950.9 | 1036.7 |
| 6 | 135.1 | 169.4 | 1132.8 |
| 7 | 52.9 | 50.2 | 432.5 |

Case 4: NASA Computational Kernels, CRAY Y-MP/832

Case 4 is a series of computational kernels from NASA/Ames that are representative of a large portion of their workload. Column 2 represents their unitasking Megaflop rates on a CRAY Y-MP. In Column 3, these same kernels have been Autotasked, showing a wide range of performance improvements. One of our benchmarkers was allowed to make a specified number of changes, and ran the codes again to get the results shown in Column 4. Again we see that the experienced programmer can often make a significant difference in overall performance. In these cases, the benchmarker used information about inhibitors to parallelism put out by the dependence analyzer, and added his own experience and knowledge about the problem domains to generate faster run-times. These NASA kernels seem representative of what one can expect from Autotasking, and Autotasking augmented by a knowledgeable programmer. Sometimes the automatic system can find little or no parallelism, and other times it is very good at finding and exploiting parallelism. Usually the programmer can improve the performance of a code, often by knowing information like tripcounts for important DO-Loops, looking beyond subroutine boundaries, and algorithm rewriting.

| Before Autotasking | 242 Megaflops |
|---|---|
| After Autotasking | 1.9+ Gigaflops |

Case 5: Utrecht Benchmark, CRAY Y-MP/832

This program implements a "Black and Red Elliptical Differential Equation Solver". This benchmark, brought in by the University of Utrecht in Holland, is well above 99% parallel. These results were generated completely automatically. It is a shame that not all codes are as parallel as this one.

## 5. Summary

This paper has discussed software for parallel processing available on Cray supercomputer systems. The three generations of Multitasking software (macrotasking, microtasking and Autotasking) have been described, showing how they evolved from the needs of the scientific and engineering applications computing community and from the strengths of their predecessors. All three generations provide some special services not available from the others, and all three will be supported by Cray in the future. Macrotasking, microtasking, and Autotasking are part of a single integrated parallel processing package, and can be used in any combination in a single program.

Particular attention has been paid in this paper to Autotasking, with discussions of its internal features and some results with production programs.

## 6. References

[1] David Callahan, Jack Dongarra, and David Levine. *Vectorizing Compilers: A Test Suite and Results.* Proceedings: Supercomputing '88.

[2] Constantine Polychronopoulus and David Kuck. *Guided Self Scheduling: A Practical Scheduling Scheme for Parallel Supercomputers.* IEEE Transactions on Computers, December 1987.

# Inherent Parallelism in Numerical Weather Prediction Algorithms

DAVID F. SNELLING

Department of Computing Studies, University of Leicester, Leicester, LE1 7RH, U.K.

## Abstract

As the number of parallel processors in massively parallel systems increases, the need for concurrent tasks or processes also increases. In this study the author investigates the total parallelism inherent in NWP algorithms. The technique employed is to implement a well known, highly simplified algorithm for solving the shallow water equations in one of the languages of dataflow systems, SISAL. In this form all extent parallelism can be detected by the compiler and made accessible to the dataflow computational environment. In this case the environment is a dataflow simulator which produces detailed diagnostics about the parallelism in the algorithm. These results are then compared to similar values obtained from a conventional SIMD system, a Cray X/MP.

## 1.    Introduction

Numerical Weather Prediction (NWP) is one of many fields for which significant increases in resolution can provide immediate, tangible, qualitative, and even financial benefits. These increases in resolution require commensurate increases in the memory and computational requirements of NWP models. As a result of this increased need for resources, computer system design has become a Sisyphean or Herculean labour. One approach adopted by machine architects is to build systems from numerous, very simple processing elements. In this case it is generally assumed that the target application will have enough parallelism to efficiently utilize all the processing elements. The question remains: just how much parallel computation is there in NWP models?

At present NWP models are split in two ways to achieve parallelism: 1) the data is organized into vectors (40 to 10000 way parallelism) and 2) the solution space is divided into sections, usually along latitude lines (2 to 200 way parallelism). The latter is frequently called spatial decomposition. In many cases an increase in one type of parallelism is accompanied by decreases in the other.

Topics in Atmospheric and Oceanic Sciences
© Springer-Verlag Berlin Heidelberg 1990

The intent of this study is to analyze the parallelism the in a standard algorithm from NWP, a simple shallow water equation model. The approach taken here is to translate this algorithm into SISAL, (Streams and Iteration in a Single Assignment Language) a language designed originally for dataflow computers. This program is then run on a dataflow simulator to determine the exact degree of parallelism inherent in the algorithm. This, and additional information on resource utilization, can provide insight into the feasibility of solving a very high resolution NWP model using many thousands of processors. Lastly, as this algorithm is well known, and data are available for it on current super computers, some comparative analysis can also be undertaken.

This paper will include three background sections and an analysis section. In the background sections the shallow water wave equation model, the dataflow environment as a tool for analysing inherent parallelism, and a collection of metrics used in this approach are described. In the final section an analysis of this algorithm is presented including some comparative results obtained on a Cray X/MP vector processor.

## 2.    The Algorithm

In this section only a very brief summary of the shallow water equations will be presented. A more detailed discussion of the formulation of these can be found in [HOFF88, SADO78]. In this model a single layer of fluid is modeled in a square Cartesian domain. These equations represent only the types of calculation that are present in an atmospheric model and are them selves only a model of the very primitive, but fundamental, equations of fluid flow.

The terms included in this implementation of the shallow water equations are the velocities in the x and y directions u an v; the pressure P; a field height term H; the mass fluxes U and V; and the potential velocity Z. These equations, as formulated in the model are given below.

$$\partial u/\partial t - ZV + \partial H/\partial x = 0$$
$$\partial v/\partial t + ZU + \partial H/\partial y = 0$$
$$\partial P/\partial t + \partial U/\partial x + \partial V/\partial y = 0$$

In the simplification of these equations, both the north/south and the east/west boundaries have been taken to be periodic, see figure 1. This structure would be replaced by more sophisticated techniques in a actual model. For example: boundary conditions interpolated from a global model could be used in a limited area model solved on a Cartesian domain.

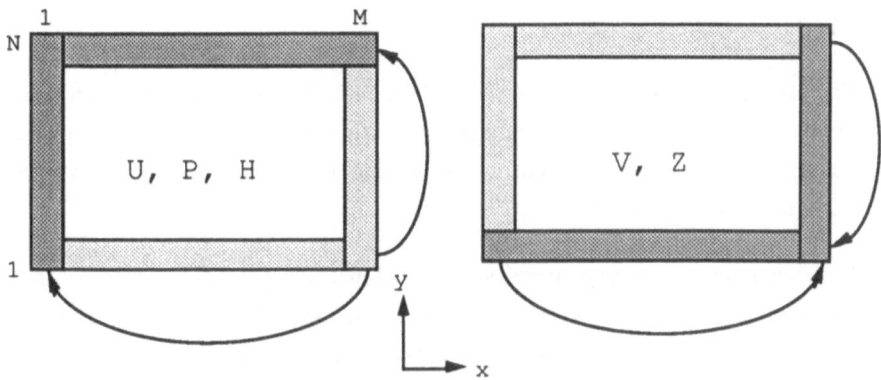

Figure 1. The solution domain.
Cartesian domain with periodic boundaries.

In earlier implementations the process of continuing these boundary condition has created an explicit global synchronization each time any values are continued. In the code outlined below periodic continuation occurs twice during each time step. In effect this restricts the available parallelism to the spatial domain and prohibits the exploitation of any parallelism that might exist between stages of the computation or in the time domain. This is certainly true in past experience, where this feature of the code was used to study the impact of synchronization on large multiprocessor systems [HOFF88, SNEL88].

> **For** each time step **Do**
>   Compute U, V, Z, & H from u, v, & P.
>   Periodic continuation of U, V, Z, & H.
>   Compute new_u, new_v, & new_P.
>   Update time filtered values.
>   Copy new values to current values.
> **End_For**

## 3. The Data Flow Analysis Environment

In this section the primary features of the dataflow environment as a tool for studying parallel applications will be discussed. This will include a summary of the dataflow model of execution, how it contrasts with the conventional parallel Von Neumann environment, a brief summary of the dataflow language SISAL, and a short description of the dataflow simulator used in this study.

*The Dataflow Execution Model*

The primary features of the conventional parallel Von Neumann machine, that have dramatically different counterparts in a dataflow machine, are instruction stream management, processes synchronization, and techniques for the detection and exploitation of parallelism.

Depending on the type of conventional parallel machine, parallel processes (instruction streams) are managed in a variety of ways. In most cases each process will have associated with it an instruction or program counter. These processes are then allocated to processors within the system and execute sequence of instructions which modify memory resident data. They will typically execute on the assigned processor until termination. It should be noted however, that on some interactive systems these processes may migrate around the processors of the system. By contrast the dataflow system has no program control counters; programs do not progress from one instruction to another. Rather, instructions wait in readiness for their operands to arrive, at which time they execute. It is this feature that gives the dataflow model its name. This model of computation can be represented graphically, see figure 2.

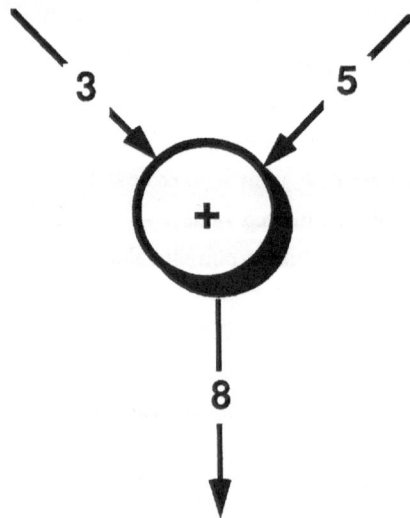

Figure 2. A Simple Dataflow Graph
of a Single Instruction.

In terms of parallel processing, it is synchronization that highlights the key differences between these two approaches. Whereas in conventional parallel processors either the user or the compiler must provide explicit information about the synchronization of parallel processes or the communication between them, in the dataflow environment all synchronization is provided implicitly in the flow of data from where it is created to where it is referenced. It is this mechanism that will be exploited to determine the total parallelism in an application, since the granularity of this parallelism is at the single instruction level.

As noted above, it is usually the user that must identify the points in an application where it would be appropriate to exploit parallelism. Even in systems where there exists some form of automatic detection of parallelism, directives to the compiler are frequently required to assist the compiler. This difficulty becomes even more evident in systems exploiting a large number of parallel processors using some form of communication. In these systems very little, if any, automatic detection of parallelism is possible.

As the goal of this study is to detect and measure parallelism inherent in an application, it is imperative to circumvent this software barrier. In the dataflow model, all applications are compiled to dataflow graphs which contain all the information necessary to extract all parallelism present in a given program. This is in fact the same technique used by existing vectorizing and parallelizing compilers except that it functions on the whole of a program rather than on a single block. In order to accomplish this, an unconventional programming language is used. In the case of this study the language is SISAL. Because the SISAL compiler performs complete data dependency analysis at all levels in a program, it is able to detect and extract parallel constructs as small as single instructions. It thus provides a vehicle for extracting virtually all parallelism in a given application. More complete discussions of the dataflow execution model are available in [GURD88, GURD85, IEEE82].

*SISAL*

SISAL (Streams and Iteration in a Single Assignment Language) can be described as a first order functional language which provides constructs familiar to the scientific programmer. It supports, in a well defined and optimized way, array and record constructs, and a very powerful iterative construct, all features essential to the scientific community. The term "single assignment" refers to the philosophy that no variable may appear more than once on the left hand side of an assignment statement. Since, in a loop, each iteration is in essence another assignment of all variables in the loop body, the language interprets each loop iteration as if new values are created during each iteration, rather than changing existing ones. Access to values created within one loop iteration are available in the next by using the "old" operator. In this way recurrence relations can be coded as well as other iterative and time stepping algorithms.

Since SISAL supports functions and must be fully analyzed by compilers, all functions must be side-effect free. Therefore all arguments to a function are accessed by value only, and the function result, which may include several variables, is assigned as part of the function operation. This feature is extended to all constructs in the language, e.g. an if statement will return a result. The example in appendix A is the main function of the shallow water model and highlights many of the key features of SISAL. More information about SISAL is available in [GRUD88, MCGR85]

*Simulation*

The dataflow simulator used in this study simulates an ideal dataflow machine in the sense that all instructions take exactly one machine cycle, the sizes of matching store and structure store can be increased indefinitely, and there exists an infinite number of parallel processing elements in the machine. Since there are an infinite number of processing elements in the simulator, all instructions that can execute on a given machine cycle will, thus exposing the maximum parallelism in the algorithm. The simulator also maintains complete statistics on the parallel behaviour of the program. These statistics are discussed in more detail in the next section.

## 4. The Metrics

The metrics produced by the dataflow simulator are generally measured in instructions or machine cycles since the two are assumed to be equivalent in the simulator. These statistics can be produced at predetermined intervals throughout the simulation and as a summary of aggregates at the end of a simulation. In this study only the summary aggregates are used. A brief definition of each of the metrics used is given below.

S1    is the total number of instructions issued during the execution of the program. For the Cray figures, each vector instruction counts as only one even though it performs many operations. This is in accordance with the dataflow "tuplicate" instruction which replicates an operand an arbitrary number of times in a single instruction.

Sinf    is the critical path for the execution of the program measured in instructions. For the measurements on the Cray, the following formulation was used:

Sinf = (Total number of cycles required to execute the program)/(Average scalar instruction length in cycles weighted by the frequency that each type of instruction occurred in this program)

This produces a reasonable approximation to the time for completion measured in Cray instructions.

Pi   is the average degree of parallelism in the algorithm as a whole. It is also measured in instructions and is simply S1/Sinf.

Pmax   is the maximum degree of parallelism encountered during the simulation measured in instructions. For measurements on the Cray Pmax is assumes to be the minimum of the vector length used in the algorithm and the combined length of all the pipelines in the hardware (i.e. 14(memory) + 7(Multiply) + 6(adder chained) = 27. This is clearly an overestimate, but in light of the comparisons made this is not a major problem.

MMR   is the MIPs to MFlops ratio. It is a measure of the efficiency of the instruction set architecture and the compiler, rather than the efficiency of hardware architecture. MMR = the number of instructions required for each floating point operation.

*Example: Complex Multiply*

In figure 3 a complex multiply is diagramed using the standard directed dataflow graph notation. Accompanying this example are the above metrics as they would be computed by the simulator. Note that the "Tup" instruction is necessary as the operands are required in several places.

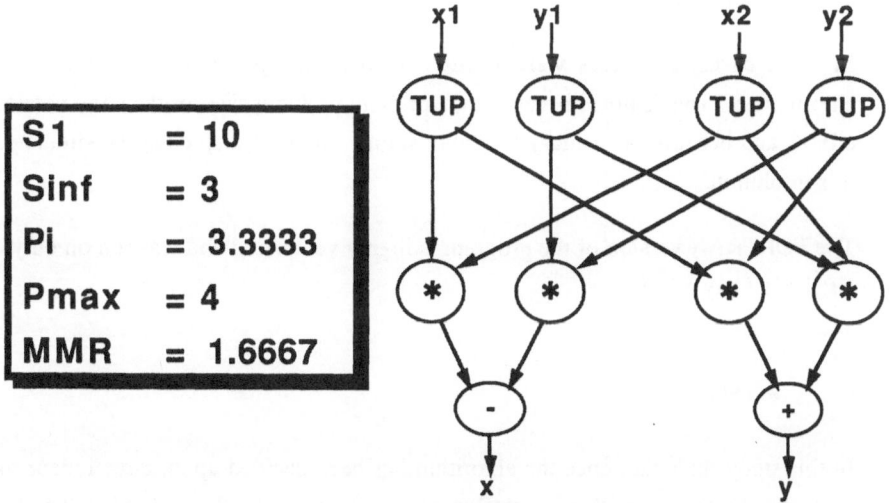

| S1 | = 10 |
| Sinf | = 3 |
| Pi | = 3.3333 |
| Pmax | = 4 |
| MMR | = 1.6667 |

Figure 3. A Complex Multiply and Associated Metrics.

This set of metrics, including the Cray approximations, provide a useful environment in which to study, comparatively, parallelism in algorithms. In the dataflow simulator all these metrics are independent of the hardware on which the program might be running, whereas the Cray results are highly dependent on the hardware in question and are, in fact, computed using the hardware monitoring facility provided as part of the system. Therefore, it should be noted that the two environments are different enough that to attribute any significance to small numerical differences between the two situations would be ill advised. However, it is the author's opinion that large (i.e. order of magnitude) differences and comparative trends are worthy of serious attention.

## 5. Experimental Assumptions

As with any investigation there are a collection of experimental assumptions within the framework of which all conclusions must be considered. In this case these assumptions cause very little confusion in any of the effects seen but are included for completeness.

As stated earlier, this application is a vastly simplified version of the shallow water equations based on the original FORTRAN source by Hoffmann, Swarztrauber, & Sweet [HOFF88].

The simulations of the SISAL version were run on SUN/3 work stations using the dataflow simulator produced by the Computer Science Department at the University of Manchester.[SARG86]

In some cases, test sizes were limited by the configuration of the workstation environment. This is noted because the optimum problem size for the Cray version is 64x64, and because of memory limitations, only one time step could be simulated at this resolution.

The FORTRAN version of the program is highly vectorized and was run on only one CPU of a Cray X/MP.

## 6. Analysis

In this study there are, once the algorithm has been decided upon, three independent variables: 1) the environment, dataflow or vector processor, 2) the system size, and 3) the number of time steps. Against each of these variables it is possible to plot all of the metrics described earlier and then discuss any enlightening comparisons. A complete comparison of all these factors would be tedious at best, and therefore I have highlighted only a few of the more interesting comparisons found with respect to each of the independent variables. Firstly, a single problem size (64x64) was run for 1 time step on both the Cray and the dataflow simulator; secondly, the problem size was varied for a fixed number of time steps; and thirdly, a fixed problem size was run for increasing numbers of time steps. In the last two cases only dataflow simulations were performed.

*Comparisons to Cray for fixed time step and system size*

| System | S 1 | Sinf | Pi | Pmax | MMR |
|--------|-----|------|------|------|-----|
| Cray | 162836 | 121845 | 1.34 | 27 | 0.20 |
| Dataflow | 720151 | 256 | 2813.10 | 29574 | 2.62 |

Table 1. S1, Sinf, Pi, Pmax, & MMR for the
Cray and the Dataflow Systems
System size = 64 & run for 1 time step.

It is clear from table 1 that the vector processing environment is far more efficient than the dataflow environment in terms of the number of instructions that must be fetched, decoded, and executed in order to perform a given calculation. This is indicated by the raw number of instructions (S1) and the MIPs/MFlops ratio (MMR). In the current dataflow environment there is only one vector instruction, the "tuplicate" instruction. Greater efficiencies could be obtained if other vector instructions were added. This would further complicate the already complex structure of a dataflow machine and erode the total degree of parallelism detectable, since vectors would then be treated as autonomous units.

Notice that, in spite of the larger number of instructions executed in the dataflow environment, the time to solution, measured in instructions, is approximately a factor of 475 smaller than the vector environment. This is because the vector machine only exploits a very small fraction of the available parallelism in the problem, whereas this ideal, infinite, dataflow machine can take advantage of all parallelism detectable in the application. This effect is also indicated by the average parallelism (Pi) and the peak parallelism (Pmax) in each case.

*Fixed time step and varying system size*

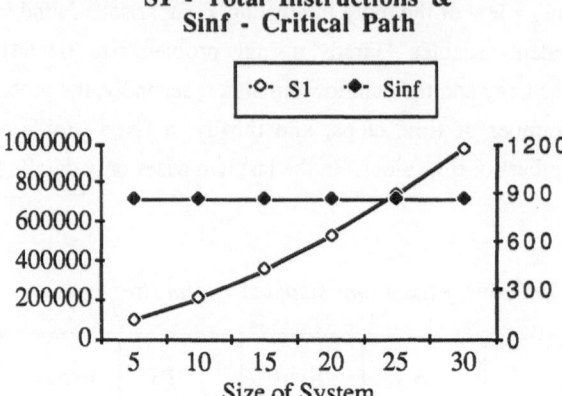

### S1 - Total Instructions &
### Sinf - Critical Path

Figure 4. S1 and Sinf vs. System size.
Simulation run for 5 time steps.

In figure 4 the quadratic growth in the total number of instructions is in line with what would be expected for a problem defined on a grid. Also, with an infinite machine we would expect that an increase in the problem size, which does not create any additional dependencies, should not increase the time to solution. In a real machine several factors, such network latency and actual machine size, would prevent realization of this effect.

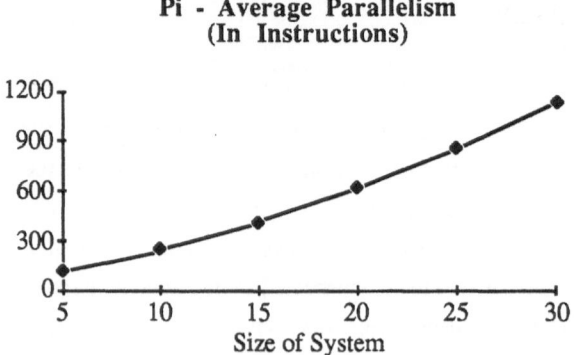

### Pi - Average Parallelism
### (In Instructions)

Figure 5. Pi vs. System size.
Simulation run for 5 time steps.

**Pmax - Peak Parallelism
(In Instructions)**

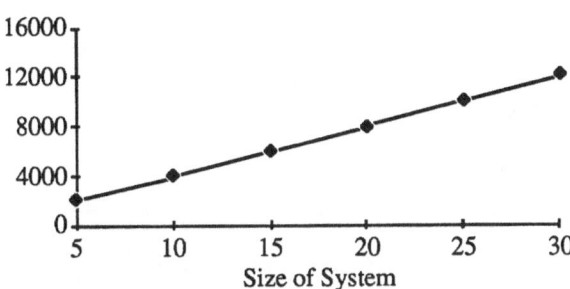

Figure 6. Pmax vs. System size.
Simulation run for 5 time steps.

Likewise, there are few surprises in the graphs of Pi and Pmax. Pi increases quadraticly as the system size changes, which is as we would expect, although Pmax appears to be linear. It is believed that if the system size were increased significantly, a quadratic effect would be observed in Pmax as well. It is also worth noting, that the absolute magnitude of the parallelism is consistent with a system in which every grid point can be treated independently, and that there is a very high degree of parallelism available even in this very small demonstration program. Real world weather models have approximately 1000 times as many grid points and several hundred more calculations per grid point, many of which are independent of each other, than the largest system studied here.

*Fixed system size and varying time step*

Critical path (Sinf) and total instructions (S1) both show simple linear growth against the number of time steps. This is not surprising and therefore no graphs have been presented.

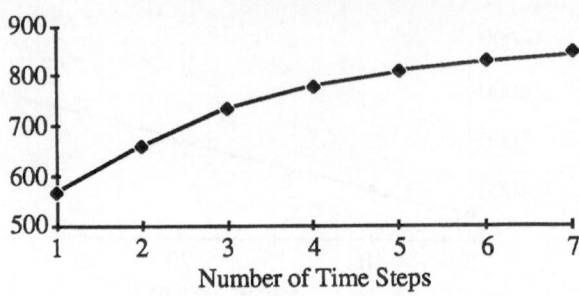

Figure 7. Pi vs. Number of time steps,.
Simulation run for System size = 24.

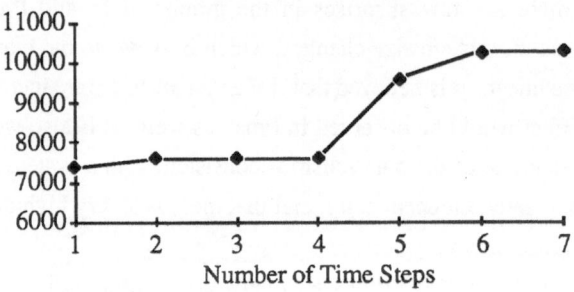

Figure 8. Pmax vs. Number of time steps,.
Simulation run for System size = 24.

The behaviour of the parallelism metrics, Pi and Pmax, is far more intriguing. Initially surprise might be registered at the increase in available parallelism of a factor of 1.5 in the seven time step case as compared to the one time step case. Particularly, as in the formulation of this application there are two separate implicit global synchronizations per time step. Yet, when all the available parallelism is detected, as in a dataflow compiler, these synchronizations are found to be spurious. A careful analysis of the time stepping mechanism, which is a two stage time filtering scheme, will reveal that for a large number of time steps it is possible that nearly two entire time steps can be in some stage of computation all the time. This effect is approached gradually because the periodic boundary conditions prevent complete parallelism across two time steps. If an explicit time stepping algorithm is employed, this inter-time-step parallelism would wave-front across the entire solution space. This means that, if the spatial size of a problem grows quadraticly, the degree of parallelism grows with forth power, for a large number of time steps.

The sharp increase in Pmax between the fourth and fifth time steps was, at the time of publication, unexplained, although it may have to do with the start up phase of the time filtering algorithm.

## 7. Conclusions

This study has made it clear that there is indeed a great deal of parallelism in even the simplest of weather models, certainly more than is exploited by our current parallel processing techniques. Much of this parallelism could be exploited by current techniques, but in order to exploit parallelism such as what was found across time steps, either very complex programming techniques or the use of implicit parallel languages will be needed. Another advantage of the later approach is the easy management of irregular calculations such as conditional computation. Note also that any attempt to exploit inter-time-step parallelism for other time stepping algorithms in a conventional environment would require repeated effort on the part of the programmer. Whereas, if implicit parallel detection techniques are employed, the time stepping algorithm can be altered freely.

There is, however, no crisis in available parallelism at the present time. As indicated above, present hardware only exploits a small portion of the available parallelism. Although only a vector machine was considered here, it is the author's belief that even current massively parallel machines, such as the CM2 from Thinking Machines, exploit only a fraction of the available parallelism in problems large enough to effectively utilize the hardware. Additionally, even this degree of parallelism is only available in an SIMD environment. It will be some time before the degree of MIMD parallelism detected in this study, and inferred in real world weather models, is accessible to commercially available hardware.

## 8. References

[HOFF88]   Geerd-R. Hoffmann, P. N. Swarztrauber, & R. A. Sweet, "Aspects of Using Multiprocessors for Meteorological Modelling," in Multiprocessing in Meteorological Models, Hoffmann, Geerd-R. and D. F. Snelling eds. (Springer-Verlag, Berlin: 1988).

[SADO78]   R. Sadourney, "The Dynamics of Finite Difference Models of the Shallow Water Equations," Journal of Atmospheric Sciences No. 32 (1975) pp.680-689

[SNEL88]   Snelling, D. F. and D. A. Tanqueray, "Performance Measurements of the Shallow Water Equations on the FPS T Series," submitted to CONPAR 88.

[GURD88]   J. Gurd, "Dataflow Architectures and Implicit Parallel Programming," in Multiprocessing in Meteorological Models, Hoffmann, Geerd-R. and D. F. Snelling eds. (Springer-Verlag, Berlin: 1988).

[GURD85]   J. Gurd et al., "The Manchester Prototype Dataflow Computer," in Communications of the ACM, 28:1, (1985) pp. 34-52.

[IEEE82]   Special issue on Dataflow computing, IEEE Computer 15:2, (1982).

[MCGR85]   J. R. McGraw et al. SISAL - Streams and Iteration in a Single Assignment Language, Language Reference Manual V1.2, Lawrence Livermore National Laboratory, (1983)

[SARG86]   J. Sargent, Simulator Users Guide, Dataflow Research Group, Department of Computer Science, University of Manchester (1986)

**Appendix A**

```
function   Shallow (Minput:    integer;
                    lastiter:   integer
        returns     stream [real])

    for initial
        P        : Real2DArray;   % Pressure
        U        : Real2DArray;   % East/west Velocity.
        V        : Real2DArray;   % North/south Vel.
        Psmooth  : Real2DArray;   % Time smoothed P.
        Usmooth  : Real2DArray;   % Time smoothed U.
        Vsmooth  : Real2DArray;   % Time smoothed V.
        M        : integer;       % Dimension of system.
        iter     : integer;       % Iteration count.
        deltat   : real;          % Time step size.
        delta    : real;          % Grid spacing.

        M        := Minput;
        iter     := 0;
        deltat   := 90.0;
        delta    := 1.0E5;

        P, U, V := Initialize (M, delta);
        Psmooth := P;
        Usmooth := U;
        Vsmooth := V
    while
        (iter < lastiter)
    repeat
        P, Psmooth, U, Usmooth, V, Vsmooth :=
        let
            Uflux    : Real2DArray;   % E/W Mass flux.
            Vflux    : Real2DArray;   % N/S Mass flux.
            H        : Real2DArray;   % Height Term.
            PotVel   : Real2DArray;   % Potential vel.
            Pnext    : Real2DArray;   % Temp results.
            Unext    : Real2DArray;   % Temp results.
            Vnext    : Real2DArray;   % Temp results.

            Uflux, Vflux
                     := Fluxes    (M, old P, old U, old V);
            H        := Height    (M, old P, old U, old V);
            PotVel   := Potential (M, old P, old U, old V,
                                        delta);
            Pnext, Unext, Vnext := TimeStep
```

```
                    (M, old deltat, delta,
                     old Psmooth, old Usmooth, old Vsmooth,
                     Uflux, Vflux, H, PotVel)
        in
            if (old iter = 0) then
                Pnext,
                old P,
                Unext,
                old U,
                Vnext,
                old V
            else
                Pnext,
                Smooth (M, old P, old Psmooth, Pnext),
                Unext,
                Smooth (M, old U, old Usmooth, Unext),
                Vnext,
                Smooth (M, old V, old Vsmooth, Vnext)
            end if
        end let;

        deltat := if (old iter = 0) then
            old deltat * 2.0
        else
            old deltat
        end if;
        iter     := old iter + 1;
    returns
        stream of  P[2,2]   % Checksum computation.
    end for
end function
```

# Report of Application of the Parallel Algorithm in NWP at SMA*

LIAO DONGXIAN and HUANGFU XUEGUAN

National Meteorological Center, SMA, 46 Baishigiaolu, Beijing, China

The report consists of two parts, namely, a barotropic experiment and a scheme for integrating a global spectral multilevel primitive equation model.

In the first part, an experiment on the parallel algorithm with the shallow water equation model has been carried out on the computer IBM-4381-p03. The results show that around 1/3 of the total time spent can be saved against the single-tasking time.

In the second part, in order to utilize the parallel algorithm well, a global multilevel spectral primitive equation model has been reduced to several sets of new equations by means of vertical normal modes. Then a time integration scheme with the algorithm is proposed and a corresponding flow graph is given.

## 1.   INTRODUCTION

Since the parallel/pipeline processors appeared in the 1960's, great changes have been brought about in numerical mathematics, and many new methods and algorithms have been proposed such as synchronous parallel algorithms and asynchronous parallel algorithms and so on. They made substantial contributions to raising the computational power and efficiency.

At the State Meteorological Administration in China, the study of the parallel algorithm started in 1981. In a few years from then Wan (1981, 1982) undertook theoretical investigations on the algorithm in consideration of the fact that the atmosphere is inseperable in some aspects, but separable in the other aspects. In 1988 his student, Jianbo, carried out a numerical experiment on the algorithm on IBM-4381-p01 with the shallow water equation model. Although those works are few in number, they are favourable to the development of the algorithm in China in the future.

In the following sections we shall describe Jianbo's experiment and the results he obtained, and then the scheme for integrating a global multilevel spectral primitive equation model on the computer with two processors.

---

*Abbreviation of the State Meteorological Administration

Topics in Atmospheric and Oceanic Sciences
© Springer-Verlag Berlin Heidelberg 1990

## 2. AN EXPERIMENT WITH THE SHALLOW WATER EQUATION MODEL

### 2.1 Separation of the equations

The computer IBM-4381-p03 has two processors. In order to exploit its potential power fully, we should allocate the job required for computation to two parts as relatively independent and equal in quantity as possible. For this reason the shallow water equations used in the experiment are separated as follows.

In the rectangual coordinates the set of shallow water equations may be written in matrix form

$$W_t + A W_x + B W_y + F W = 0 \tag{1}$$

where $W = (u, v, 2c)^T$, u and v are the wind components in x and y directions respectively, $c = \sqrt{\varphi}$, $\varphi$ is the geopotential height, f Coriolis parameter, t time, the superscript T a transpose, $W_s = (u_s, v_s, 2c_s)^T$, s = x, y or t.

$$A = \begin{pmatrix} u & 0 & c \\ 0 & u & 0 \\ c & 0 & u \end{pmatrix}, \quad B = \begin{pmatrix} v & 0 & 0 \\ 0 & v & c \\ 0 & c & v \end{pmatrix}, \quad F = \begin{pmatrix} 0 & -f & 0 \\ f & 0 & 0 \\ 0 & 0 & 0 \end{pmatrix}$$

Replacing the time derivatives by the central difference quotients gives

$$W^{\tau H} = W^{\tau-1} - (A W_x^\tau + B W_y^\tau + F W^\tau) \cdot 2\Delta t \tag{2}$$

where $\tau = t/\Delta t$, $\tau = 0, 1, \ldots$ Let

$$F = F_1 + F_2 \tag{3}$$

$$L_1(W^\tau) = -4 (A W_x^\tau + F_1 W^\tau) \Delta t \tag{4}$$

and

$$L_2(W^\tau) = -4 (B W_y^\tau + F_2 W^\tau) \Delta t \tag{5}$$

Then Eq. (2) is equivalent to the equations below

$$W_1^{\tau+1} = W^{\tau-1} + L_1 (W^\tau) \tag{6}$$

$$W_2^{\tau+1} = W^{\tau-1} + L_2 (W^\tau) \tag{7}$$

and

$$W^{\tau+1} = \frac{1}{2} (W_1^{\tau+1} + W_2^{\tau+1}) \tag{8}$$

where

$$F_1 = \begin{pmatrix} 0 & -f & 0 \\ 0 & 0 & 0 \\ 0 & 0 & 0 \end{pmatrix} \qquad F_2 = \begin{pmatrix} 0 & 0 & 0 \\ f & 0 & 0 \\ 0 & 0 & 0 \end{pmatrix}$$

Thus in one time step the integration of Eq. (6) can be carried out on one processor and that of Eq. (7) can be carried out on the other processor. Repeating this process time step by time step, the entire time integration expected beforehand will be completed.

## 2.2 Numerical experiment

At the initial instant the geopotential heights were assumed to be distributed like the Haurwitz wave. Based upon them the wind field was derived from the geostrophic wind relationship. Thus a series of hemispheric forecasts up to three days were made. Comparing the forecasts with those made with the serial algorithm under the same conditions, we can see that those forecasts are quite similar to each other (the figures are omitted). It indicates that the computations are reliable. Furthermore, in the case of highest priority to the user, around 1/3 of the total time can be saved with the parallel algorithm against the single-tasking time.

## 3. A GLOBAL MULTILEVEL SPECTRAL PRIMITIVE EQUATION MODEL

### 3.1 Vertical discretization of the atmosphere and vertical boundary conditions

In the vertical the $\sigma$-coordinates defined as

$$\sigma = p/p_s \tag{9}$$

178

is adopted, where p is the pressure, $p_s$ the pressure at the earth surface. The atmosphere is divided into K layers (see Fig.1). At each integer level the temperature T, $\varphi$, u and v are predicted, while the vertical velocity $\dot{\sigma}$ is evaluated at each half level. The thickness of each layer $\Delta\sigma_k (= \sigma_{k+\frac{1}{2}} - \sigma_{k-\frac{1}{2}})$ may vary with k.

At $\sigma=1$ and 0, the homogeneous boundary condition

$$\dot{\sigma} = 0 \tag{10}$$

is taken.

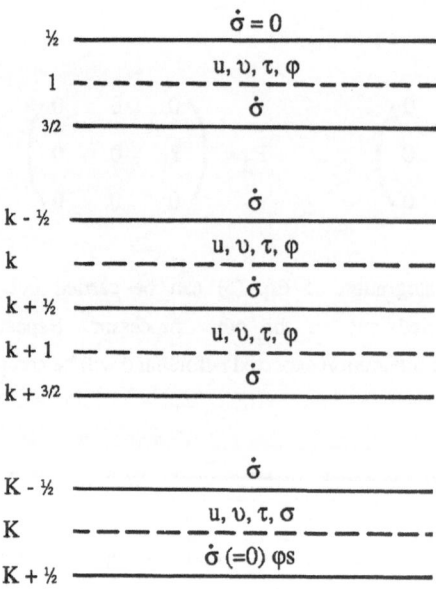

Fig. 1 Vertical discretization of the atmosphere

## 3.2 The hydro-and-thermodynamic equations

Without loss of generality, in the spherical coordinates the discrete hydro- and thermodynamic equations of the atmosphere may be written as follows.

$$\frac{\partial \zeta_k}{\partial t} = \frac{1}{a(1-\mu^2)} \frac{\partial}{\partial \lambda} (N_v)_k - \frac{1}{a}(N_u)_k \tag{11}$$

$$\frac{\partial \delta_k}{\partial t} = \frac{1}{a(1-\mu^2)} \frac{\partial}{\partial \lambda} (N_u)_k + \frac{1}{a}\frac{\partial}{\partial \mu}(N_v)_k - \nabla^2 E_k - \nabla^2 G_k \tag{12}$$

$$\frac{\partial T_k'}{\partial t} = -\frac{R\overline{T}}{c_p} \left(\sigma \frac{\partial}{\partial \sigma} \frac{\dot{\sigma}_L}{\sigma} + \delta\right)_k + (N_T)_k \tag{13}$$

$$\frac{\partial H}{\partial t} = -\sum_{j=1}^{k} \delta_j \Delta\sigma_j + \sum_{j=1}^{k} A_j \Delta\sigma_j \tag{14}$$

$$\frac{\partial \varphi_k}{\partial \ell n \sigma} = -RT_k \tag{15}$$

otherwise by taking dissipation and heating as parts of the nonlinear terms in the corresponding equations, the forms of the equations are still the same as the original. Here $\zeta$ is the absolute vorticity, $\delta = \nabla \cdot \vec{v}$, $\vec{v}$ is the horizontal wind velocity, $T = \overline{T} + T'$, $\overline{T} = \overline{T}(\sigma)$, $G = \varphi + R\overline{T}H$, $H = \ell n\,(p_s/p_0)$, $p_0$ is the standard pressure at sea level, $E = (u^2 + v^2)/2(1-\mu^2)$, $U = u\cos\varphi$, $V = v\cos\varphi$, $\mu = \sin\varphi$, $\varphi$ is the latitude, $A = -\vec{v} \cdot \nabla H$,

$$N_u = \zeta V \dot{\sigma} \frac{\partial U}{\partial \sigma} - R\frac{T'}{a} \frac{\partial H}{\partial \lambda} \;\;,$$

$$N_v = -\zeta U - \dot{\sigma} \frac{\partial V}{\partial \sigma} - R\frac{T'}{a} (1-\mu^2) \frac{\partial H}{\partial \mu} \;\;,$$

$$N_T = \left[-\frac{1}{a(1-\mu^2)} \frac{\partial}{\partial \lambda} (T'U) - \frac{1}{a} \frac{\partial}{\partial \mu} (T'V)\right] + N_{T2} \;\;,$$

$$N_{T2} = \delta T' - \dot{\sigma} \frac{\partial T'}{\partial \sigma} - \dot{\sigma}_N \frac{\partial \overline{T}}{\partial \sigma} + \frac{R\overline{T}}{c_p} \left(-\sigma \frac{\partial}{\partial \sigma} \frac{\dot{\sigma}_N}{\sigma}\right) + \frac{RT'}{c_p} \left(-\sigma \frac{\partial}{\partial \sigma} \frac{\dot{\sigma}_L}{\sigma} - \delta\right) \;\;,$$

$$\dot{\sigma}_{k+\frac{1}{2}} = (\dot{\sigma}_L)_{k+\frac{1}{2}} + (\dot{\sigma}_N)_{k+\frac{1}{2}} \;\;,$$

$$(\dot{\sigma}_L)_{k+\frac{1}{2}} = \sigma_{k+\frac{1}{2}} \sum_{j=1}^{k} \delta_j \Delta\sigma_j - \sum_{j=1}^{k} \delta_j \Delta\sigma_j$$

and

$$(\dot{\sigma}_N)_{k+\frac{1}{2}} = -\sigma_{k+\frac{1}{2}} \sum_{j=1}^{k} A_j \Delta\sigma_j + \sum_{j=1}^{k} A_j \Delta\sigma_j$$

The hydrostatic equation (14) also needs discretization. In this paper we adopt the difference sheme employed by ECMWF (1978), namely,

$$\varphi_k = \varphi_s + \sum_{\ell=k}^{k} B_{k\ell} T_\ell \tag{16}$$

where the coefficient $B_{k\ell}$ is only related to R, $\ln(\sigma_{k+\frac{1}{2}}/\sigma_{k-\frac{1}{2}})$, $\ln(\sigma_{\ell+\frac{1}{2}}/\sigma_{\ell-\frac{1}{2}})$, $\sigma_1$ and 0.

### 3.3 The spectral equations

Let F represent $\zeta$, $\delta$, T', H, $\varphi$, U, V, Z, D, A, E or $\varphi_s$ and express it in terms of spherical harmonics. Then Eqs. (11) - (14) and (16) in spectral form become

$$\frac{\partial \zeta_n^m}{\partial t} = Z_n^m \tag{17}$$

$$\frac{\partial \delta_n^m}{\partial t} = \underline{D}_n^m + \frac{n(n+1)}{a^2} (\underline{G}_n^m + \underline{E}_n^m) \tag{18}$$

$$\frac{\partial \underline{T}_n^m}{\partial t} = -\underline{L}\,\underline{\delta}_n^m + (\underline{N}_T)_n^m \tag{19}$$

$$\frac{\partial H_n^m}{\partial t} = -\Delta^T \underline{\delta}_n^m + \Delta^T \underline{A}_n^m \tag{20}$$

and

$$\underline{\varphi}_n^m = \underline{\varphi}_{s,n}^m + \underline{B}\,\underline{T}_n^m \tag{21}$$

where $\underline{F}_n^m = [F_n^m(1), F_n^m(2), \dots F_n^m(k)]^T$, $\Delta^T = (\Delta\sigma_1, \Delta\sigma_2, \dots \Delta\sigma_k)$, $\underline{L}$ is a matrix whose elements are $L_{k\ell} = R\overline{T}\,\hat{\ell}_{k\ell}/c_p$,

$$\frac{\Delta\sigma_\ell}{2\Delta\sigma_k} [\sigma_{k+\frac{1}{2}}(\overline{T}_{k-1} - \overline{T}_k) + \sigma_{k-\frac{1}{2}}(\overline{T}_k - \overline{T}_{k-1})]\frac{c_p}{R\overline{T}} , \qquad \ell > k;$$

$$\hat{\ell}_{k\ell} = \quad 1 - \frac{\sigma_k}{\sigma_{k+\frac{1}{2}}} + \frac{1}{2}[\sigma_{k+\frac{1}{2}}-1)(\overline{T}_{k+1} - \overline{T}_k) + \sigma_{k-\frac{1}{2}}(\overline{T}_k - \overline{T}_{k-1})]\frac{c_p}{R\overline{T}} , \qquad \ell = k;$$

$$\sigma_k \frac{\Delta \sigma_\ell}{\Delta \sigma_k} \{ \frac{1}{\sigma_{k-\frac{1}{2}}} - \frac{1}{\sigma_{k+\frac{1}{2}}} + \frac{1}{2\sigma_k} \left[ (\sigma_{k+\frac{1}{2}} - 1) (\overline{T}_{k+1} - \overline{T}_k) \right.$$

$$\left. + (\sigma_{k-\frac{1}{2}} - 1) (\overline{T}_k - \overline{T}_{k-1}) \right] \frac{c_p}{R\overline{T}} \} , \qquad\qquad \ell \le k-1$$

$$Z_n^m = \int_{-1}^{1} \{ \frac{im}{a(1-\mu^2)} (N_v)^m P_n^m(\mu) + \frac{(N_u)^m}{a} \frac{d}{d\mu} P_n^m(\mu) \} \, d\mu ,$$

$$D_n^m = \int_{-1}^{1} \{ \frac{im}{a(1-\mu^2)} (N_u)^m P_n^m(\mu) + \frac{(N_v)^m}{a} \frac{d}{d\mu} P_n^m(\mu) \} \, d\mu ,$$

$$N_{T,n}^m = \int_{-1}^{1} \{ \frac{-im}{a(1-\mu^2)} (T'_U)^m P_n^m(\mu) + \frac{(T'_V)^m}{a} \frac{d}{d\mu} P_n^m(\mu) \} \, d\mu + (N_T 2)_n^m$$

$P_n^m$ is the associated Legendre polynomial, $(N_u)^m$ and $(N_v)^m$ are Fourier coefficients of $N_u$ and of $N_v$ respectively, $\underline{B}$ is a matrix whose elements are $B_{k\ell}$.

It should be noted that the approximation

$$(\dot{\sigma} \frac{\partial F}{\partial \sigma})_k \approx \frac{1}{2\Delta\sigma_k} [\dot{\sigma}_{k+\frac{1}{2}} (F_{k+1} - F_k) + \dot{\sigma}_{k-\frac{1}{2}} (F_k - F_{k-1})] \qquad\qquad (22)$$

has been applied to all the quantities or terms invovling vertical advection.

## 4. NEW EQUATIONS

### 4.1 The necessity of reducing Eqs. (17) - (21) to new equations

In principle Eqs. (17) - (21) can be used to carry out time integration. However, if we want to take larger time steps than the ones required for the explicit scheme, it is most simple to adopt the semi-implicit scheme. Thus a three-dimensional Helmholtz equation would be solved. Although the computational amount for solving it is not very much for the spectral model, owing to the computation of the right hand side terms and the inseperability in some aspects, in general, the computation required for solving the equation can only be carried out on one processor, and the other has no work to do. Thus the computer will not be fully operated. Therefore, the method for integrating the above model should be greatly altered.

## 4.2 The new equations

As is known, the Helmholtz equation is of elliptic type, and the coefficient matrix of its discrete equation is allowed to be partitioned. Therefore, the synchronous or asynchronous parallel algorithm can be used and the computational amount required can be allocated to two processors. Thus the difficulties mentioned above can be overcome. However, the difficulties can be overcome in another way.

It is known that a multilevel primitive equation model can be reduced to sets of shallow water equations by means of vertical normal modes. Thus, a three-dimensional problem can be taken as two-dimensional problem formally under certain conditions. Therefore, such a reduction is favourable to the usage of multiprocessors. In the following we shall use a reduction to reestablish our model.

The resulting spectral equations are

$$\frac{\partial \tilde{\zeta}_n^m}{\partial t} = \tilde{Z}_n^m \tag{23}$$

$$\frac{\partial \tilde{\delta}_n^m}{\partial t} = \tilde{D}*_n^m \tag{24}$$

and

$$\frac{\partial \tilde{G}_n^m}{\partial t} = -\Lambda \tilde{\delta}_n^m + \tilde{N}_{G,n}^m \tag{25}$$

where $\tilde{F}_n^m = \underline{W}^{-1} \underline{F}_n^m$; $\underline{W} = (W_1, W_2, \dots W_k)$, W's are the characteristic vectors corresponding to the latent values nf the matrix $\underline{C}$, $\lambda_1, \lambda_2, \dots \lambda_k$, ordered from the largest to the smallest; $\underline{C} = \underline{B} \ \underline{L} + R\overline{\underline{T}} \ \Delta^T$; $\underline{G}_n^m = \Phi_n^m + RH_n^m \overline{\underline{T}}$; $\underline{N}_{G,n}^m = \underline{B} \ \underline{N}_{T,n}^m + R\overline{\underline{T}} \Delta^T$; $\underline{A}_n^m \ \Lambda = \text{diag}(\lambda_1, \lambda_2, \dots \lambda_k)$; $\tilde{D}*_n^m = \underline{W}^{-1} (\underline{D}_n^m + \frac{n(n+1)}{a^2} \cdot (\underline{G}_n^m + \underline{E}_n^m)]$.

Eqs. (23) - (25) are the new equations characterized by various latent values $\lambda_\ell$ ($\ell = 1, 2 \dots k$).

As for $H_n^m$ and $T_n^m$, we have

$$\frac{\partial H_n^m}{\partial t} = -\Delta^T (W \ \tilde{\delta}_-^m - A \ _-^m) \tag{26}$$

and

$$\frac{\partial I_n^m}{\partial t} = -\underline{L}\ \underline{W}\ \tilde{\delta}_n^m + \underline{N}_{T,n}^{\ m} = \underline{B}^{-1}\ (\underline{W}\ \frac{\partial \tilde{G}_n^m}{\partial t} - R\underline{\overline{T}}\ \frac{\partial\ H_n^m}{Mt})$$  (27)

## 5.  TIME INTEGRATION

According to many author's computations (Kasahara, 1976; Temperton et al, 1981), except the characteristic velocities of the first few normal modes, those of the others amount to usual wind velocities. Therefore, we may adopt the integration technique similar to that used by Burridge (1975), namely, using the explicity scheme for Eq. (23) for all $\ell$'s and for the other equations when $\ell > \ell^*$, and using the semi-implicit scheme for all the equations when $\ell \leq \ell^*$ except Eq. (23), where $\ell^*$ is a small positive integer. For convenience, however, the time step is taken the same for both the schemes.

For each vertical normal mode the explicit and semi-implicit schemes of Eqs. (23) - (27) are as follows.

1.  The explicit scheme

$$\Delta_t \tilde{\zeta}_n^m(\ell) = 2\ (\tilde{Z}_n^m(\ell))^\tau\ \Delta t$$  (28)

$$\Delta_t \tilde{\delta}_n^m(\ell) = 2\ (\tilde{D}*_n^m(\ell))^\tau\ \Delta t$$  (29)

$$\Delta_t \tilde{G}_n^m(\ell) = -2\lambda_\ell\ (\tilde{\delta}_n^m(\ell))^\tau\ \Delta t + 2\ (\tilde{N}_{G,n}^{\ m}(\ell))^\tau\ \Delta t$$  (30)

$$\Delta_t H_n^m = -\Delta^T\ (\underline{W}\tilde{\delta}_n^m - \underline{A}_n^{\ m})^\tau .\ 2\Delta t$$  (31)

and

$$\Delta_t\ T_n^m(k) = \underline{B}^{-1}\ [(\underline{W}\ \Delta_t\ \tilde{G}_n^m)\ (k) = R\underline{\overline{T}}\ \Delta_t\ H_n^m]$$  (32)

where

$$\Delta_t (\ ) = (\ )^{\tau+1} - (\ )^{\tau-1}.$$

## 2. The semi-implicit scheme

$$\Delta_t \tilde{\zeta}_n^m(\ell) = 2\,(\tilde{Z}_n^m(\ell))^\tau \Delta t \tag{33}$$

$$[1 + \frac{n(n+1)}{a^2}\, \lambda_\ell \Delta t]\, (\tilde{\delta}_n^m(\ell))^{\tau+1} = \frac{2n(n+1)}{a^2}\, \Delta t\, [(\tilde{G}_n^m(\ell))^{\tau-1} + (\tilde{E}_n^m(\ell))^\tau]$$

$$+ 2(\tilde{N}_{G,n}^{\;\;m}(\ell))^\tau \Delta t + [1 - \frac{n(n+1)}{a^2}\, \lambda_\ell\, \Delta t]\, (\tilde{\delta}_n^m(\ell))^{\tau-1} + 2(\tilde{D}_n^m(\ell))^\tau \Delta t \tag{34}$$

$$\Delta_t \tilde{G}_n^m(\ell) = -2\lambda_\ell\, \Delta t\, \overline{\tilde{\delta}_n^m(\ell)}^{\,t} + 2\Delta t (\tilde{N}_{G,n}^{\;\;m}(\ell))^\tau \tag{35}$$

$$\Delta_t \tilde{H}_n^m = -2\Delta^T\, [\underline{W}\, \overline{\tilde{\delta}_n^m}^{\,t} - (\underline{A}\,_n^m)^\tau]\, \Delta t \tag{36}$$

and

$$\Delta_t T_n^m(k) = \underline{B}^{-1}\, [\underline{W}\, \Delta_t \tilde{G}_n^m)\,(k) - R\overline{\underline{T}}\, \Delta_t H_n^m] \tag{37}$$

## 6. PARALLEL COMPUTATION

In time integration the steps below suitable for both the explicit scheme and the semi-implicit scheme can be taken.

1. Find $\zeta_n^m(k)$ and $\delta_n^m(k)$ from $\tilde{\zeta}_n^m(\ell)$ and $\tilde{\delta}_n^m(\ell)$, and $N_u(k)$, $N_v(k)$, $N_T(k)$, $A(k)$ and $E(k)$ from $\zeta_n^m(k)$, $\delta_n^m(k)$, $T_n^m(k)$ and $H_n^m$ at the instant $\tau$.

2. Find $[Z_n^m(k)]^\tau$, $[D*_n^m(k)]^\tau$, $[A_n^m(k)]^\tau$ and $[N_{G,n}^{\;\;m}(k)]^\tau$, and then $[\tilde{Z}_n^m(\ell)]^\tau$, $[\tilde{D}*_n^m(\ell)]^\tau$ and $[\tilde{N}_{G,n}^{\;\;m}(\ell)]^\tau$.

3. Carry out time integration for one time step by using (28) - (32) or (33) - (37), and then find $\tilde{\zeta}_n^m(\ell)$, $\tilde{\delta}_n^m(\ell)$, $\tilde{G}_n^m(\ell)$, $T_n^m(k)$ and $H_n^m$ at the instant $\tau+1$.

The above steps may be illustrated in the diagram below.

$$[\tilde{\zeta}_n^m(\ell)]^\tau \text{ etc. } \xrightarrow{\text{W}} [\zeta_n^m(k)]^\tau \text{ etc. } \rightarrow [N_u(k)]^\tau \text{ etc. } \xrightarrow{\text{Gauss integral}}$$

$$[Z_n^m(k)]^\tau \text{ etc. } \xrightarrow{\text{W-1}} [\tilde{Z}_n^m(\ell)]^\tau \text{ etc. } \rightarrow (\tilde{\zeta}_n^m(\ell))^{\tau+1} \text{ etc.}$$

However, the steps below also may be employed.

$$[\tilde{\zeta}_n^m(\ell)]^\tau \text{ etc. } \xrightarrow{\text{W}} [\zeta_n^m(k)]^\tau \text{ etc. } \rightarrow [N_u(k)]^\tau \text{ etc. } \xrightarrow{\text{W-1}} [\tilde{Z}(\ell)]^\tau \text{ etc.}$$

$$\xrightarrow{\text{Gauss integral}} [\tilde{Z}_n^m(\ell)]^\tau \text{ etc. } \rightarrow (\tilde{\zeta}_n^m(\ell))^{\tau+1}$$

It can readily be seen that the two diagrams can lead to the same results, irrespective of round-off errors. However, we prefer the second diagram to the first one.

For convenience of discussion we make the agreement in the below that all the quantities in the spectral space are stored in memory as horizontal slices of data, and all the quantities in the grid space are stored in memory as vertical slices of data.

Having the above agreement and considering the main storage of IBM-4381-p03 to have 16 MB (about 2.09 million 64-bit words), we can illustrate the parallel computation of the above second steps with the flow graphs (Fig. 2) under the conditions that $(\tilde{\zeta}_n^m, \tilde{\delta}_n^m, \tilde{G}_n^m, T_n^m, H_n^m)^\tau$ has been stored in core, and $(\tilde{\zeta}_n^m, \tilde{\delta}_n^m, \tilde{G}_n^m, T_n^m, H_n^m)^{\tau-1}$ has been stored on disk.

If triangular truncation is taken, the maximum wave number is 63, and $\Delta\lambda=\Delta\varphi=1.875°$, then for a 9-level model, the total memory needed is about $111 \times 10^4$ words. Hence, the main storage is sufficient for the time integration of the model.

However, if the physical processes are included in the model and the number of levels greatly increases, the in core memory may not be enough. In such a case, only by using I/0, desk or other peripheral devices can the difficulty be overcome.

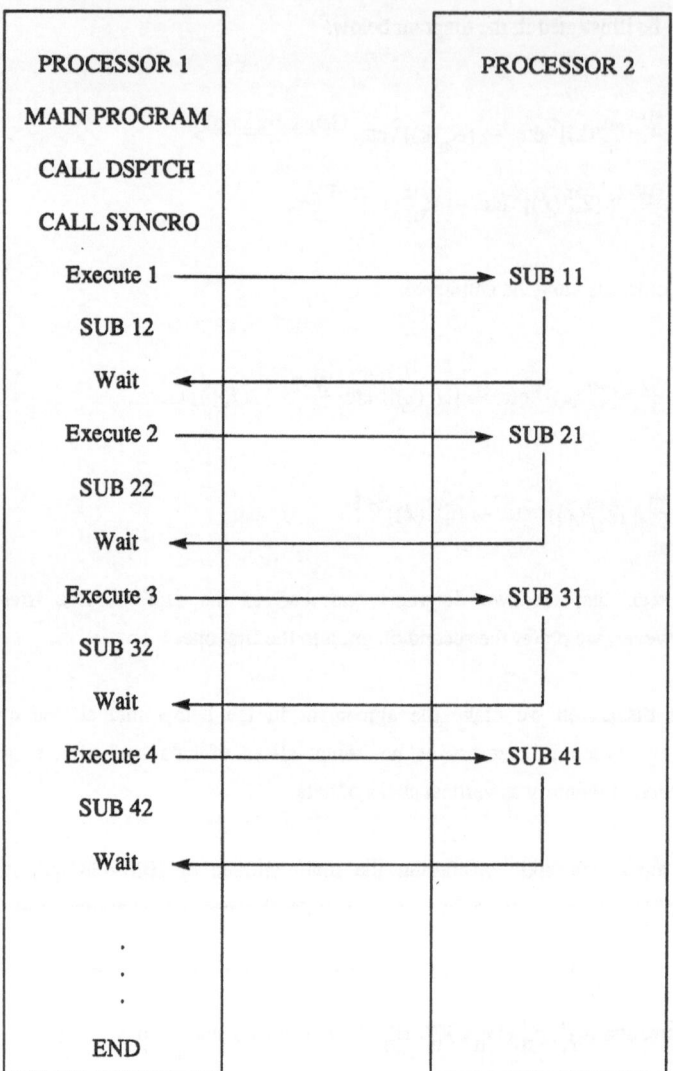

Fig. 2 FLow graph of parallel computation in one time step

SUB11: Compute $\zeta_n^m$, $\delta_n^m$ and $H_n^m$ at the instant $\tau$ when n-m is even, and then compute $(\delta, H_\lambda, H_\mu)_s$, where the subscript "s" denotes the symetric part.

SUB12: Compute $\zeta_n^m$, $\delta_n^m$ and $H_n^m$ at the instant $\tau$ when n-m is odd, and then compute $(\delta, H_\lambda, H_\mu)_A$, where the subscript "A" denotes the antisymmetric part.

SUB21: Compute $\delta$, $\dot{\sigma}_i$; u, v, A, $H_\lambda$, $H_\mu$ and $\dot{\sigma}_N$ in the Northern Hemisphere.

SUB22: The same as in SUB21 but in the Southern Hemisphere.

SUB31: Compute $\zeta$, T', $N_u$, $N_v$, E, T'U, T'V and $(N_T)_2$, and then $\tilde{N}_u$, $\tilde{N}_v$, $\tilde{E}$, $\tilde{T}$'U,

$\tilde{T}$'V and $\tilde{(N_T)}_2$ in the Northern Hemisphere.

SUB32: The same as in SUB31 but in the Southern Hemisphere.

SUB41: Compute $(Z_n^m, \tilde{D}_n^{*m}, \tilde{N}_{G,n}^{m}, A_n^m)_s$ and then $(\tilde{\zeta}_n^m, \tilde{\delta}_n^m, \tilde{G}_n^m, T_n^m, H_n^m)_s^{\tau+1}$.

SUB42: The same as in SUB41 but the antisymmetric part.

## References

Arakawa, A. and V.R. Lamb, 1976: Methods in computational physics, 17, J. Chang, Ed., Academic Press, 173-265.

Baede, A.P.M. et al., 1979: Adiabatic formulation and organization of ECMWF's spectral model. Tech.Report 15.

Burridge, D.M., 1975: A split semi-implicit reformulation of the bushy-Timpson 10-level model. Q.J.R.M.S., 101, 777-792.

Hoffman, G-R. et al., 1984: Aspects of using multiprocessors for meteorological modeling. Workshop on using multiprocessors in meteorological models, 3-6 December 1984, 270-358.

Jiangbo, 1988: Application of IBM-4381-p03 and numerical experiments. Master thesis (unpublished).

Kasahara, A., 1976: Normal modes of ultralong waves in the atmosphere. Mon.Wea.Rev., 104, 669-690.

Temperton, C. and D.L. Williamson, 1981: Normal mode initialization for a multilevel grid-point model. Part I: Linear aspects. Mon.Wea.Rev., 109, 729-743.

Wang Zonghao, 1981: A theoretical model of the system engineering task of the weather analysis and forecasting. Collected papers of medium-range numerical weather prediction, Beijing, Meteorological Press, 111-122.

Wang Zonghao et al., 1982: Editor program of meteorological built-up-model. Science Exploration, 2, 1-14.

# Operating Systems and Strategies for Highly Concurrent Systems

CHRIS JESSHOPE, GAJINDER PANESAR and JELIO YANTCHEV

Dept. of Electronics and Computer Science, The University, Southampton SO9 5NH, U.K.

## 1. FUTURE HARDWARE SYSTEMS

We now face an era in which large scale parallelism in computer equipment will become increasingly common. This paper addresses that issue; it notes the drive that arises from VLSI technology and sets out the requirements for systems support. Some of the factors that are be considered include the support for heterogeneous systems, or systems that comprise many processor types, including vector or SIMD components; the provision of multiple programming paradigms; and the use of object orientated methodologies.

### 1.1 Introduction

The transputer (figure 1) is one of the first of a new generation of microprocessors, that is designed to support concurrency in operation. It has led, in the UK at least, to a revolution in the way in which system design has been considered. To some extent, the transputer represents a timely convergence of two threads of research, both going back to the 1960s. The first of these threads has been the development of the microprocessor, which of course, has itself been closely linked with the major advances made in MOS integrated circuit technology. This research has led to the VLSI components we see today, commonly containing around 1 Million active devices. In retrospect, such devices must have seemed impossible to conceive of in 1959, when the first working integrated circuit was fabricated by Jack Kilby of Texas Instruments (This IC

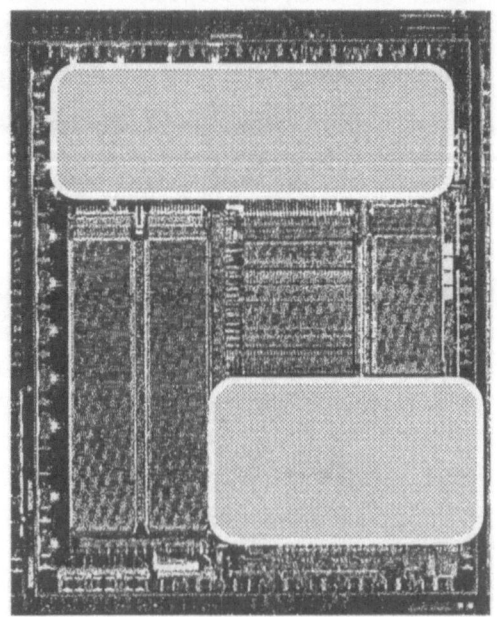

Figure 1. The INMOS T800 Chip, a 1.5 MFlop dual processor chip. The floating point unit (top) and support for communication channels (bottom right) are highlighted for comparison.

Topics in Atmospheric and Oceanic Sciences
© Springer-Verlag Berlin Heidelberg 1990

contained 1 transistor, 1 capacitor and several resistors). This represents an increase of approximately 6 orders of magnitude in complexity in just three decades, sufficient now to implement a complete computer system onto a single chip. At this rate, in another decade, we must know what to do with 100 complete computer systems on a single chip.

The answer to this will no doubt come from the second of the threads mentioned above, that of the development of parallel computer architectures and the methods by which they can be utilised. This thread has been motivated by the inability of the developments in technology to produce ever faster serial computers, and hence increase processing performance. The major barrier to this goal is the limitation of the propagation of signals by the speed of light (approximately 1-2 nsec per foot in a good transmission line). This dilemma can be illustrated by considering the Cray range of supercomputers, which have evolved out of the CDC 7600 range. In two decades, since the launch of the CDC 7600, clock rates in this type of computer have not even increased by a single order of magnitude (CDC 7600: 27.5 nsec; Cray 2: 4 nsec). Fast technology dissipates a great deal of power and to minimise signal propagation delays, circuit modules must be packed closely together; these factors combine to produce large power densities which mandate expensive cooling technology.

The cost effectiveness of parallel computers can be directly attributable to the development of processing in breadth to achieve performance, rather than squeezing increasingly smaller clock periods from diminishing returns in technological speed. This breadth is, of course, exactly what the VLSI technology can deliver, moreover it can deliver it in large and economic quantities, through the economies of scale.

## 1.1 Non-homogeneous hardware

It is clear that the transputer is not the only computer technology to exploit the density of CMOS VLSI, there are other microprocessors, with other techniques for combining them into single systems, there are SIMD machines like the Connection Machine and the AMT DAP, there are even vector supercomputers like the ETA 10, all of which use high density CMOS VLSI circuits. Moreover, active code development is proceeding on all of these machines, using different languages and programming paradigms. It seems clear that with so many different varieties of parallel machines, each of which with its own advantages, then the obvious solution is to integrate those models into a heterogeneous whole. Then, if two components of an application naturally map onto two different hardware models, this may then be accommodated. An example of this can be found in vision applications, where processing in pixel and object domain may be more appropriate for fine grain SIMD and more coarse grain MIMD hardware models respectively.

This has already been implemented, to some extent, in MIMD/SIMD systems (Baba 1987, Falk 1976, Jesshope 1987b, Lea 1988). The major motivation for this, has been in maintaining synchronisation over an otherwise indefinitely-large SIMD array. The argument being that, if at some point for electronic reasons, it becomes necessary to duplicate control processors, then the computational model need not be restricted to purely SIMD, but should exploit the multiple threads of control.

Other forms of non-homogeneity can be found within a single hardware model. For example the Esprit project P1085 (Harp 1986, Harp 1987,Jesshope 1987a) has developed a switched transputer architecture (The Supernode), which may be extended to thousands of processors. Although comprising only transputers, the machine is not homogeneous; memory and other services are not distributed evenly within the architecture. The hardware system is modular, and although any given configuration can be considered as being static, any failure of components can easily be identified and then mapped around, giving a dynamic set of non-homogeneous hardware components. More severely, the mapping realised by the switching network is user defined and may change dynamically during program execution.

With such the P1085 Supernode already in commercial production, we have had to consider how the system software models these resources. We have elected to model and control these this computer using an entity-relation model (Jesshope and Panesar 1988). We will be returning to and expanding on this model later.

## 1.2 Active-memory as a building block

It is increasingly clear in VLSI technology, that design must account more and more for communication between integrated circuits and less and less to computation or functional density. This arises from the following trends in CMOS IC technology. As line widths shrink, then the gate-Hertz product, which measures functional density, increases by a cube law in the inverse line width. Correspondingly, the pin-Hertz product, which measures addressability or communications density, grows far more slowly. This will depend on the assumptions made about packaging etc, but typically the communications density will only show sub-linear dependence on inverse line-width. This divergence favours the processor-memory chip with optimised communications channels, of which the transputer is a first example.

Let us consider this further; the transputer's maximum external memory bandwidth is 32 bits in three processor cycles (150nsec). However three cycle memory systems will require the more expensive RAM chips, and common large memory systems will have 4 or even 5 cycle memories. At three cycles, a bandwidth of 213Mbits/sec is achieved, using a total of 47 pins and giving a figure of 4.5 Mbit/pin/sec. On the other hand, the INMOS links each give 20Mbit/pin/sec, as each link requires 2 pins (Linkin and Linkout) and provides 20Mbit/sec in each direction. To be fair, in fully interleaved traffic, only 8 out of every 13 bits contains user data, the rest being related to protocol. A realistic bandwidth per pin for the INMOS link is therefore 12.3 Mbit/sec.

Thus the communication density of the 'specialised' link is nearly three times that of the conventional bus structure. Moreover it is more likely that link bandwidths will increase faster than memory bandwidths. A case in point is the proposed two cycle T800 transputer memory. This requires an additional 32 pins to increase the memory bandwidth by a factor of 1.5, yielding 4.0 Mbits/pin/sec, an intrinsically less efficient interface. On the other hand, link circuits will be able to exploit on-chip circuit speed improvements, without compromising pin-count further. Figures of 100MHz have been mentioned for future transputer products. On inspection of the transputer chip floorplan (figure 1), it is evident that the silicon support for the 'specialised' links is considerable, occupying an area similar to that used by floating point co-processor for example, or to that used by the on-chip RAM. However, this is in line with the divergence between computation and communication identified above. On chip, where the number of connections is not so restricted, and where the loads to be driven are less, it is possible to implement very fast memory systems, and to optimise their interface to the processing unit (also on chip). For example, whereas the maximum bandwidth for external memory on the transputer is 213 Mbits/sec, the internal bandwidth on the 4 Kbyte memory on the T800 chip is 640 Mbits/sec and could easily be improved 1.3 Gbits/sec by providing 64 bit parallel access to floating point operands for example.

This development of the processor-memory chip thus seems to be the direction most sympathetic to the rapidly advancing technology. We should certainly expect more developments along these lines.

### 1.3 Shared or distributed memory

Another issue, and one that may be used as an argument against the above proposition, is that many applications require the implementation of a global rather than a distributed memory model. Clearly on-chip memory supports only a distributed model. This is not a sound argument however, and any message passing system may be programmed to implement a global memory model.

Any request to a memory system requires an address at which the data is located. In a conventional sequential (shared) memory system, this address must be output by the requestor, usually onto a bus. It must then be used to steer a signal to the appropriate cell of memory, which then dumps its contents onto the bus for the requestor to sample.

This is exactly the same mechanism that can be used to implement a global parallel memory system from a distributed memory hardware configuration. Consider the extreme case, of one memory cell per processor; the address now required is the address of that processor in the system that contains the data. To complete the model, busses in the conventional system are replaced by message routing over the inter-processor links. Moreover, because the cells have intelligence, the strict and dumb memory model of a sequential, shared-memory machine may be improved upon. For example, when adding a value to the contents of a memory cell, in the dumb memory model, this requires a memory read followed by a memory write (assuming the value to be added is already at the processor). In the distributed global memory model, this operation can be modified by extending the addressing model to include an operation tag and an operand (a more object orientated view of the operation). The value is now updated in place, with a single decode or communication latency. This technique leads to an active-data model of concurrency described in (Jesshope 1988) and outlined in section 3.4 below.

In general, a distributed memory model must support an address partitioned into processor selector and cell selector. Neither of which need be absolute.

The major issues in memory design however, whatever its implementation, are:

bandwidth between source and destination, and latency beteen request and completion of access.

In a bus-based memory system, the bandwidth is constant, and as processors are added to the system, to a first approximation, the bandwidth per processor in inversely proportional to number of processors added. In a distributed memory system, the memory bandwidth per processor is constant, however there is still a bandwidth degradation in a global memory model as processors are added, but this can be minimised by good design. For example a switched architecture, see below, will not suffer any degradation in bandwidth, although latency may become unacceptably high. Also in fixed networks, the inverse dependency may be limited to the k-th root of the number of processors added, for a network of degree k. This follows as the global bandwidth is measured by the number of virtual channels shared by the same physical link, which is $n/2$ for n processors in a line (assuming all virtual channels are active); $n^{1/2}/2$ for n processors in a plane (again assuming that all virtual links are active; etc.

Latency is the second major factor in any memory system and currently this is the area where bus-based memory systems score over distributed memory systems. However this is due entirely to implementation, the former uses hardware decoding and gating, while the latter, usually uses slower software techniques. The use of hardware techniques for decoding message propagation will show equal benefits, as will be demonstrated later. Latency will eventually be reduced to two factors, signal propagation and fanout. It will also be shown that ultra-low latency routing techniques can achieve these optimal limits.

In both types of memory system, locality is important in maximising the perceived bandwidth and in minimising the perceived latency. In a bus system, perceived bandwidth at each processor can be improved by adding cache memory to the processor (actually distributing a 'shadow' of the global store). For repeated access to regularly stored data, this will improve both latency and bandwidth. However, there are problems of cache integrity which will nullify the additional bandwidth despite the additional hardware, in applications that require true global updates of store. In the distributed memory system, advantage is always gained by locality, as this implicitly reduces the number of virtual channels per physical link, without additional overhead (to be fair, the overhead has already been paid). Latency will depend on the implementation of routing strategy, network topology and the matching of form to function. However for all reasonable implementations locality will indeed reduce the latency in delivering a message.

## 2. SYSTEMS SOFTWARE ISSUES

There is a pressing need for operating systems and languages to support massive parallelism effectively. This is true for a variety of applications, many of which exhibit a **natural** parallelism. These applications will generally require diverse models for their support. Requirements for such operating systems and languages are therefore:

- Portability;
- Support for multiple hardware paradigms;
- Support for multiple software models;
- Support for existing languages and systems;
- Efficiency of implementation.

These goals are not yet met, indeed it is indicative of the state of the industry, that in a recent report concerning the commercial impact of parallel processing commissioned by Ovum (Johnson 1986), a report of some 450 pages, only 1% discusses the requirements for parallel operating systems. Even this small concession to the problem facing the industry is centered largely on the role of Unix a wholly inappropriate choice of operating system for a massively parallel computer, when implementing the fine grain concurrency which will be required, to fully exploit the hardware on the most general application programs.

Unix may best be considered as a programming paradigm, or at least should be, as the kernel or manager program supports entries or primitives for the generation and control of rather heavyweight processes. It is however, a prime candidate for the first of the priorities listed above, that of portability. It should be recognised however, that this paradigm is only suitable for applications yielding coarse grained concurrency, and should only be one of a number of paradigms available to the programmer and/or compiler on the system.

Other programming paradigms supported would include: embedded system languages and their compilers, for example occam and C languages supporting the transputers native communication model; active-data managers (see section 5), which are being developed by the authors within a collaborative Alvey project for DAP and transputer systems; Linda augmented languages Carriero(1988); occam objects managers (Thomas 1989); data-flow compilers; etc. Given this diversity of requirements, where many models will require their own dedicated run-time system (e.g Unix kernel), perhaps it is time to radically review the role of an operating systems in such highly parallel and probably heterogeneous environments.

### 2.1 Resource allocation

Perhaps the first old chestnut that should be shucked is that of resource allocation and management. Conventional operating systems provide for the sharing of resources between the various software objects that are being executed concurrently, whether it be simulated or real concurrency. In a highly concurrent environment, it is desirable to allocate whole resources or sets of whole resources to software objects, for their sole and expert supervision, similarly de-allocation would return those resources to the resource pool, for re-allocation to other software objects.

Static software objects, such as configured occam programs would need to be allocated resources on a for-life basis, while more sophisticated systems should be able to manage a dynamic set of resources. As an example, we propose active-data managers based on transputers being able to adapt to system load and acquire or relinquish resources based on some protocol, between the active-data and resource managers in the system. The following is an example of the kind of transaction that might take place between them:

**On active-data manager instantiation,**

   **Active-data manager to Resource manager...**

      give me up to 1024 floating point processors, each with at least 1 Mbyte of memory.

**Some short time later:**

   **Resource manager to Active-data manager...**

      here are 64 processors, you can call them Fast 1..64,

      each with: comms. rating: 4 by 100 mbits/sec,fl. pt. rating: 5 Mflops.

   **Active-data manager to Resource manager...**

      load active-data queue manager on processors Fast 1..64

      configure them as a broken, skewed, torroidal mesh.

   **Resource manager to Active-data manager...**

      done.

**A discrete time later (as they say in the movies)...**

   **Resource manager to Active-data manager...**

      you can now have the 1024 processors you requested earlier, you can call them Slow 1..1024

     each with: comms. rating: 7 by 10 mbits/sec, fl. pt. rating: 2 Mflops.

     please return processors Fast 1..64.

   **Active-data manager to Resource manager...**

      load active-data queue manager on the Slow 1..1024

      configure them as a binary tree superimposed on a torroidal mesh.

   **Resource manager to Active-data manager...**

      done.

**After restarting code on new resources,**

    **Active-data manager to  Resource manager...**

        eturning Fast 1..64.

**Resource manager to Active-data manager...**

        thank you.

It should be noted that the active data manager has a single point of control, and takes full responsibility for the resources allocated, including the possibility of belligerently retaining the originally allocated resources. One can postulate intelligent responses, based on the quality of request. For example the qualifier 'please' might have been omitted for a stronger command. The managers may thus act as autonomous agents, with detailed knowledge of the exploitation of a set of resources for a particular paradigm.

Notice also that the resource manager abstracts the resources, in this case no processor type is mentioned, but ratings allow the active-data manager to tune itself. Also no explicit code files are referenced, a code manager within the resource manager would be able to load code appropriate to the processor type, thus maintaining portability within the active-data manager implementation. These generic operations are a feature of the object orientated approach. In this system, the objects are the managers, which may themselves control other user objects and resources, which are bid for and acquired in totality.

There will of course be resources in the system that must be shared. Peripherals or specialised devices such as large RAM servers are examples of these. A strategy for the use of such shared resources is to provide a protocol and instantiate the device as an object in its right, that serves its resource to other distributed objects wishing to share its facilities.

## 2.2 Memory management

Memory management is one area where criticism has been applied to the transputer as a microprocessor. It is not clear however, that devices intended for parallel processing require address translation. By its very nature, address translation will add additional latency to the access of data from memory, although buffering will allow bandwidth to be maintained. Consider first the requirements for memory management:

to provide easy relocation of software objects; but we have seen above proposals that programming models and resource allocation are moving towards abstraction in the processor domain;

to provide an abstraction over two level stores; but we have seen above proposals to implement shared resources as distributed server objects; and finally

to provide domains of memory for protection purposes, such that one user or software object will not cripple another when it falls over.

This last point is a key requirement, and must be satisfied in a multi-user environment, or where concurrency is additionally used for extending the reliability of a system. Again the way in which to provide this protection is to decompose software objects in the processor domain, building the firewalls and any relocation required into the communications managers, or their guardians if the hardware does not support it.

### 2.3 A footnote on storage transmutation

One final issue that should be noted concerns the transmutation between computation and storage which has a major impact on communication. As a simple example, consider a distributed program which requires a computed object in every processor. One solution is to compute the value of the object once using a single resource, store it and distribute it when and where it is required. Alternatively, it can be computed where it is required and stored for later use. Finally if the object was large it may even be computed everywhere whenever it is required, despite the apparent redundancy in computation. Table 1 summarises the complexity of each option.

## Table 1. Generation of a distributed object on n processors

|  | Computation | Storage | Communication |
|---|---|---|---|
| Calculate and store global | 1 | 1 | n |
| Calculate and store local | n | n | 0 |
| Calculate k times locally | k*m | 0 | 0 |

In a sequential computer this trade-off almost invariably favours the storage of data-values if they are used repeatedly, indeed this is common compiler optimisation. However, in the distribution of a loop containing an invariant, the choice will more likely be to compute values locally. Given the trends of technology referred to above this strategy makes sense. We should not forget that this option exists.

## 3. ANOTHER LOOK AT THE HARDWARE

### 3.1 Switches as resource allocators

Having now outlined a strategy for distributed systems operation, We would like to reconsider some hardware issues, looking first at the role of circuit switches in parallel computer design. One example of the use of this kind of resource is in the Supernode computer which has been developed in Esprit project P1085. The Supernode has the ability to dynamically re-configure the connection between any two transputers in the system. The circuit that supports this is shown in figure 2. This facility is supported by a system bus to provide the global control. There are now many variants of this architecture, and the basic components of such 'Supernode' architectures are:

- switched links to allow arbitrary programmer-defined networks to be established;
- non-homogenous design to allow customisation of machine to application;
- scaleable computer by replication of clusters, either by fixed topology replication, or by further levels of switches.

On the latter issue, in the P1085 Supernode, it can be seen from figure 2 that inputs to the switch are grouped, with only half coming from the transputers within the Supernode; the other half of the links are available as input and output to the node, thus maintaining a scaled processing to communications ratio for the Supernode. Thus it is possible to have all four links from all transputers within the supernode being

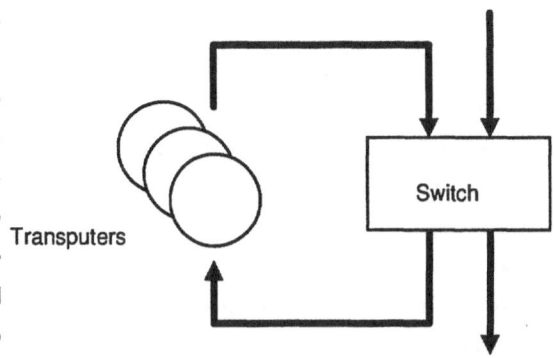

Figure 2. The Supernode switch, showing input and output channels.

switched to external sources of data. It is clear that with such a large number of external links, very rich fixed networks of interconnections between supernodes may be achieved. However, like the transputer, the supernode is itself a node which can be switched. One construction strategy for maximum flexibility is therefore to recursively define the structure, i.e. to form a supernode of supernodes.

For example, with a Supernode containing 36 transputers say, it will have 4 x 36 links external links, one for each face of each transputer. For a 32 Supernode machine, we will therefore require four 1152 way crossbar switches. This is simply realised by interconnecting the intra-supernode switches of figure 2 with thirty six 32 way cross-bar switches, as illustrated in figure 3, which establishes a three stage Clos network between each face of all of the transputers in the system. The first stage of the network is the intra-supernode switch (going out), the second stage of the network is the inter-supernode switchand the final stage is the intra-supernode switch (coming in).

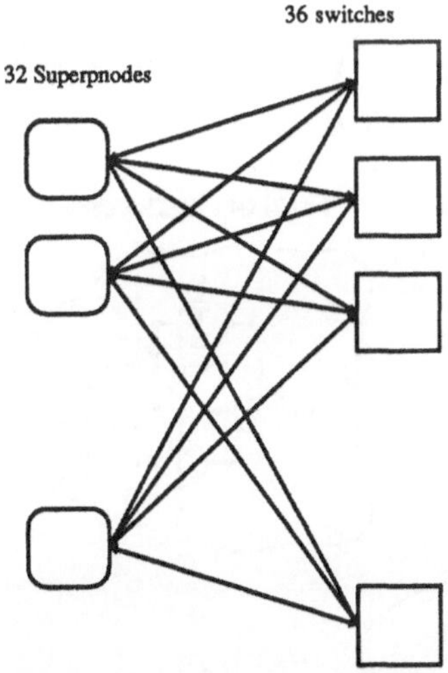

The key feature of the generic supernode architecture is that it allows for arbitrary software defined program networks, where the resources are drawn from a pool of available transputers. System software should support this model, as circuit switches provide an ideal way of allocating resources to software objects, while maintaining locality of reference. This issue is discussed further in section 3.2 below.

## 3.2 Processor fragmentation

It is surprising how many analogies can be taken from sequential computer experience when looking at con-

Figure 3. Internode supernode switch, showing 32 supernodes and 36 32 way switches.

198

current machines, but the analogy is usually transferred from the memory to the processor domain. Fragmentation of resources through repeated allocation (e.g. memory allocation) is one such example.

Consider a computer comprising many fixed-interconnected processors, and an allocation strategy as described in section 2. Moreover assume that we have a message passing scheme that supports a virtual communication from any processor to any other processor . This is a flexible resource set. If we now consider the repeated allocation and return of 'blocks' of processors to software objects, then if the size of the block varies, after a while fragmentation will occur, and although allocation will always be possible (we have arbitrary virtual channels), locality will eventually be lost. A solution to this, adopted from the sequential/memory analogue, is to allocate and deallocate in in large blocks. This however may be wasteful. Alternative solutions are to migrate processors between software objects based on the perceived traffic on that processor.

For example, if a processor clearly forwards more messages for an external object than it forwards within its owning object, then it more rightfully belongs to that external object. The migration of processors however, may require considerable effort; it may involve the for example the transmission of many Mbytes of code and data (one can begin to see advantages of fine grain memory systems here) from one processor to another. Also the traffic density may be very dynamic and hence difficult to analyse in a distributed manner.

Circuit switching provides the optimal solution to this problem, although not without cost. In the supernode architecture, for example, this model literally provides a bag of processors from which any

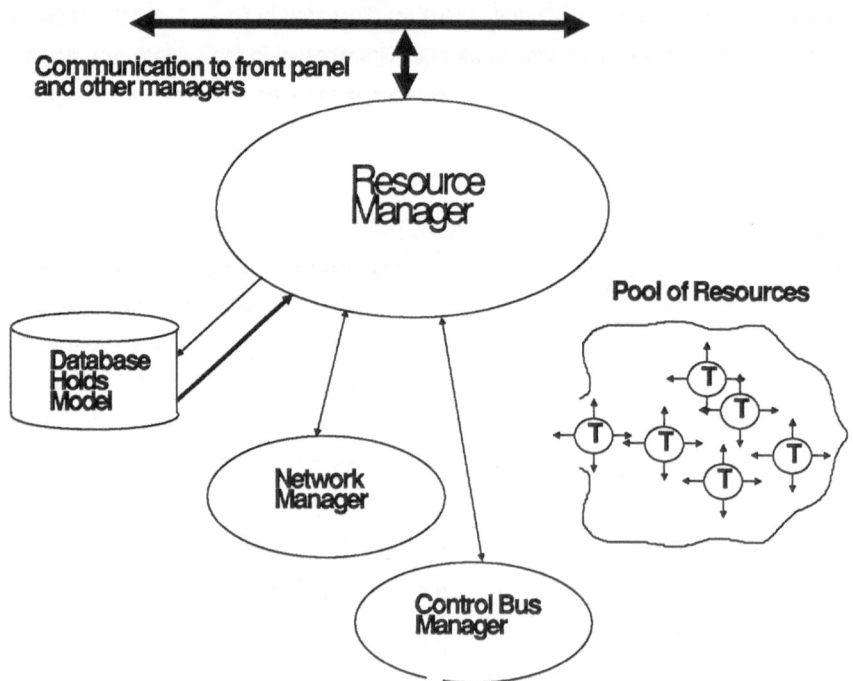

Figure 4. Software model of resource allocation base on pools of resources

software object may be given resources. The software view of this model is shown in figure 4.

## 3.3 Packet routing as an alternative to circuit switching

The design of an efficient general communication network is one of the most important and difficult issues in the design of a computer system. Ideally it should permit any processor to communicate with any other. One choice to achieve this is a network of reconfigurable topology as described above. However, even a fully reconfigurable network has limitations, imposed by its static nature, fixed valency and central control; it is also very expensive to implement.

In order to provide a fully dynamic arbitrary connectivity in a fixed valence network of processors, a transport mechanism must be implemented, which provides the propagation of data from processor to processor, based on addresses contained within a packet of data. In such a network each processor may send a message to any other. Such a data routing mechanism must satisfy a number of requirements depending on the particular application. Among the most demanding ones are:

- the routing protocol must be deadlock free;
- no packet is infinitely delayed in the network;
- a packet always takes the shortest route to its destination;
- the routing mechanism must adapt to traffic conditions and exploit the full available communications bandwidth (no restriction on the routing must be imposed);
- a node must not be able to refuse the input of a message from its user for ever;
- the highest possible throughput must be achieved (possibly only limited by the bandwidth of the internode links);
- the lowest possible latency must be achieved.

Clearly software implementations are possible, but these do not satisfy all of these requirements, in particular, as noted above, bandwidth and latency are compromised.

The requirements for high throughput and low latency in communication time are extremely important if the technology is to be exploited to gain breadth and if the ratio of computation to communication costs are to remain well balanced. This ratio scales rather badly in replicated designs because of increases in the network diameter or because of wiring costs.

Deadlock freedom is a natural requirement unless one wants the routing network to have the potential of being non-deterministically brought to a complete halt until some external intervention brings it back into life. The requirement for fair allocation of routing resources ensures that no packet is indefinitely delayed in the routing network before being delivered at its destination, which has an obvious bearing on load balancing.

## 3.4 Overview of existing packet routing mechanisms

Many deadlock-free routing algorithms have been developed for *store-and-forward* computer communications networks (Galertner 1981, Merlin 1980 and Roscoe 1987). In store-and-forward, each packet is stored completely in a node and then transmitted to the next node. The main disadvantage of this method is the very high latency in delivering an individual packet; it is proportional to the product of the packet length and the number of channels traversed. It also requires significant buffer space in each node, as a node must be able to accommodate a number of full length packets.

Instead of storing a packet completely in a node and then transmitting it to the next node, *wormhole* routing (Seitz 1985) operates by advancing the head of the of a packet directly from incoming to outgoing channels. Only a few flits are buffered at each node. A *flit* is the smallest unit of information that a queue or a channel can accept or refuse, it will usually contain sufficient information to address the remainder of the message.

As soon as a node examines the header flit of a message, it selects the next channel on the route and begins forwarding flits down that channel. As flits are forwarded, the message becomes spread out across the channels between the source and the destination. Because most flits contain no routing information, the flits in a message must not be interleaved with the flits of other messages. Thus, when the header flit of a message is blocked, all of the flits of a message stop advancing and block the progress of any other message requiring the channels they occupy.

A solution to the problem of deadlock freedom in wormhole routing has been proposed in (Dally 1987). It is based on preventing cycles in a channel dependency graph. This virtual channels approach restricts the routing of packets. each packet must follow a unique path, instead of being able to adapt its route when it encounters a hot spot in the network.

*Virtual cut-through*, described in Kermani (1979), is a method similar to wormhole routing. It differs from wormhole routing in that it buffers messages when they block, removing them from the network; this improves the performance of the network when heavily loaded, as message transport will tend towards store-and-forward routing, as the network traffic build up.

### 3.4 Low latency packet routing

In order to solve this problem, we must first find less restrictive solution to deadlock free packet routing, than that proposed by Dally for wormhole or virtual cut through techniques.

Deadlock occurs in a concurrent network when no further action can take place. This is usually because, even though each component process is in a state in which it can communicate, its potential communications are blocked by its neighbours. Three conditions are required to provide the potential for deadlock; at least one must therefore be avoided to prevent it. Firstly there must be cycles of requests for communication within a network, and at first sight this seems difficult to avoid, with packets of data being sourced at any processor and addressed to any other processor. Secondly there must be convergence of data, which is impossible to avoid as all nodes in a network could address data to the same processor. Finally any buffering must be finite, which is difficult if not impossible to avoid as naively sufficient buffering for all messages being delivered would be required at all processors in the system. The best strategy for avoiding deadlock is therefore to avoid cycles communications within the network. An acyclic network of buffer processes must therefore be deadlock free. These arguments are presented more rigorously in Yantchev (1989).

We have already indicated, that for the complete network this can not be guaranteed. For a subset of the messages or a partitioning of the network however, this becomes a possibility. More formally, if a physical network can be split into a set of independent virtual networks, all of which are acyclic networks of buffer processes, then the whole network will be deadlock free. To illustrate this consider the 2-D array case.

**The 2-D array.** Although each node of the physical grid must route messages in all four directions, this provides for a large amount of unnecessary connectivity, as far any one packet is concerned. If we constrain packets to always move towards their destination, then any one packet will need to be routed in at most two directions. Thus all packets can be divided into four classes according to their directions of travel. These four classes correspond to the four quadrants of a 2-D plane:

- Class I (+X,+Y)
- Class II (-X,+Y)
- Class III (-X,-Y)
- Class IV (+X,-Y)

Four independent virtual networks, each routing messages in one of the four quadrants, can provide for the necessary routing paths given that each packet is initially injected into the appropriate network.

**Application to arbitrary networks.** It is easy to see that this method can be applied to arbitrary networks. In the extreme case, each message class will consist of one message only and the number of virtual subnetworks will be equal to the number of classes. Although possible, in such extreme cases the method may not be of great use. For example, some ill conditioned networks may require excessive storage requirements, but then one can always reduce the packet size or indeed use wormhole routing. In most practical cases, however, the number of virtual networks need not be large.

For arbitrary networks the procedure below must be followed:

- subdivide all messages into classes according to their direction of routing;
- split the whole network into directed-cycle free virtual subnetworks, each of which provides all possible paths (if needed) for any of the message classes above;
- implement a buffer process at each node of the virtual subnetworks and map them, together with all virtual channels, onto the physical network.

A distinct advantage of this method is that it has reduced the redundancy of choice in each virtual network, which can in turn be used to optimise latency. In a 2-dimensional network, a one-of-four choice for routing is replaced by a one-of-two choice in each virtual network, but only after an initial one-of-four choice in selecting that virtual network.

Latency is defined here, as elsewhere (Dally 1987), as the delay in delivering a single message in isolation. In order to provide comparison with a routing strategy, known as the mad postman, and presented in (Yantchev 1989), latencies for previously known routing strategies need to be defined. Thus assuming bit serial communication, if:

- L is the packet length in words;
- W is the word length;
- T is the time to transmit one bit;
- D is the number of channels traversed;

then the latencies for the routing strategies introduced above are:

- store-and-forward routing, latency = LWTD;
- iwormhole routing, latency = WTD + LWT;
- virtual cut-through, latency = WTD + LWT.

Store-and-forward routing gives a latency that depends on the *product* of the message length L, and the number of communication channels traversed D. Wormhole routing results in a latency which depends on the *sum* of the two terms. Virtual cut-through provides for the same latency as wormhole but eliminates the degradation in bandwidth due to blocked messages.

A good strategy for minimising latency is to always route the packet along the dimension of the leading address digit, until it has been fully routed in a that dimension. The associated address digit is then stripped off. It will also increase susceptability to hot spots, if when a packet cannot be routed in the first dimension then it can either be routed in some other dimension or if that too is blocked, wait util either channel is available.

A further observation is that if dedicated routing hardware is available, then there is nothing inherently wrong in using it for any purpose if it would otherwise stay idle. Furthermore, there is nothing wrong if some (even all) of the additional work turns out to be useless, if this helps us achieve a latency far less than the minimum latency achievable otherwise. This is the strategy adopted in mad postman routing. In order to reduce latency, without changing any of the terms L,W,D,T, the only choice left is to output every packet along the same dimension it arrived from from, but to do this immediately it arrives, i.e. without buffering sufficient information to make a routing choice. The routing decision is then performed later. The latency then becomes:

- mad postman routing, latency = kTD + LWT

where k is a coefficient between 0 and 1 and is implementation dependent. In case of synchronous communication k is most likely to be 1, and in case of asynchronous communication kT will be the time to restore the pulse waveform (usually far less than T). When the whole leading address digit has been examined, one of the following will be true:

The leading address digit indicates that the packet, indeed, had to be forwarded further along the same dimension; then simply continue transmitting the packet, thus achieving minimum latency.

The leading address digit indicates that the packet has been fully routed in this dimension, and now has to be routed in the next dimension; then stop the transmission along the first dimension, strip off the first address digit and start transmitting the second digit along the next routing dimension. The second address digit need not be delayed for more than kT time and again minimum latency is achieved! The penalty is that a 'dead' address digit has been transmitted along the first routing dimension.

The packet has arrived at its destination; then stop transmission and store the rest of the packet into memory. Again a 'dead' address digit has been transmitted.

What then, if any, are the costs of this reduction in latency. The mad postman only works for the class of directed cycle-free networks described above. The only additional requirement to obtain this low latency is that the reverse acknowledge signals introducing discipline in the processing of packets be implemented in hardware, as they determine the smallness of kT.

The mad postman works properly in any traffic condition and can be made to adapt to traffic density. Of course, in conditions of increased traffic, some blocking is inevitable due to the contentions for output channels. If because of this a packet cannot be routed in the dimension of the leading address digit, then it can either wait for the corresponding output channel to be freed or be routed along some other dimension by swapping the first address digit with any other depending on the traffic conditions (i.e. adapting to traffic density). Note that no dead address digit is generated when a packet changes its dimension of routing due to traffic conditions.

The maximum number of dead address digits that will ever be generated is equal the degree of the network. They will also never reduce the performance of the network beyond that of virtual cut-through routing. A dead address digit can always be identified by comparing it with the local address, it can always therefore, be recognised and ignored by the nodes through which it passes. It will either quickly reach the boundary of the associated virtual network or will be blocked at some intermediate node. In both cases it will be immediately discarded from the network. Of course, a dead address digit may affect some other packet if there is a contention for the output channel of the associated routing dimension. However, the delay will never be greater than the duration of one address digit and this is precisely the minimum delay introduced at each node by virtual cut-through or wormhole routing. This is evident by considering the situation where the channel, instead of being blocked by a dead flit, is empty. In this situation virtual cut-through would in any case delay the incoming message by one flit transmission time.

If well implemented therefore, the mad postman routing strategy will provide a message latency which is the absolute minimum achievable for its class. In an empty network, the speed of propagation of a packet need only be limited by the physical constraints of propagating a signal pulse between the source and the destination (e.g. synchronous, asynchronous, electrical or optical). Thus from the discussion in section 1.2 above, and the methods presented here, message passing need not provide inferior performance to bus-based systems for implementing global memory functionality.

## 4. DATABASE SYSTEM MODELS

Returning again to system software models and resource allocation, ask yourself how best should we represent a collection of processors, the allocation of those resources to software objects, the connection of those resources in an appropriate network, be it switched or packet routed or indeed both? How do we provide the further abstractions necessary to make the software models portable, in a heterogeneous environment, and represent code objects in a resource independent manner.

Clearly from the section heading we believe the solution is found in databases. In this sense we consider a database to be an abstraction over stored data, i.e. one that does not take account of a physical filing system, but which provides a model and access protocol to the stored data. The model adopted however, is very critical, and most systems use the simplest database model to model the machine configuration, namely fixed tables. For a dynamically evolving system, a relational approach to the stored data is more suitable, as this provides access to data as sets of tuples. These sets are called relations and each tuple provides an element of that mapping; mathematically a relation is a one to many mapping. These may best be thought of as tables accessed by content on one or more columns, where each row represents a tuple. Moreover the model allows data to be joined from many different relations. In this way relations can define attributes of the real world objects, which may be extracted and joined with other attributes from other relations. The result of a query to a relational database is itself another relation.

In our model for example, we may have a relation concerning processors' memory, another concerning their connectivity, and others defining their status, allocated or not. A query may then select all processors with a given memory requirement, which are unallocated, and an update may well change the status and connected_to relations of the processors allocated.

Within an operating system model, there is function involved as well as storage, and from software engineering practice we know that function + storage = abstract data types. A better model of stored data would therefore be to have methods or functions associated with classes of data. For example, we store connected_to relations which define networks and ideally the operating system would like to deal with network objects, i.e establishing, copying, updating, etc, but does not wish to be concerned with network implementations (i.e. terminal, switch drivers etc.) To provide this abstraction we have implemented a layer of software on top of the data server. This however, would be more flexibly implemented if the data model itself supported this. Such database models are called object orientated databases, and although this is a relatively new model(Baroody 1981), commercial systems supporting this abstraction are now beginning to appear.

Our database is called the SEP(Panesar 1988) and is described in more detail in section 4.1 below. It holds information concerning the static configuration of the hardware, and also dynamic information concerning the switch settings, the process to processor placement, processor usage and in fact any other information the system or user may require. This experiment uses a small data engine driven using the SQL language.

Although not implemented in this experiment, databases can also provide support for the storage of data structures even in a long term (filestore) manner, giving persistence of data objects. Databases have already been used in sequential operating systems to provide a more flexible operating environment, although usually for commercial database applications. However it has been recognised elsewhere that the abstraction over stored data has more general applicability for use in operating systems (Balzer 1986).

## 4.1 The SEP structure and implementation

**Level 0 -** The database engine is illustrated by the lowest layer in figure 5; it comprises a small relational data-base, supporting a subset of functions that would normally be expected in a commercial, relational-data-base product. The hardware is modelled as a collection of tables or relations in this database. Operations are performed on this data at the set level, and any operation results in the creation of a new relation, so the system is closed. This level of the SEP is implemented in the C programming language, with request arbitration being handled by a small occam harness. All communication to and from the SEP database engine is therefore over occam channels. The interface

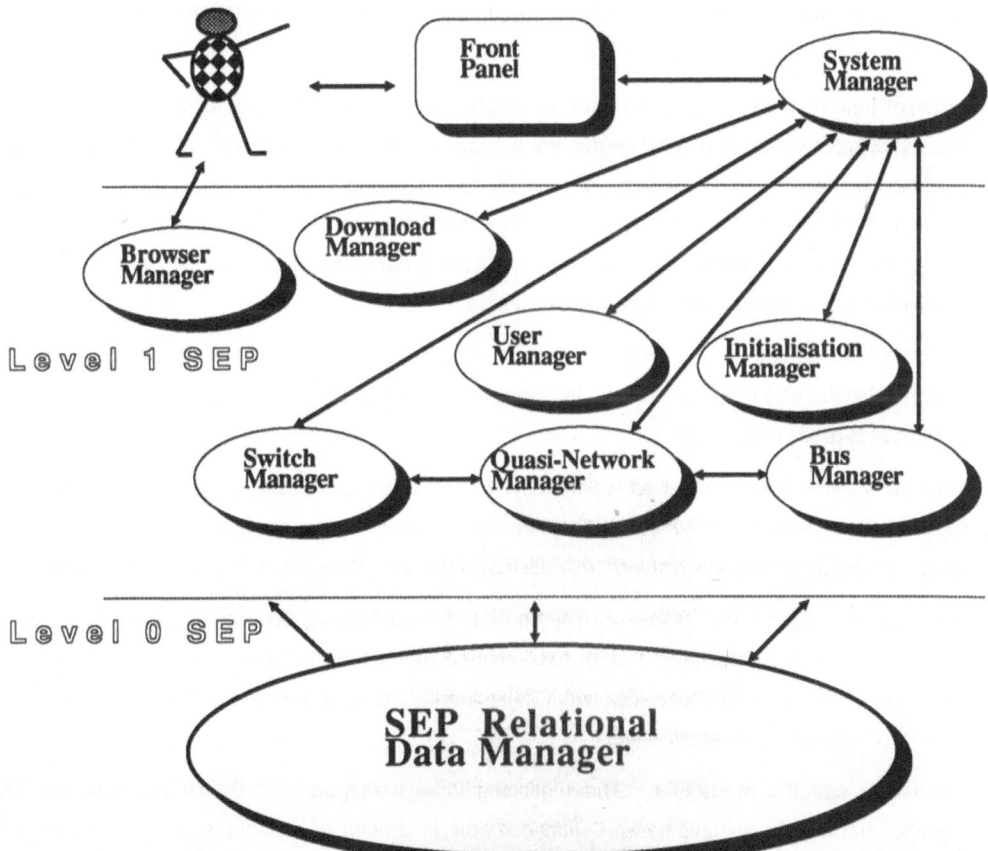

Figure 5. The layered SEP software structure.

to this data base is the SQL language and again only a subset of this language is supported. A typical transaction with the SEP consists of creating a SQL-tokenised message and passing it onto the data-base and then receiving the result.

The subset of operations supported is the set that is necessary for basic data storage and manipulation. These are:

- creating Base tables;
- deleting Base tables; altering Base tables;
- inserting into Base tables;
- updating Base tables;
- Locking and unlocking tables;
- killing Base tables.

Only two data types are currently supported, integer and strings of characters, but knowledge of relations and abstractions is added to this data by higher level code. **Level 1** - The applications manager form the outer layer of code, which embed the knowledge of the architecture and programming models supported. Each of the managers (See figure 5) communicates with the database engine and with the system manager to receive requests for action. This set comprises:

**Browser manager** - This gives the user an interface to the SEP. It allows for example the display, creation or modification of relations within the database (with of course appropriate ownership rights). It would typically be used for system configuration, querying resource usage, and other similar functions. Users may also make use of the SEP for application specific purposes and this would provide an interface to browse that usage. Programs may dump debug or monitoring information here, which may be interrogated during program execution.

**Initialisation manager** - This manager is responsible talking to any file system and initialising the various relations and managers. When this is complete, it indicates to the system manager that the supernode is ready to run code.

**User manager** - This manager allows the user to download application specific code, which would bypass systems management function. From this 'code slot', the user may have complete control of all resources and devices in the system. No protection on the use of resources is provided at this level.

**Download manager** - This manager is responsible for establishing code on a set of resources. In addition to a standard bootable file, this will require a run time relation generated by a modified configurer program, which specifies such relations as user_to_supernode processor mappings, network or connected_to relations etc.

**The target specific managers** - The remaining three managers, the Switch manager, the Bus manager and the Quasi-static network manager work in concert to control the physical resources in the system. The Bus and Switch managers contain drivers for the control bus and switch chips respectively. These in turn will provide synchronisation information and establish network configurations within the system. The quasi-static network manager has knowledge of switch settings, reconfiguration points and reconfiguration sets, which define when and between what processors a given network will be established. Currently only a static configuration is possible (a configurer constraint), although the quasi-static network manager actually provides hooks for group reconfigurations at any time.

### Above level 1 - The system manager

Above level 1 is currently a simple run time system, implemented in the form of a load-and-go system manager. This provides orchestration of the level 1 managers for a single-user multiple-load environment. The SEP would be booted up with the system on power on and would remain resident until powerdown or system failure. The user may load or kill jobs using simple window driven commands (in stand alone) or through the normal TDS interfaces. Program termination or user soft break will cause reset of the slave transputers but maintain the current state of the control transputer. This level would in a more general system, be replaced with protocols to the various language paradigm managers, each competing for resources.

### 4.4 Multiple SEP systems

An SEP controls a set of resources which are in its domain. To propose a single SEP for all resources would eventually lead to a bottleneck in resource allocation. We currently have the SEP running on a single Supernode machine, comprising a memory server transputer with 16 MBytes of RAM, a control transputer with 1/2 a MByte of RAM and 16 worker transputers each with 1/4 MByte of RAM. Larger Esprit Supernode machines will be built by replicating these clusters, as outlined above. In such multi-Supernode systems each controller transputer within a supernode will contain a database of its own local resources.

Logically the distributed SEPs will comprise a model of the overall system, and each manager will be able to make global enquiries. This logical model will be provided by a transaction protocol between SEPs at the level 1 layer. As will happen with any other agent, like minded objects will have a special relationship (protocol).

## 5. PROGRAMMING MODELS

### 5.1 Portable models for heterogeneous machines

Portable models for software paradigms (paradigm managers) can be implemented on multiple hardware to achieve single model management of concurrency in heterogeneous environments. An example of this is being investigated in an Alvey project, ARCH/001 . This project is defining a data-concurrent virtual systems architecture, VSA (Davis 1988), to support the active-data model. This virtual machine is being implemented on machines as diverse as AMT DAPs and transputers. Other examples of portable programming paradigms are the virtual objects can be found in (lucc87) and in occam objects (Thomas 1989) and the Linda model (Carriero 1988).

We will study a selection of these models in this section, but for obvious reasons we concentrate on the models being implemented here at Southampton.

## 5.2 Occam object management environment

This project is a research project which makes independent use of the SEP described above, to maintain and action support for a more dynamic programming environment based on the occam language. It is a research student project, which currently provides for the dynamic creation and binding of occam based processes. It does more than this for it has investigated the issues in distributing objects and providing the features normally associated with sequential objected-orientated programming systems, such as object classes and code sharing through inheritance.

The occam object manager is a class of objects that provide an environment through which occam programs, augmented with communication protocols to talk to manager, may be dynamically instantiated and bound to other occam objects. Unlike Actors, the sphere of influence of an occam object is defined by the channels it has to other occam objects (including the occam object manager). However through this set of acquaintances, it can 'get to know' other occam objects.

Objects may be referenced by class or by instance, therefore avoiding some of the restrictions imposed by the static point-to-point referencing found in the occam programming language. For example generic servers can be defined and instantiated as required by the unfolding structure of the computation.

The SEP provides a means by which the occam object manager can model the universe it knows about. Again the principal of owned resources defines the extent of this universe, or at least the extent of the universe over which the object manager has influence. Occam objects must be spawned by an object manager, so there is, in this model, the the concept of a single point of control, which may manage multiple threads of computation. Dynamic behaviour of instances of objects, their environment and their connection topology, which may be through simulated concurrency, are all stored as relations in the SEP. The load on the data engine is far greater in this application than when providing an environment for statically configured occam programs. This project has therefore provided a severe test of the SEP system.

## 5.3 The Linda model

The Linda programming model (Carriero 1988) has been developed at Yale University and is gaining widespread acceptance as an effective portable programming environment. It is a CSP like model, but is based on the concept of a tuple pool, rather than CSP's point-to-point communication. Like the relational database described above, tuples do not have addresses; to find one you match field values. Linda languages consist of a base language, C for example, augmented by a few simple operations. These operations interface to the tuple pool. Concurrent operation is provided by a process generation mechanism which generates a 'live tuple'. Processes are just like any other tuples in the pool, with the exception that some of their components are evaluated by programs, rather than being data values.

The operations provided over the tuple space are as follows:

- out(t), which causes a tuple to be added to the tuple space;
- in(s), which causes some tuple t that matches the template S to be withdrawn from the tuple space. This process binds values to values to parameter in the template s, and the inputting process then continues;
- rd(s), is the same as in(s), but the tuple remains in the tuple space; and finally
- eval(t), which is the same as out t, except that t is evaluated after rather than before it enters the tuple space.

Much emphasis is placed on compile time optimisation in implementing Linda programs, the reasons are obvious, for without optimisation, the penalty implicit in associatively matching all communications and process instantiations would be unacceptable. For example a pair of tuples commands.

- out(channel,A+B)
- in(channel,?x)

is equivalent to an occam communication if there exists only one tuple requiring input, it acts as an occam ALT if there are multiple tuple inputs of this form. Thus if these can be recognised by compilation, efficient transputer implementations could be implemented. The model is obviously far less restricted than that of occam, for example a single out and multiple rds provides for broadcast to a set of processes.

Knowledge of resources on which tuple space is implemented is clearly a requirement of this compilation process, so again one can conceive of an allocation of resources to a Linda environment. However, it is not clear to the authors, whether a run-time set of resources can be exploited by a Linda system.

### 5.4 The VSA implementation of active-data

The VSA model model being defined in Alvey project Arch/001 allows programs to be expressed in terms of element-by-element operations over whole data structures. In this model, the whole data structures are treated as single objects, with reference to subsets of elements identified not by a control structure but rather by an activation-set, which is also a parallel data object. Clearly scaler objects are allowed in the model, as a special case of a parallel data object with extent 1, but their use is discouraged. Object space manipulations will have minimal overhead in a large concurrent object, but a large impact on a scaler object, unless carefully optimised by a compiler. Moreover, unlike some other models, parallelism based on active-data thrives on massive concurrency. The active-data, i.e the elements of a parallel data object to be updated, are distributed across the processors in the system and it is by this data re-distribution that dynamic load balancing may be achieved (Jesshope 1988).

This illustrates that load balancing is not just an operating system responsibility, but may and indeed should be delegated to paradigm managers. We are designing the VSA to be serviced by an SEP like object for resource management.

The VSA can be considered as a virtual machine architecture, in which we stretch the processor space in order to map onto it the concurrency of the user's data objects. In this way, code can be written (by code-generators of compilers), as if each element of a data structure, or each element which has been defined for updating, had its own processor. The VSA will then hide the explicit mapping operations that will be required in order to map a given data structure, or subset of that structure, onto the actual processors in a real system. Both data structure and resources (as we have seen) may be dynamic.

The operational model of the VSA is that the data structures are created and updated as if by executing the same instruction in each (virtual) processor. If the extent of the real processor space is smaller than the concurrency required, the VSA will provide the appropriate sequence of instructions to map the virtual to the real resources. This is a simple operation if the elements in the data space are independent, as for example in the Mandlebrot set calculations. However, when there are dependencies across the data space, then data communication is required and the form of the mapping of the virtual to real processors becomes important.

A VSA object or instance name provide a mapping onto the values that define the resources associated with that object, onto the values that define the **rank, extent, precision** and **type** of that object and also a mapping onto the set of values which comprise the elements of the data-structure object. There is no direct identification of individual elements of a data object, although elements or sets of elements may be obtained using the *Select* function or by the use of Activation sets.

Activation sets are the most general reference mechanism to the individual elements of a data object. Operationally they define the active virtual processors associated with the data structure object, which are required to perform calculations in order to update the operation during the execution of a VSA operation. Within the model, they are boolean data objects that conform to the structure being updated.

Thus the VSA supports a model of synchronous concurrency. However, we feel that if the model is to have a wide applicability, it must be flexible enough to support multiple threads of control between global synchronisations and we are implementing this by allowing the concurrent data structure definition to be extended to include a process type. For example given a data structure S, whose elements $s_i$ (i in {o...n} ) are data, and a conforming data structure P, whose elements $p_i$ (i in {0...n}) are processes, then the application of P to S will apply the process $p_i$ to $s_i$, for all i in {0...n}. In general not all of the $p_i$ will be unique, for example if the $p_i$ belong to the set of unique processes defined within the VSA operation, $P = p_i$ i in {0...m} and if m<n, then one or more of the processes will be a true SIMD concurrent data operation.

The model can therefore exploit both SIMD and MIMD target hardware. However it must be realised, that for efficient SIMD implementation, each of the process types must be associated with a sufficiently large set of data (virtual processors), when compared with the set of physical resources. The implementation of multiple threads of control within a synchronous instruction requires some form of time-slicing or multiplexing of the SIMD processor.

The model of computation described for the VSA so far is not sufficient for general computation, unless the VSA instructions allow the migration of values between virtual processors associated with the data structure elements. The data manipulation processes required will depend on the type of object and the the algorithms being implemented.

The most general remapping can be defined by a triple of arrays, the source, the destination and the pointer array. In one case the pointer array specifies, for each element of the source array, where in a resulting array data from the source array, is to be placed. In the other case, the pointer array specifies, for each element of the resulting array, from which element the data is to be selected from in the source array. In the first case the pointer array must conform to the source array, in the second case, the pointer array must conform to the destination array. In the first case, the values of the pointer elements must be bounded by the extent of the destination structure, and only those elements of the destination structure are updated. In the second case the pointer element values must be bounded by the extent of the destination array. These correspond to an indirected parallel write and an indirected parallel read operation respectively.

The indirected parallel write operation is in fact non-deterministic, as multiple data elements may be sent to the same address, specified at more than one source location. There must exist some constructor function or reduction operation associated with the remapping, to fully define the semantics of the operation. The most general constructor is to generate a set of data objects at that "location". This operation is directly provided by the packet routing abstraction described above.

This provides the one exception to the referencing of individual elements of a data object. In practice, the model holds together in the absence of direct, data-object-element references due to the manner in which operations are performed on the data objects. Data objects are updated and operations performed on them between corresponding data elements of the objects involved in the computation (i.e. dimension for dimension, and index position for index position). Thus there is an implicit but arbitrary ordering applied to the elements. This is not true however for certain of the routing operations, described above, where a data structure element or virtual processor may be referenced by the VSA indirected read and write operations.

Other VSA instructions are the "order code" of the virtual system. VSA instructions are represented by a function interface, which can be interpreted as being either a procedure invocation, which performs the instruction with appropriate arguments replacing the formal parameters, or may be interpreted as a message containing the arguments together with an instruction selector, which is delivered to the object, where it is actioned.

VSA instructions are indivisible, i.e external synchronisation between concurrent VSAs is only possible at the start and finish of a VSA operation. If they require more than a single data object as an operand, then those operands must conform, that is, they have equal **rank** and an equal **extent** in each dimension. The VSA assignment will also normally require the source and destination objects to conform. A scalar data object (**rank** 0) is deemed to conform with all arrays.

To demonstrate the suitability of this model as a programming paradigm for heterogeneous environments, the VSA is currently being implemented within the Alvey project ARCH/001, both on transputer and AMT DAP systems. We are interfacing the VSA as a manager to the SEP defined within section 4 and as a consequence are implementing a relational database manager in FORTRAN PLUS on the AMT DAP.

## 6. CONCLUSIONS

This paper has presented a coherent strategy for the implementation of large scale computational systems based on collections of heterogeneous elements, be they microprocessor chip level devices, or complete systems such as the AMT DAP. This strategy adopts an object oriented approach to the system definition, where manager objects allocate resources, or act as expert kernels for a particular programming paradigm. This model has been argued against a background of the rapidly changing VLSI technology, and well suits the next generations of that technology. We give examples of experiments being undertaken within two collaborative projects, one funded by Esprit, implementing flexible transputer architectures, the other funded by Alvey, implementing an experimental portable active-data manager for transputers and the AMT DAP.

## 7. REFERENCES

Baba, T (1987) *Microprogrammable Parallel Computer MUNAP and its applications* MIT Press.

Balzer, R M (1987) Living in the next-generation operating system, *IEEE Software*, pp.77-85

Baroody, A J and DeWitt D J (1981) An Object-Oriented Approach to Database System Implementation, *ACM Trans Database Systems*, **6**, pp 576-601

Carriero, N and Gelernner, D (1988) Applications and experience with Linda, *Proc. ACM/SIGPLAN Symp. on Parallel programming* pp 173-187.

Dally, W J and Seitz, C (1987) Deadlock-Free Message Routing in Multiprocessor Interconnection Networks, *IEEE Trans*, **C-36**, pp 547-553.

Davis, P, Flanders, P, Jesshope C R and Laventhol, J (1988) The VSA Definition, Alvey/Arch001 internal document (Contact: Pavlin C, Project Manager, AMT Ltd, Reading, UK) Falk, H (1976) Reaching for the Gigaflop *IEE Spectrum 13 (13) 65-70.*

Gelernter, D (1981) A DAG-Based Algorithm for Prevention of Store-and-Forward Deadlock in Packet Networks *IEEE Trans* **C-30** pp 709-715.

Harp J G, Jesshope, Muntean T and Whitby-Strevens C (1986) Phase 1 of the development and application of a low cost high performance multiprocessor machine *Esprit 86: Results and achievements* 551-562(North Holland)

Harp J G (1987) Phase 2 of the reconfigurable transputer project (p1085), *ESPRIT 87 Achievements and Impact Part 1* (North Holland) pp583-591.

Jesshope C R (1987a) Transputers and switches as objects in occam, Proc VAPP III, Liverpool University, to be published in Parallel Computing.

Jesshope, C R (1987b) The RPA as an intelligent transputer memory system, *Systolic Arrays* 283-294 (Adam Hilger, Moore and McCabe Eds).

Jesshope, C R (1988) A dynamic, load-balanced, active-data model of parallel processing for vision, *Parallel Architectures and Vision* 315-329 (Oxford Science Publications, Page Ed.).

Jesshope, C R and Panesar, G S (1988) P1085: Somebody Else's problem, *Proc CONPAR 88*, To be Published by CUP.

Johnson, T and Durham, T (1986) *Parallel processing: the challenge of new computer architectures*, (Ovum Ltd.)

Kermani, Parviz and Kleinrock, L (1979) Virtual Cut-Through: A New Computer Communication Switching Technique *Computer Networks* 3 pp 267-286

Lea, M and Bolouri, H S (1988) Fault tolerance: step towards WSI, *IEE Proc E*, **135** 289-297.

Lucco, S E (1987) Parallel programming in a virtual object space, *Proc OOPPSLA 87* pp26-34.

Merlin, P M and Schweitzer, P J (1980) Deadlock Avoidance in Store-and-Forward Networks-I: Store-and-Forward Deadlock *IEEE Trans* **COM-28** pp 345-354.

Panesar G and Jesshope C R (1988) Distributed operating systems based on databases, *Proc 1988 SCS Computer Simulation Conference*, Seatle 9-13

Roscoe, A W (1987) Routing Messages Through Networks: An Exercise in Deadlock Avoidance, Oxford University Computing Laboratory Report.

Seitz, C L (1985) The Cosmic cube *Comm ACM* **28(1)** pp22-23

Thomas, I (1989) A support system for occam objects on transputers, To be published *Microprocessors and Microsystems* March 1989 (special issue on transputers and their application)

Yantchev, J and Jesshope, C R (1989) Adaptive low latency, deadlock-free packet routing for networks of processors, to be published *IEE Proc E*.

# The SUPRENUM Architecture and Its Application to Computational Fluid Dynamics*

KARL SOLCHENBACH[1] and CLEMENS-AUGUST THOLE[2]

[1] SUPRENUM GmbH, Hohe Str. 73, D-5300 Bonn, FRG
[2] Gesellschaft für Mathematik und Datenverarbeitung, D-5205 St. Augustin, FRG

## 1   Introduction

SUPRENUM is the German supercomputer project aiming at the development and construction of a distributed-memory (dm) multiprocessor system.

In this paper we give a short overview on the SUPRENUM system and we demonstrate its use for a particular CFD application.

In Section 2 we look at the architecture and the hardware components of SUPRENUM. The system consists of up to 256 nodes which are connected via a two-level interconnection network of buses. Each *node* consists of a CPU, private memory, a fast floating point vector unit, and dedicated communication hardware. 16 nodes and a local disk are combined to a *cluster*. The clusters are connected by a matrix of serial high-speed buses.

The programmer's view of the architecture is defined by the *abstract SUPRENUM architecture* which is described in Section 3. It is based on a dynamic process system and message passing communication. The program language for numerical applications is SUPRENUM-Fortran. The programming environment provides a lot of tools which support the programmer in developing parallel software.

The development of SUPRENUM is characterized by the simultaneous construction of the hardware and the adaptation of many application software packages. The software packages available are briefly listed in Section 4. We also explain the fundamental parallelization technique for CFD applications which is based on grid algorithms and grid partitioning.

One of the CFD codes is considered in more detail in Section 5. Although it is a code for the direct simulation of turbulence it is very similar to codes used for weather forecast simulations. The parallelization strategy is outlined and some estimation of performance results are given.

---

*This work was funded by the Bundesminister für Forschung und Technologie der Bundesrepublik Deutschland under project number ITR 8601 9 and the Minister für Wirtschaft, Mittelstand und Technologie der Landes Nordrhein-Westfalen under project number 323-860 5200 as part of the SUPRENUM supercomputer project.

---

Topics in Atmospheric and Oceanic Sciences
© Springer-Verlag Berlin Heidelberg 1990

# 2 The SUPRENUM system

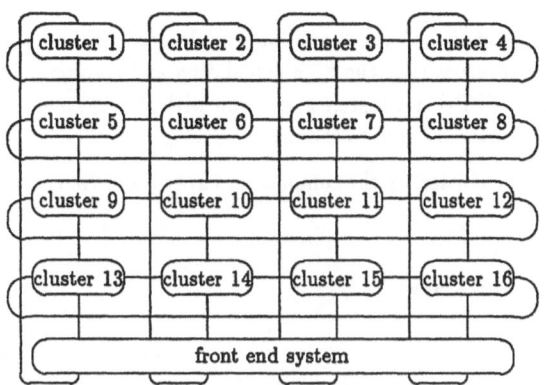

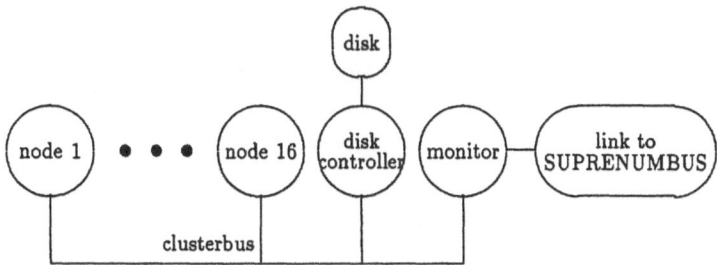

Figure 1. Structure of the SUPRENUM system with 256 computing nodes in 16 clusters (top) and a cluster (bottom)

Figure 1 shows the overall structure of the SUPRENUM-prototype hardware as it was designed by Giloi [3]. In the SUPRENUM-prototype 256 nodes are connected via a two-level interconnection network of buses.

## 2.1 The node

Each *node* consists of a CPU (MC 68020), private memory (8 MB), and three coprocessors (all running with 20 MHz):

1. A scalar arithmetic coprocessor (MC 68882) for certain scalar floating point calculations.

2. A micro-programmable communication coprocessor which can transfer data dircetly from the memory (via a DMA unit) to the cluster bus.

3. A very fast vector floating point unit (VFPU). The VFPU uses Weitek processors (WTL 2264/2265), is micro-programmable, and has its own static memory (64 KB). The VFPU can load/store data directly from/to the main memory with constant increment

via a DMA unit. The peak performance of the VFPU is 10 Mflops for vector additions and multiplications and 20 Mflops for chained operations like axpy and dotproduct. The $n_{1/2}$ is about 20. For more details on the VFPU cf. [6].

All coprocessors are highly intgerated and the whole node is located on one board. The integrated node architetcure guarantees that the high calculation and communication rates can be obtained practically. The VFPU, for instance, can run with its peak rate even if one of the operands has to be loaded/stored from/to the main memory.

The coprocessor instructions are generated by the compiler and are passed via the coproceesor interface of the MC 68020.

## 2.2 The interconnection network

Up to 16 computing nodes are combined to a *cluster* using the *clusterbus* (320 Mbyte/s). Each cluster also contains a local disk, a disk controller node, a monitor node which supports performance measurements, a communication node for the connection to the second bus level (the SUPRENUM-bus).

As shown in Figure 1, 16 of these clusters are connected by a matrix of serial high-speed SUPRENUM-buses (200 Mbit/s) and form the *high performance kernel*. The SUPRENUM system is completed by a front end machine which is used for operating and maintaining the high performance kernel as well as for software developement.

## 3   The system software

The software concept for SUPRENUM is based on a *process* system and on a *message-passing communication* handling. The process concept (the so-called *Abstract SUPRENUM architecture*) is a dynamic one which is characterized by the following elements:

- Processes are autonomous program units which run in parallel.

- Processes can terminate themselves and can create but not terminate other processes.

- Processes communicate only by exchange of messages, and no shared memory is available.

- Applications are started by one initial (or host) process.

- In arithmetic expressions and communication instructions, array constructs are especially supported.

- The user defined process system is homogeneous and independent from the actual hardware configuration. The two-level architecture (cluster structure) is not reflected in the Abstract SUPRENUM architecture and is completely transparent to the user. The processes are mapped to the clusters and nodes at run-time.

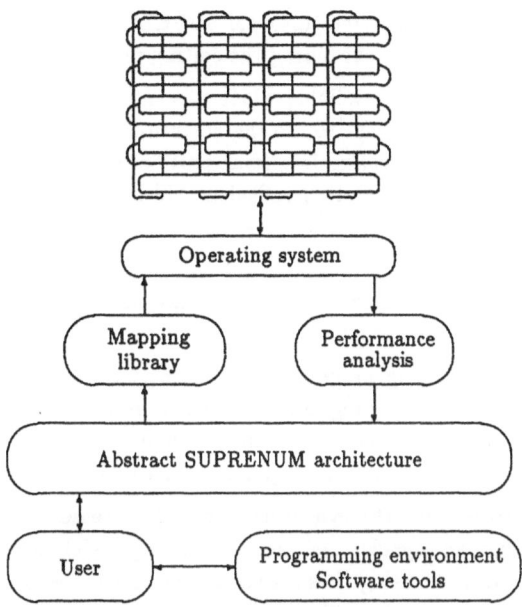

Figure 2. The SUPRENUM system software

Figure 2 shows, that this Abstract SUPRENUM architecture is the central model in the system software. The user should write his codes only in terms of processes.

The mapping of processes to nodes is supported by the *mapping-library*. It provides optimal mapping startegies for some standard process systems (like trees, rings, grids) and uses heuristical strategies for irregular process structures.

The SUPRENUM *operating system* consists of two components residing on the front end system and on the node level. The front end system is operated under UNIX V[1]. In each node a small operating system (PEACE) is responsible for the process scheduling and the message handling. Special servers of the operating system support the local disk, the performance analysis and the collection of diagnosis information.

The *programming language* for numerical computations is SUPRENUM-Fortran, an extended Fortran 77. The extensions include special process handling and message-passing constructs and an array syntax formulation according to the proposed Fortran-8X standard. Additionally Concurrent Modula-2 and a parallel version of C will be available.

*Performance analysis* tools collect performance data from each cluster for graphical presentation. This enables the user to analyze the utilization of nodes and buses and to tune his parallel programs.

The SUPRENUM *programming environment* provides a lot of tools which support the programmer in developing parallel software.

---

[1]registered trademark of AT&T

The set of tools contains a syntax-directed editor for SUPRENUM-Fortran which checks the syntactical correctness of a program text during the editing phase.

A state-of-the-art auto-vectorizer is available which automatically generates code for the vector nodes – if the user does not want to use the Fortran-8X array notation. This allows porting of existing "dusty deck" software since large parts of the code can be kept without modifications. Later on, a semi-automtatic parallelizer will be available which supports the user in constructing host and node processes and which automatically distributes arrays over the nodes [17].

A communications library for grid applications allows easy, safe and portable programming for regular and block-structured grids [4].

A simulator is available supporting program development for the Abstract SUPRENUM architecture. Usually the complete program development is done on workstations where the simulator checks the logical correctness of the communication and gives performance estimates. Only the final tuning of the code requires the access to the parallel hardware.

The communication between and the synchronization of processes is done via message-passing. SUPRENUM offers an *asynchronous* message-passing model, i.e. the sending process does not have to wait until the message has arrived at the receiver and can continue immediately. The message arrives in the mailbox of the receiving process and can be selected by the application program. The use of the SUPRENUM message passing constructs is demonstrated in [10].

# 4   The application software

The availability of practical relevant application software is decisive for the scientific and commercial success of a new computer architecture. This software must, of course, make use of the specific advantages of the architecture and translate these advantages into gains of speed.

The basic numerical software available on SUPRENUM consists of a parallel Linear Algebra package (similar to LINPACK), some solvers for basic partial differential equations (PDEs) using highly efficient parallel multigrid (MG) algorithms, and a package for ordinary differential equations.

The emphasis within the application software development lies on computational fluid dynamics (CFD) codes. These cover most of the standard CFD models used today and include solvers for the potential, the Euler, and the Navier-Stokes equations as well as grid generation systems, all for 2 and 3 dimensions.

Besides the CFD applications many different application software packages are currently under development for SUPRENUM during the project (Finite Element package, climate research, oil reservoir modelling, nuclear safety, etc.).

## 4.1   Grid algorithms

Although the typical applications in scientific supercomputing belong to very different scientific and technological fields and seem to be wide spread, most of them are characterized by a remarkably uniform type of mathematical models and - as a consequence of that - by very typical uniform data structures: Grids and grid-like data structures.

The grid structures themselves may be very simple but may also be very complex data structures, depending on the application. We here consider only regular grids although many applications on SUPRENUM are based on more complex grid structures (like block-structured, locally refined, or even completely irregular grids).

An important class of grid algorithms are those iterative methods in which the new values at any given grid point are calculated using the values at certain neighbor grid points. Here two principal approaches are used: Jacobi methods (using only old values) and Gauss-Seidel methods (using also new values from neighboring points if available). Explicit time stepping methods correspond in this respect to a Jacobi method.

For a regular grid with $N$ grid points the degree of parallelism of a Jacobi method clearly is (proportional to) $N$. With respect to convergence and smoothing properties, often Gauss-Seidel methods are preferred. By using a *red-black* order of the grid points rather than the lexicographical order each step of Gauss-Seidel consists of two half Jacobi steps, thus giving a degree of parallelism of $N/2$.

In the case of larger difference operators or larger neighboring formulae, one always can find an ordering – by coloring with $F$ colors – that yields an degree of parallelsim of $N/F$.

Whereas the asymptotic convergence properties typically are independent of the ordering of the grid points, the *smoothing* properties which are important in a multigrid context may be essentially improved by the red-black (multi-color) ordering.

## 4.2  Grid partitioning

According to the considerations above, the parallelization of the typical grid algorithms can, in principle, be achieved. The next question is, how such parallel versions of grid algorithms are to be mapped (and are to be implemented practically) on MIMD multiprocessor systems like SUPRENUM. Details about the grid partitioning can be found in [11].

The grid is divided into rectangular (box-shaped) subgrids. Each subgrid is assigned to a node in such a way that neighboring subgrids are mapped on neighboring nodes. (Whether such a homomorphic mapping exists depends on the topology of the multiprocessor system. It exists for hypercubes but not for ring and tree computers.)

The grid variables in each subgrid are exclusively computed by its associated node. At the inner boundaries of the subgrid the nodes need values at points which are contained in the neighboring subgrid and node. Instead of transfering these values exactly at the time when they are needed – this would prevent vector processing – they are stored in so-called *overlap areas*. After each iteration the values in the overlap areas are exchanged and actualized via the message-passing communication mechanism. The introduction of overlap areas does not change the calculation (i.e. it leads to an equivalent algorithm) since a strict synchronization following to each iteration step (e.g. by employing a blocking receive instruction) is ensured.

The way how the geometric partitioning of the grid should exactly look like depends among other things also on the communication costs of the hardware. For example, let us have a

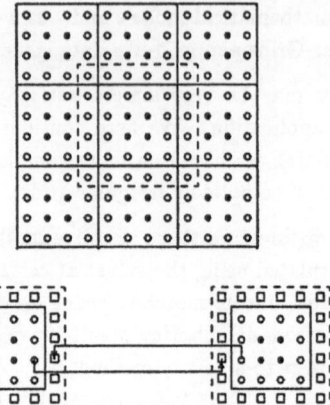

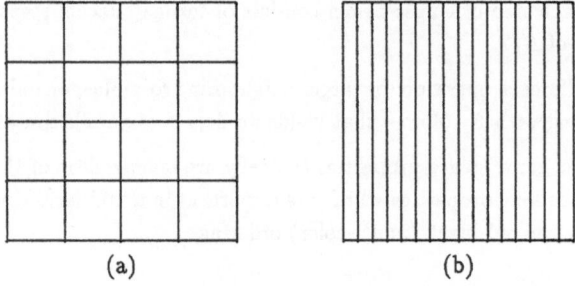

Figure 3. Overlap areas and their exchange.

(a)

(b)

Figure 4. (a) Minimum length of messages, (b) minimum number of messages.

look at a regular 2D structure (Figure 4). If the start-up time, which is needed to initiate the message-passing process, is very high, a stripewise partitioning (with a minimum number of inner boundaries) will be advantageous. If the start-up time is low compared to the transfer time per data word, a square partitioning is preferred in order to minimize the length of the inner boundaries.

## 4.3 Programming aspects

The organization, control, and synchronization of hundreds of parallel processes is often claimed to be an enormously difficult – or even unsolvable – problem inherent to parallel processing on dm-multiprocessors. The experience gained so far in the practical work on different distributed memory multiprocessors shows that this bias is completely wrong. Because of their regular communication structure the implementation of the grid partitioning method for regular grids is extremely simple and straightforward. For a more detailed description of the implementation we refer to [10]. On SUPRENUM, the programming of grid applications is supported by various tools like the SUPRENUM communications library for regular [4] and block-structured grids.

# 5 An example: Direct numerical simulation of turbulent flows

A code of Gerz, Schumann and Elghobashi [2] for the simulation of stratified homogeneous turbulent shear flow has been choosen to demonstrate the performance of CFD codes on SUP-RENUM and the parallelization strategies. As shown in Section 5.2 the code characteristics are very similar to the ECMWF spectral model. On the other hand, the formulae describing the physical properties are much simpler. This results in a substantial reduction of the computational work per gridpoint for our example. Due to the better calculation/communication ratio the ECMWF codes should behave better on SUPRENUM as the performance estimates in Section 5.6 indicate.

## 5.1 Formulation of the problem and discretization

In this part we summarize a description of the problem given in [2, Section 1 and 2].

The objective of the code is to study the effects of stable stratification and shear on the behaviour and the evolution of homogeneous turbulent flows using the method of direct simulation. The work considers only homogeneous turbulence in which the mean velocity and temperature have a uniform gradient in the vertical direction. The ultimate goal is to use the results of the simulations to calibrate the parameters of existing second-order turbulence closure-models.

For the simulation a finite cubical domain with side-length $L$ is considered. The mean horizontal velocity $U(z)$ and mean temperature $T_0(z)$ possess uniform and constant gradients relative to the vertical coordinate $z$. The fluid is assumed to have constant diffusitives $\nu$ and $\gamma$ for momentum and heat, respectively. The density $\rho_0$ is assumed to be constant except for small density fluctuations due to temperature fluctuations affecting buoyancy accelerations. All fields are expressed nondimensionally using $\rho_0$, $L$, $\Delta U = |\frac{dU}{dz}|L$, and $\Delta T = |\frac{dT_0}{dz}|L$ as reference scales for density, length, velocity, and temperature, respectively. The normalized momentum, heat, and contuniuty equations are

$$\frac{\partial u_i}{\partial t} + \frac{\partial(u_j u_i)}{\partial x_j} + Sx_3\frac{\partial u_i}{\partial x_1} + Su_3\delta_{i1} = \frac{1}{Re}\frac{\partial^2 u_i}{\partial x_j^2} - \frac{\partial p}{\partial x_i} + |Ri|\,T\,\delta_{i3}, \quad i = 1,2,3; \quad (1)$$

$$\frac{\partial T}{\partial t} + \frac{\partial(u_j T)}{\partial x_j} + Sx_3\frac{\partial T}{\partial x_1} + su_3 = \frac{1}{RePr}\frac{\partial^2 T}{\partial x_j^2}, \quad (2)$$

$$\frac{\partial u_j}{\partial x_j} = 0. \quad (3)$$

where the summing convention is used and $u_i$, $T$ and $p$ are the derivations of the velocity, temperature and pressure from their respective mean profiles. $S = \{0,1\}$ and $s = \{-1,0,1\}$ distinguish between cases with and without shear and with unstable, neutral and stable stratification, respectively. The Reynolds number $Re$, the absolute value of the Richardson number $Ri$, and the Prandtl number $Pr$ characterize the flow. The values of the dimensionless spatial coordinates $x_i$ are within the range $-0.5 \leq x_i \leq 0.5$ and $x_3$ points vertical upwards. The boundary conditions are perdiodic in $x_1$- and $x_2$-direction and shear periodic in $x_3$-direction with offset in $x_1$-direction.

The equations are discretized in an Eulerian framework using finite-difference technique on a staggered grid for all terms in the equations except the mean advection in $x_1$-direction where pseudo-spectral (Fourier) approximation is used. The Adams-Bashforth scheme is used to integrate the equations in time. Pressure is treated implicitly. The timestep is choosen to be $\Delta t = \Delta x/2$.

Starting from $u_i^k$ and $T^k$ the approximations for the next time step $k+1$ are calculated in three steps

- STEP-1: Evaluate intermediate values $\tilde{u}_i, p^*$, and $T^{k+1}$ using the Adams-Bashforth scheme and discrete Fourier interpolation in $x_1$-direction. The boundary conditions are applied to the intermediate velocities. The basic operations in this part are evaluations of finite difference operators (cf. [2] for details).

- STEP-2: Solve the Poisson equation

$$\partial_i \partial_i p^{k+1} = p^*$$

  with a fast solver [8] which applies Fast Fourier transforms (FFTs) in $x_1$- and $x_2$-direction and Gaussian elimination in $x_3$-direction in this sequence.

- STEP-3: Obtain the new velocities

$$u_i^{k+1} = \tilde{u}_i - \Delta t \partial_i p^{k+1} \tag{4}$$

  and adjusted them to the boundary conditions using Fourier interpolation in $x_1$-direction at the boundaries, where necessary.

## 5.2 Code characteristics

The code consists of about 16,000 Fortran lines and is is highly vectorized with a usual vector length of $n = \frac{1}{\Delta x}$. The data area of the code requires in total 9 3D-fields. That means that for a typical problem size of $n = 128$ gridcells in each direction 18.6 MWords are necessary. This data area does not fit into the main memory of todays CRAY-XMP's. As a consequence, each 3D-field is split into n 2D-slices with respect to the $x_2$-direction. A memory management system [16] allows the optimization of the workspace available of in the main memory of the CRAY. The programmer ensures via subroutine calls that the slices necessary for the

|        | add | mult  | axpy  | other |
|--------|-----|-------|-------|-------|
| STEP-1 | 121 | 56    | 51    | 40    |
| STEP-2 | 34  | 18.25 | 10.75 | 5     |
| STEP-3 | 3   |       | 3     |       |

Table 1. Number of floating point operations per gridpoint for the different parts of the integration ($n = 128$).

computations are present in the workspace, while most of the data resides in an SSD file.

| $n$ | 128 | 256 | 512 |
|---|---|---|---|
| CPU-times (secs) | | | |
| Initialization | 75 | 600 | 4800 |
| Integration | 25200 | 400000 | 6560000 |
| Statistics | 5100 | 41000 | 328000 |
| Total CPU-time (d) | 0.35 d | 5.2 d | 79 d |
| Memory (MWords) | | | |
| Incore | 2.1 | 16.4 | 130 |
| Outcore | 18.6 | 146.2 | 1160 |

Table 2. CPU-times and memory resources required on CRAY-XMP using one processor. ($n = 256$, 512 estimated)

After an initialization step the integration is performed in an loop over the time steps. At certain time steps a statistic package is called, to analyse the current solution. The initialization, statistics, STEP-1, and STEP-3 steps are executed on the basis of 2D-slices. For the treatment of the shear periodic boundary conditions each 2D-slice contains one line of dummy cells at each boundary. The dummy cells contain copies of the appropriate values at the opposite boundary. At the $x_2, x_3$-boundaries these values are generated by Fourier-interpolation.

For the STEP-2 part (Poisson equation) an FFT of one 3D field is required. This field lies completely in the workspace of the main memory and the FFT is performed inplace. This requires in total 2.1 MWords of workspace ($n = 128$).

Table 1 shows the number of floating point operations necessary for the different basic parts. Table 2 summerizes the computing times needed for the different parts of the program on a CRAY-XMP 216, if only one processor is used. A performance of approximately 83 MFlops is obtained for the integration part.

## 5.3  Parallelism of the integration

The description of the integration part at the end of Section 5.1 shows that four different parts of the algorithm have to be distinguished:

- Gridtype calculations with local relations in the STEP-1 and STEP-3 part,

- 1D-FFTs in $x_1$-direction for the Fourier interpolation in the STEP-1 part,

- 1D-FFTs in $x_1$-directions at the boundaries for Fourier interpolation and

- the fast Poisson solver.

The *gridtype calculations* are fully explicit. As for Jacobi relaxations (cf. Section 4.1) only local information is required. In each computational step in the STEP-1 and STEP-3 part of the algorithm the values at distinct grid points can be treated in parallel.

In the *fast Poisson solver* 1D-FFTs are applied first in $x_1$-direction and than in $x_2$-direction at each $x_1$- or $x_2$-line, respectively. Then in each $x_3$-line Gaussian elimination is used to solve the remaining tridiagonal systems. Finally the results are transformed back into grid point space using 1D-FFTs in the same manner as described before. In each step at least $\frac{n^2}{2}$ lines can be treated in parallel. While the Gaussian elimination itself is sequential, in the FFT itself all values can be treated in parallel [12,1] in each basic step (i.e. a radix-2 transformation). This parallelism, however, is not exploited in our implementation.

Similar considerations can be made for the treatment of the boundary and the Fourier interpolation.

## 5.4   Data distribution by grid partitioning

As it was shown in the previous section the code offers a high degree $n^2/2$ of data parallelism (i.e. the same code can execute on different data in parallel). The basic data structure used by the codes are grids. Therefore, using grid partitioning as explained in Section 4.2 will give good speed-up rates on SUPRENUM.

Several ways are possible how the grid can be split into subgrids. In order to get optimal performance on dm architectures, the computational load must be well balanced and the communication should be minimized.

An analysis of relaxation methods (Section 4.1) and multigrid methods (cf. [13],[10]) shows that

- most of the communication is local (except on very coarse grids);

- the amount of data which is communicated is small compared to the calculation work (typically a surface/volume ratio).

FFT algorithms are different: They are nonlocal, i.e. the coefficient of one specific wavenumber depends on all input values. Furthermore, the amount of data that has to be exchanged between the processes is at least proportional to the number of input values. A lower bound for the amount of information might be $\frac{p-1}{p}n$, where $p$ is the number of processes and $n$ is the number of input values. (This shows clearly the superiority of multigrid methods over FFTs for the solution of Poisson's equation on dm computers: While the number of operations is nearly similar for FFT's and multigrid methods, the "communication complexity" of multigrid methods is much lower than that of FFT's. Global information is transported through the grid on coarser meshsizes and therefore less information has to be transported between the different processes. Details on the parallel implementation of multigrid methods on SUPRENUM and performance evaluation are reported in [7].)

A convenient way to implement multidimensional FFTs on a dm architecture is the following [1, Chapter 11-9]:

The grid is splitted in such a way, that the FFT in the first dimension can be performed without any communication in the local memory of the processes. This means, that the grid is not partitioned in the dimension the transformation is applied to. After this step the data are globally rearranged in such a way, that the next transformation can be treated locally.

This rearrangement of data is equivalent to the global transpose of a multidimensional matrix. If $p$ processes are involved in this transpose $\frac{p-1}{p}n^2$ data have to be exchanged in the case of a 2D-problem of size $n \times n$.

In the 3D case, the transpose has to be applied for each dimension which is initially splitted.

This basic treatment of the FFT fits also nicely to the sequential nature of the Gaussian elimination used for tridiagonal systems in the fast Poisson solver used in this code. For the whole STEP-2 solver the grid has to be rearranged 4 times in total, if two dimensions are partitioned.

We summarize the possible ways of partioning

- Partitioning in $x_1$-direction:
  A lot of overhead is introduced due to the parallel treatment of the FFTs in the Fourier interpolation of STEP-1 and for the $x_3$-boundaries.

- Partitioning in $x_2$-direction:
  Splitting in this direction means that some parallel features have to be added to the memory management system.

- Partitioning in $x_3$-direction:
  Splitting in this direction is convenient. The overhead for the memory management is not distributed amoung the processes. The boundary condition at the $x_3$ boundary causes some load imbalance. This can be handled by assigning differently many points to the processes and using only a moderate number in this direction. Using only a moderate number of processes also reduces the amount of additional workspace needed for intermediate results in the integration.

As a consequence, the $x_2$- and $x_3$-directions of the problem have to be partitioned in such a way that 256 nodes can be used.

## 5.5  Code changes

In order to be executed on SUPRENUM, the code has to be adapted to the process concept and the message passing model. In the grid computation parts of the integration (STEP-1, STEP-3) the code has to be changed only at two places, namely

1. The boundary update routine has to be altered, to cover the boundary update in the $x_2$-direction.

2. The boundary exchange in the $x_3$-direction is handled by the data management system. Each time, when a process needs a local copy of a specific slice that belongs to the data space of an other process, a message is send to this process and the data is send back by the data management of the requested process.

In each integration step 4.5 updates of the boundaries in each direction are necessary to provide actual information for the gridfunctions.

The FFT has to be replaced by a parallel FFT which is provided by the scientific library on SUPRENUM.

| | FLOP | INST | $r_\infty$ | $f$ |
|---------|------|------|------------|-----|
| NEWJ    | 9    | 13   | 6.9        | 1.0 |
| VTJ     | 33   | 39   | 8.5        | 1.6 |
| RADIX-4 | 34   | 38   | 8.9        | 1.8 |

Table 3. Single node performance of selected program kernel. FLOP: Number of floating point operations per grid point. INST: Number of vector instructions per gridpoint. $r_\infty$: Asymptotic performance. $f$: Relation between floating point operations and load/store operations to memory

## 5.6 Performance evaluation

### 5.6.1 Performance on one single node

As mentioned in Section 2.1 each SUPRENUM node contains a vector unit which requires a more detailed analysis than a scalar floating point unit.

For performance estimates three characteristic routines were evaluated. NEWJ computes equation (4), VTJ computes some intermediate values in STEP-1, and RADIX-4 a radix-4 step of an FFT. All routines are fully vectorized. Table 3 shows the estimated performance (using Hockney's notation, c.f. [5]). Since $n_{\frac{1}{2}}$ is $\approx 20$ we assume an overall performance of 7 MFlops.

### 5.6.2 Parallel performance

Kolp and Mierendorff introduced a simulation model for SUPRENUM systems [7]. They introduce five types of system components (cf. Section 2) namely

i=0: the vector floationg point unit,

i=1: the transfer facility of the computing node,

i=2: the cluster bus,

i=3: the SUPRENUM-bus,

i=4: the communication node in each cluster.

Each of this component is modelled by a linear cost function of the form:

$$T_i = \sum_w (a_{i1} x(w) + a_{i2}) \tag{5}$$

$T_i$ is the time cost in $\mu$sec of an algorithm on one of the components, $a_{i1}$ and $a_{i2}$ are coefficients of the linear cost model for each of the components, $w$ is the number of vectors handled by the specific components and $x(w)$ is the length of the vectors in 64-bit words.

The overall execution time is then estimated as

$$T_{calc} = T_0 \tag{6}$$

$$T_{comm} = \max(T_1, T_2, T_3, T_4) \tag{7}$$

$$T = T_{calc} + T_{comm} \tag{8}$$

Reasonable values for the cost coefficients for SUPRENUM-1 are [7]

$$a_{11} = 0.6 \qquad a_{12} = 600$$
$$a_{21} = 0.031 \qquad a_{22} = 1.25$$
$$a_{31} = 0.4 \qquad a_{32} = 30$$
$$a_{41} = 0.4 \qquad a_{42} = 30$$

Although the two-level structure of SUPRENUM is transparent to the user, the performance usually depends on the particular mapping of the data structure (process system) to the nodes. We assume that each process is mapped to one node and vice versa and that the corresponding subdomains are of the same shape and size. Furthermore the subdomains treated by the nodes of one cluster form again a regular subdomain of the whole grid. The mapping is characterized by the numbers

$c_2 \times c_3$  logical configuration of the clusters in $x_2$- and $x_3$-direction

$p_2 \times p_3$  logical configuration of the nodes in $x_2$- and $x_3$-direction

If $n^3$ is the total number of grid points

$$n \times \frac{n}{p_2} \times \frac{n}{p_3} \qquad (9)$$

is the size of the subdomain treated by each node. The size of the subdomain handled by all nodes of one cluster together is

$$n \times \frac{n}{c_2} \times \frac{n}{c_3}. \qquad (10)$$

| | time- | CRAY-XMP | | SUPRENUM | | | | | |
|---|---|---|---|---|---|---|---|---|---|
| $n$ | steps | $T$ | Mflops | $c_2 \times c_3$ | $p_2 \times p_3$ | $T_{calc}$ | $T_{comm}$ | $T$ | MFlops |
| 128 | 2560 | 25,000 | 83 | 1 | $4 \times 4$ | 18,800 | 1,200 | 20,000 | 104 |
| | | | | $2 \times 1$ | $8 \times 4$ | 9,400 | 2,600 | 12,000 | 175 |
| | | | | $4 \times 1$ | $8 \times 8$ | 4,700 | 2,700 | 7,400 | 280 |
| 256 | 5120 | 394,000 | | $4 \times 4$ | $16 \times 16$ | 20,000 | 27,800 | 47,800 | 700 |
| | | | | $16 \times 1$ | $16 \times 16$ | 20,000 | 22,000 | 42,000 | 790 |
| 320 | 6400 | 960,000 | | $16 \times 1$ | $16 \times 16$ | 49,400 | 50,300 | 99,700 | 815 |

Table 4. Performance estimation for the integration part of the turbulence simulation (times in sec)

Table 4 summarizes the performance estimations based on

- a sustained performance of 7 MFlops per node (Section 5.6.1),

- 406.75 floating point operations per gridpoint and iteration (Table 1),

- 4.5 exchanges of boundaries of the subdomains in $x_2$-direction as whole slices and in $x_3$-direction as boundaries of slices (Section 5.4),

- 2 $(x_1, x_2)$ and 2 $(x_2, x_3)$ transposes of one grid function (Section 5.4).

| $n$ | CRAY | | SUPRENUM per node | | |
|---|---|---|---|---|---|
| | main memory | SSD | $c_2 \times c_3$ | 64 bit | 64/32 bit |
| 128 | 17 | 149 | $1 \times 1$ | 10.15 | 5.73 |
| | | | $2 \times 1$ | 5.29 | 3.30 |
| | | | $4 \times 1$ | 2.87 | 1.75 |
| 256 | 132 | 1170 | $16 \times 1$ | 5.56 | 3.47 |
| 320 | 256 | 2280 | $16 \times 1$ | 10.43 | 6.27 |

Table 5. Memory requirements in MByte

Table 5 shows the memory requirements for the problem. The numbers in the column labelled 64/32 bit refer to the case that the calculations and the FFTs are performed with 64 bit data but the data are truncated by the memory management system to 32 bit.

### 5.6.3 Remarks

- A fast Poisson solver based on multgrid instead of FFTs would increase the overall performance to 1.6 GFlops on a 16 cluster system.

- Outcore calculations on SUPRENUM using the cluster disks would increase the maximum possible problem size to $n = 576$. In this case the Mflops rate would decrease since the current disk system is not fast enough to catch up with the speed of the nodes in the case of this example. This decrease could be removed by increasing the number of disk nodes and disks per cluster.

## 5.7 Comparison with ECMWF model

The information in [9] indicates that the basic structure and the computational properties of the ECMWF operational model for weather prediction are very similar to the CFD code discussed here. Both codes use similar memory management systems. Grid point calculations involving the evaluation of finite difference operators are alternating with local evaluations in the space of "wave numbers" for each time step. Although the current parallel implementation of the ECMWF model involves the partitioning along one dimension, an implementation on a multiprocessor architecture with a medium number of nodes (256-1024) would require a 2D partitioning.

On the other hand the computational work per gridpoint and iteration should be much higher in the ECMWF model than in the turbulence simulation because more physical properties are modelled and a turbulence model is incorporated. Also the Legendre transformation used instead of FFT's in one dimension is much more expensive than an FFT. Due to the better calculation/communication ratio an implementation of the ECMWF model on SUPRENUM should be much more efficient than the discussed implementation of the turbulence model and it should also be possible to use the cluster disks efficiently.

# 6 Conclusion

We have given a brief description of the SUPRENUM architecture, its hardware components and its system software. Application software packages are available and the parallelisation of grid applications has been outlined. The detailed investigation of a CFD code for turbulence simulation demonstrates that considerable performance rates of several hundred Mflops can be obtained on SUPRENUM and that only minor code changes are necessary.

# 7 Acknowledgements

The authors thank Dr. Gerz and Prof. Schumann (DFVLR, Institut für Physik der Atmosphäre) for providing their turbulence code and for their help in its analysis.

# References

[1] Fox, G.; Johnson, M.; Lyzenga, G.; Otto, S.; Salmon, J.; Walker, D.: Solving Problems on Concurrent Processors. Vol. 1. Prentice-Hall International, Englewood Cliffs, New Jersey (1988).

[2] Gerz, T., Schumann, U., Elgobashi, S.: Direct numerical simulation of stratified homogeneous turbulent shear flows. J. Fluid Mech., in press, 1989.

[3] Giloi, W.K.: SUPRENUM – a trendsetter in modern supercomputer development. In [15].

[4] Hempel, R.: The SUPRENUM communications subroutine library for grid-oriented problems. Report ANL-87-23, Argonne National Laboratory, 1987.

[5] Hockney, R.W.: Parametrization of computer performance. Parallel Computing 5, 97-104, 1987.

[6] Kammer, H.: The SUPRENUM vector floating-point unit. In [15].

[7] Kolp, O., Mierrendorff, H.: Performance estimations for SUPRENUM systems. In [15].

[8] Schumann, U.: The countergradient heat flux in turbulent stratified flows. Nucl. Engrg. Des., 100 (1987), pp. 255-262.

[9] Simmons, A.J., Dent, D.: The ECMWF multi-tasking weather prediction model. Internal report, ECMWF, 1988.

[10] Solchenbach, K.: Grid applications on distributed memory architectures: Implementation and evaluation. In [15].

[11] Solchenbach, K., Trottenberg, U.: SUPRENUM: System essentials and grid applications. In [15].

[12] Swarztrauber, P.N.: FFT algorithms for vector computers. Parallel Comput, 1 (1984), pp. 45-63.

[13] Thole, C.A., Trottenberg, U.: A short note on standard parallel multigrid algorithms for 3D-problems. In Supercomputing, A. Lichnewsky, C. Saguez (eds.), North-Holland, Amsterdam, 1987.

[14] Trottenberg, U.: The SUPRENUM project: idea and current state. Proceedings of the SPEEDUP Workshop on Vector and Parallel Computing in Berne, 15 Jan. 88, to appear.

[15] Trottenberg, U. (ed.): Proceedings of the 2nd International SUPRENUM Colloquium "Supercomputing based on parallel computer architectures". Parallel Computing 7 (1988).

[16] Volkert, H., Schumann, U.: Development of an atmospheric mesoscale model on a Cray – Experiences with vectorization and input/output. In: W. Schönauer (Ed.): Notes on Numer. Fluid Mech. Vol. 12, Vieweg, Braunschweig.

[17] Zima, H.P., Bast, H.-J., Gerndt, H.M.: SUPERB: A tool for semi-automatic MIMD/SIMD parallelization. Parallel Comput. 6, pp-1-18, North Holland, Amsterdam, 1988.

# Data Parallel Supercomputing

S. Lennart Johnsson

Thinking Machines Corporation, 245 First St., Cambridge, MA 02142 USA
and
Department of Computer Science, Yale University, New Haven, CT 06520, USA

## Abstract

Supercomputers with a performance of a trillion floating-point operations per second, or more, can be produced in state-of-the-art MOS technologies. Such computers will have tens of thousands of processors interconnected by a network of bounded degree. Reducing the required data motion through a careful choice of data allocation and computational and routing algorithms is critical for performance. The management of thousands of processors can only be accomplished through programming languages with suitable abstractions.

We use the Connection Machine as a model architecture for future supercomputers, and Fortran 8X as an example of a language with some of the abstractions suitable for programming thousands of processors. Some of the communication primitives suitable for expressing structured scientific computations are discussed, and their benefit with respect to performance illustrated. With thousands of processors engaged in the solution of a single scientific problem, several subtasks are often treated concurrently in addition to the concurrent execution of each subtask. Some issues in constructing scientific libraries for such environments are discussed. Concurrent algorithms and performance data for matrix multiplication and the Fast Fourier Transform are presented. The solution of the compressible Navier-Stokes equation in three spatial dimensions by an explicit finite difference method, and the solution of a parabolic approximation of the Helmholtz equation by an implicit method are two examples of applications for which data parallel implementations are described briefly. The Helmholtz equations models three dimensional acoustic waves in the ocean.

# 1 Introduction

In the next decade, supercomputers are expected to have a performance of at least one trillion instructions per second, and a primary storage of tens to hundreds of Gbytes [5]. At this rate of computation and memory size, the operation code, the operand addresses, and the operands require 300–400 bits for a single instruction. The storage (including registers, or caches) must deliver 300–400 trillion bits per second, or about 16 million bits per cycle at a 25 MHz clock rate. This clock rate is somewhat conservative for

Topics in Atmospheric and Oceanic Sciences
© Springer-Verlag Berlin Heidelberg 1990

MOS technologies, but it cannot be expected to become higher by more than a small constant factor. The width of the storage needs to be several million bits. Assuming each processor can deliver 50 Mflops/sec, 40,000 processors will have a nominal peak capacity of two trillion floating-point instructions per second. A system of this complexity is entirely feasible to build. In half micron technology, 40,000 chips with on-chip floating-point units and memory are projected to have a total of about 64 Gbytes of primary storage. With the required storage bandwidth and with tens of thousands of processing units, a network is the only feasible alternative for passing data between processors and storage units. Using a technology that is an order of magnitude faster than MOS technologies, such as bipolar GaAs technology (used for the CRAY-3), would still require thousands of processing units for an architecture with a performance of a trillion floating-point operations per second.

In highly concurrent network architectures, the nominal processing capability is determined by the processing speed of a single processor and the number of processors. The real processing capability is determined by how well the individual processing units can be utilized, load balance, and how well the network supports the data motion required by the computation. The capacity available for the data motion is determined by technological constraints, and the requirements determined by data placement, computational algorithms, and routing algorithms. Of the various technological constraints that determine the performance characteristics of an architecture, the ones related to data motion are the most unforgiving with respect to performance.

In this paper we first review the communication capabilities of MOS technologies, and the communication needs of a few typical scientific applications. We then briefly discuss a programming model for architectures with a large number of processing units. The Connection Machine® architecture is presented in section four which also discusses features provided for efficient communication. Two basic computational primitives, matrix multiplication and the Fast Fourier Transform, are described in section five; and two applications, the solution of the compressible Navier-Stokes equations and the solution of a parabolic equation forming an approximation to Helmholtz equation are discussed. The last problem occurs in underwater acoustics.

## 2   Communication issues

In this section, we consider the communication capabilities of architectures built out of MOS technologies, the requirements of typical scientific computations, and the potential benefits of a good data placement or *address map*. The purpose of a good address map is to reduce the need for communication resources by placing data that frequently interacts close to each other.

A suitable metric for measuring locality of reference is determined either by the topology of the data set or the communications network. In solving partial differential equation, common distance measures are of the form $(\sum_{i=0}^{d} |x_i - y_i|^p)^{\frac{1}{p}}$, where $d$ is the dimensionality of the problem domain and $p$ the type of *norm*. The 2-norm (Euclidean distance) is often used in the physical domain. The 1-norm measures the distance between two points corresponding to traversals along coordinate axes. This measure is particularly interesting for Boolean cube networks. In such a network of $n$ dimensions with $x$ and $y$ being processor addresses, and $x_i$ and $y_i$, $0 \leq i < n$ being the distances (0 or 1) along the coordinate axes, the 1-norm is equal to the *Hamming* distance between the two points. The Hamming distance is equal to the minimum number of communication links a data item must traverse to move from processor $x$ to processor $y$ in a Boolean cube network. The 1-norm is not ideal for all networks. In a completely interconnected network all points are at unit distance from each other, and the 0-norm is a relevant distance measure.

In state-of-the art MOS technologies, $10^3$ - $10^4$ wires fit across a chip. The total data motion capacity of 40,000 chips is 100 - 1,000 TBytes/sec at 25 MHz clock rate without sharing of on-chip channels between different data paths. Assuming current standard packaging technologies of 100 - 300 pins per chip, the data motion capacity at the chip boundary is about 10 TBytes/sec. The data transfer rate on a chip is one to two orders of magnitude higher than the transfer rate at the chip boundary. At the board boundary, assuming connectors with 500 pins, the data motion capacity for a 200 board system is about 0.16 TBytes/sec. The transfer rate at the chip boundary is one to two orders of magnitude higher than the rate at the board boundary. The transfer rate at the board boundary is two to three orders of magnitude below the required rate for a system with a performance in the Tflop/sec range. A sustained performance of this magnitude is not possible with current packaging technologies without locality of reference.

Many mathematical models of physical phenomena, such as (partial) differential equations, are derived from local interaction rules. The discrete approximation of the continuous models are typically also derived from local approximations. For instance, the difference stencils used to approximate derivatives in finite difference techniques are local approximations. Finite elements provide a different local approximation. The difference stencils in finite difference techniques and the elements in finite element techniques completely define the spatial data interactions in one step of an *explicit method* for the solution of the discretized equations. The data interaction is local in the physical domain. The classical iterative methods for the solution of linear systems of equations only require local data interaction in the index space used for the solution variables. The conjugate gradient method requires a global reduction operation for the computation of scaling factors, and a global copy, or broadcasting, operation for the distribution of these factors in addition to the same local communication as required by Jacobi's method. Though each step in the iterative methods only involves local communication in the physical domain, most problems require global communication to attain a correct solution. Elliptic problems are of this type [6].

The requirement for non-local, or global, communication is more apparent in direct methods. Factoring matrices by Gaussian elimination or Householder transformations can be performed as a sequence of rank-1 updates of the submatrix that remains to be factored. (Higher rank updates may yield better performance on some architectures.) For a dense matrix, the pivot row is distributed to all the rows of the remaining submatrix, and the pivot column to all the remaining columns. For sparse matrices, the rank-1 update only affects the rows having non-zero entries in the pivot column, and the columns having non-zero entries in the pivot row. In Gaussian elimination one variable at a time is eliminated from the system of equations. Depending on the topology of the graph that the sparse matrix represents, the elimination process may only require local communication in the processor network, even for networks of bounded degree.

The problem of determining an address map that takes advantage of locality of reference in the physical domain, such that local communication in the processing network is possible, is often formulated as a graph embedding problem. A network of high degree local communication may be possible also when the references in the physical domain are non-local. For instance, divide-and-conquer methods for solving linear systems of equations, such as odd-even cyclic reduction, nested dissection, and multi-grid methods perform a recursive subdivision of the physical domain. For some of the recursion steps these methods require interaction between subdomains that are not adjacent in the physical domain. But, the references may still be performed by local communication in a network, such as a Boolean cube network. Any lattice can be embedded in a lattice of higher dimensionality preserving locality, but the converse is not true. It is also possible that the distance between a pair of lattice points is reduced when the lattice is embedded in a lattice of higher dimensionality. For instance, with a binary-reflected Gray code embedding of lattices in Boolean cubes [35,28] lattice points at a distance $2^j, j > 0$ in the index space are at distance 2 in the Boolean cube [16].

The communication requirements for explicit finite difference methods are determined by the grid and the difference stencils. Similarly, the elements, their order, and domain discretization determine the communication requirements for explicit finite element methods. The communication requirements for iterative solvers are determined by the adjacency matrix. The communication requirements for direct solvers can be determined by considering the graph underlying the (sparse) matrix, and by viewing variable elimination as node elimination in the graph [34]. The communication requirements are a function of the elimination order. The communications for the Fast Fourier Transform is identical to a butterfly network. Parallel cyclic reduction requires communication in the form of a data manipulator network or a PM2I (plus-minus $2^i$) [37] network. Of regular communications, the emulation of multi-dimensional lattices, butterfly networks, data manipulator networks, pyramid networks, and various forms of trees for reduction and copy operations are the most common.

The potential benefits from exploiting locality of reference is illustrated by three frequently used operations: matrix multiplication, a 7-point symmetric difference stencil applied at each node in a three dimensional grid, and butterfly based computations (FFT, bitonic sort). Applying a symmetric, 7-point difference stencil at every point in a three dimensional grid with $k$ variables per grid point and 2 operations per variable, the number of operations per remote reference is $r = \frac{1}{2d}(\frac{M}{k})^{\frac{1}{d}}$. For $d = 3$ $r = \frac{1}{6}(\frac{M}{k})^{\frac{1}{3}}$. Tables 1 and 2 give some values of $r$ for different sizes of the local memories. In the tables, $k = 8$. If the

| Computation | Registers only | 4 Mbit chips | 256 4 Mbit chips (board) | 256 Boards |
|---|---|---|---|---|
| Mtx mpy | 0.5 | 104 | 1600 | 26000 |
| 3-d Relaxation | 0.17 | 4.27 | 26.7 | 170.7 |
| FFT | 1 | 18.8 | 28.8 | 38.8 |

Table 1. Number of operations per remote reference of a single variable.

| Computation | 4 Mbit 1 proc. 1 chip | 256 Procs. = Board | 256 boards = Machine |
|---|---|---|---|
| Mtx mpy | 1 | 10 | 160 |
| 3-d relaxation | 32 | 480 | 24600 |
| FFT | 3 | 1140 | 160000 |
| no locality | 300 | 76800 | 19660800 |

Table 2. Number of bits across the chip/board/system boundary per cycle.

local variables form matrices and the local operations imply matrix multiplications, then the number of arithmetic operations per variable is higher. Several linear algebra operations have a ratio of operations to remote references that can be modeled by the same expression as was given for the difference molecules, i.e. $\frac{1}{\alpha}(\frac{M}{\beta})^{\frac{1}{\gamma}}$ for suitable values of $\alpha$, $\beta$ and $\gamma$. In the Navier-Stokes flow computation, the local state vectors are of length 5 and local matrices typically of size $3 \times 4$, or $4 \times 5$, or some similar size [32]. For butterfly based algorithms, such as the Fast Fourier Transform (FFT) and sorting, the dependence is of the form $\alpha \log(\frac{M}{\beta})$. For the FFT the ratio is $1.25 log_2(M/2)$ real operations per remote reference using a radix-M algorithm, which is optimum [12].

Table 2 gives the number of bits that have to cross the chip, board, and system boundaries during a single cycle, assuming the optimum locality or no locality of reference. It is assumed that each chip has one processing unit, that a board has 256 processing units, and that all variables are in single precision.

Exploiting locality reduces the required communication bandwidth by a factor of 8–100 at the chip boundary for these computations, a factor of 80–5000 at the board level, and at least a factor of 125 at the I/O interface. A sustained performance in the Tflops/s range is possible with state-of-the-art technology only if locality is properly exploited.

# 3   Programming model

Architectures in which tens of thousands of operations can be performed concurrently are often referred to as *data parallel* to distinguish them from *control parallel* architectures, which offer a considerably lower degree of concurrency. Algorithms are designed based on the structure and representation of the problem domain. Objects in data parallel languages are represented by higher level data types such as the array extensions of Fortran 8X [31]. In a language with an array syntax, a number of nested loops (often equal to the number of axes in the array) disappear from the code, compared to a language without the array syntax. We illustrate this property by two examples. The first example is the implementation of a 7-point stencil in three dimensions. The second example is taken from a finite element code for stress analysis.

In the example below which defines the computation of a 7-point stencil at every point in a three dimensional grid, the operation CSHIFT defines a circular shift. The first argument is the variable to which the shift is applied, the second defines the axis along which the shift takes place, and the third argument defines the length and direction of the shift. Since there is no conditional statement in the code below, it implements periodic boundary conditions. Note that there are no explicit loops for the array axes.

```
          subroutine psolve(phi, omega, inside, n, iter)
          real phi(n, n, n), omega(n, n, n), factor
          logical inside(n, n, n)
          factor = 1.0/6.0
          do 100 i=1,iter,1
             phi = factor * (
  1             CSHIFT(phi, dim=1, shift=-1) +
  2             CSHIFT(phi, dim=2, shift=-1) +
  3             CSHIFT(phi, dim=3, shift=-1) +
  4             CSHIFT(phi, dim=1, shift=+1) +
  5             CSHIFT(phi, dim=2, shift=+1) +
  6             CSHIFT(phi, dim=3, shift=+1) ) +
  7             omega
 100      continue
          return
          end
```

In the finite element example below, the elements are brick elements of first order. There is one nodal point in each corner of an element. The state is represented by three displacements, $x = (u, v, w)$. The local interaction matrix, the elemental stiffness matrix, is a $3 \times 24$ matrix, with one row for each of the three components of the local displacement vector. The code segment also contains one compiler directive, SERIAL, which affects the data layout. The meaning will be explained later. The code fragment is from the iterative solver which requires the computation of a matrix vector product. In the particular finite element code from which the code segment is selected, the elemental stiffness matrices are not assembled into a global stiffness matrix. Instead, a matrix vector product is performed for each element, and a total product vector assembled.

```
CMF$LAYOUT K(:SERIAL, :SERIAL, , , ), R(:SERIAL, , , ), X(:SERIAL, , , )
REAL K(3,24, 32, 32, 32), R(3,32,32,32), U(3,32,32,32), V(3,32,32,32), W(3,32,32,32), X(24,32,32,32)
CALL ALL-TO-ALL-ELEMENT-BROADCAST(U,V,W,X)
R = 0.0
DO I=1,24
   DO J=1,3
      R(J,:,:,:)=R(J,:,:,:)+K(J,I, :, :, :) * X(I, :, :, :)
   END DO J
END DO I
(WHERE (.NOT. I-RIGHT-BOUNDARY)) R=R + EOSHIFT(R, 1, 1)
(WHERE (.NOT. I-LEFT-BOUNDARY)) R= EOSHIFT(R, 1, -1)
(WHERE (.NOT. J-RIGHT-BOUNDARY)) R=R + EOSHIFT(R, 2, 1)
(WHERE (.NOT. J-LEFT-BOUNDARY)) R= EOSHIFT(R, 2, -1)
(WHERE (.NOT. K-RIGHT-BOUNDARY)) R=R + EOSHIFT(R, 3, 1)
(WHERE (.NOT. K-LEFT-BOUNDARY)) R= EOSHIFT(R, 3, -1)
```

In the above code segment, I-RIGHT-BOUNDARY, I-LEFT-BOUNDARY, etc. are boolean arrays which define the right–hand and left–hand boundaries of each finite element in the three dimensions respectively.

# 4   The Connection Machine Architecture

The Connection Machine [8] is a data parallel architecture. It has a total primary storage of 512 Mbytes using 256 kbit memory chips, and 2 Gbytes with 1 Mbit memory chips. The data transfer rate to storage is approximately 45 Gbytes/s at a clock rate of 7 MHz. The primary storage has 64k ports and a simple 1-bit processor for each port. The storage per processor is 8 kbytes for a total storage of 512 Mbytes and 64k bytes with 1 Mbit memory chips. The Connection Machine model CM-2 can be equipped with hardware for floating-point arithmetic. With the floating-point option, 32 Connection Machine processors share a floating-point unit, which is an industry standard, single chip floating-point multiplier and adder with a few registers. The peak performance available from the standard instruction set and the higher level languages is in the range 1.5 Gflops/s - 2.2 Gflops/s. The higher level languages do not at the present time make efficient use of the registers in the floating-point unit for operations that vectorize. With optimum use of the registers, a performance that is one order of magnitude higher is possible. For instance, for large local matrices, a peak performance in excess of 25 Gflops/s has been measured.

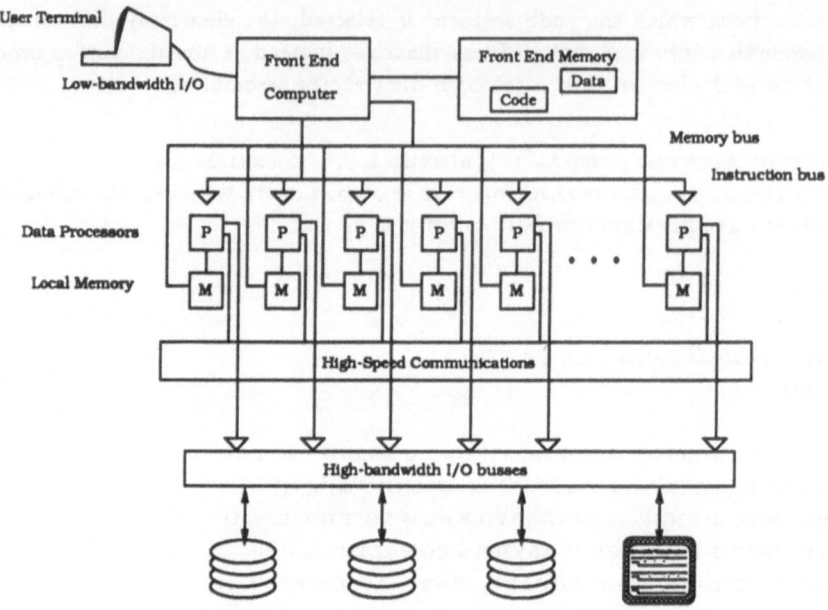

**Figure 1.** The Connection Machine System

The Connection Machine needs a host computer. Currently, three families of host architectures are supported: the VAX family with the BI-bus, SUN 4, and the Symbolics 3600 series. The Connection Machine memory is mapped into the address space of the host. The program code resides in the storage of the host. It fetches the instructions, does the complete decoding of scalar instructions, and executes them. Instructions to be applied to variables in the Connection Machine are sent to a microcontroller, which decodes and executes instructions for the Connection Machine. Variables defined by array constructs are allocated to the Connection Machine, unless allocation on the front-end is requested. The architecture is depicted in Figure 1. The Connection Machine can also be equipped with a secondary storage system known as the data vault. There exist 8 I/O channels, each with a block transfer rate of up to approximately 30 Mbytes/s. The size of the secondary storage system is in the range 5 Gbytes to 640 Gbytes. The Connection Machine can also be equipped with a frame buffer for fast high resolution graphics. An update rate of about 15 frames per second can be achieved.

The Connection Machine processors are organized with 16 processors to a chip, and the chips interconnected as a 12-dimensional Boolean cube. The communication is bit-serial and pipelined. Concurrent communication on all ports is possible. Through the bit-serial pipelined operation of the communication system, remote processor references require no more time than nearest neighbor references provided there is no contention for communication channels. For communication in arbitrary patterns, the Connection Machine is equipped with a router which selects one of the shortest paths between source and destination, unless all of these paths are occupied. The router has several options for resolving contention for communication channels.

## 4.1 Configuring the address space

The address field of the Connection Machine is divided into three parts: (off-chip|on-chip|memory). The off-chip field consists of 12 bits that encode the Connection Machine processor chips, the on-chip field encodes the 16 processors on each Connection Machine processor chip, and the lower order bits encode the memory addresses local to a processor. The lowest order off-chip bit encodes pairs of processor chips sharing a floating-point unit. On-chip communication is considerably faster than inter-chip communication. On-chip communication is a local memory reference. Off-chip communication is slower due to the limited bandwidth at the chip boundary. With the current chip (VLSI MOS) and interconnection (metal wire) technologies, such a characteristic is expected. The non-uniformity in access time impacts the optimum data allocation [15,17].

The default data allocation scheme on the Connection Machine first determines how many data elements need to be stored in each processor for an equal number of elements per processor, then stores that many successive elements in each processor, *consecutive storage* [15]. With the $n$ highest order bits encoding the processors and the lower order bits encoding memory addresses in each processor, the consecutive assignment can be illustrated as follows.

$$\textit{Consecutive assignment:} (\underbrace{x_m x_{m-1} \cdots x_{m-n+1}}_{rp} \underbrace{x_{m-n} x_{m-n-1} \cdots x_0}_{vp})$$

The field denoted $rp$ encodes *real processor* addresses as opposed to local *memory* addresses $vp$. For a data set of $M = 2^m$ complex points, $m + 1$ address bits are required, $n$ of which are processor address bits. There are $m - n + 1$ local storage address bits. Another frequently used address form is *cyclic* assignment, for which the lowest order address bits determine the *real processor* address.

$$\textit{Cyclic assignment:} (\underbrace{x_m x_{m-1} \cdots x_n}_{vp} \underbrace{x_{n-1} x_{n-2} \cdots x_0}_{rp}).$$

In the cyclic assignment, all data elements in a processor have the same $n$ low order bits. In the consecutive assignment, the elements in a processor have the same $n$ high order bits. The cyclic allocation scheme currently is not supported on the Connection Machine system. However, for some computations, it offers certain performance advantages [15,27].

Current implementations of the Connection Machine languages encode each axis of a multi-dimensional array separately. Each axis is extended to a length that is equal to some power of two. For an axis length $P$, $\lceil log_2 P \rceil$ address bits are assigned to the encoding of the elements along that axis. The consecutive allocation scheme is used for each axis. The encoding of the axes in the total address space attempts to configure each part of the address space (off-chip, on-chip, and memory) to conform with the array. To the extent possible, all axes have a segment of each address field, and the ratio of the lengths of segments for different axes is the same as that of the length of the axes.

The default allocation of axes to off-chip, on-chip, and memory bits may not always be the preferred allocation. For a computation in which the interaction between virtual processors is equally frequent in each direction, the total amount of communication is minimized if the virtual processors assigned to a physical processor, or actually a processor chip, forms a single subdomain with an aspect ratio as close to one as possible [17]. The different Connection Machine languages provide different means for user controlled data allocation. In CM-Fortran compiler directives allow a user to specify an axis as SERIAL, which implies that the axis is allocated to a single processor. In PARIS (PARallel Instruction Set), the Connection Machine native language, a user has full control over what dimensions of the address space an axis occupies. But, only consecutive allocation of data to processors is supported.

If an array has fewer elements than the number of real processors in the configuration, the array is extended such that there is one element per real processor. In CM-Fortran an axis is added to the array with a length equal to the number of instances of the specified array that matches the number of real processors.

Array elements in the Connection Machine programming languages are often referred to as *virtual processors* [8,1]. In general, several virtual processors (array elements) are mapped to the storage of each physical processor. The number of virtual processors per physical processor is called the *virtual processor ratio* [1]. The storage of a physical processor is divided between as many virtual processors as is given by the virtual processor ratio. That many virtual processors time-share a physical processor.

## 4.2   Encoding of array axes

In the common binary, encoding successive integers may differ in an arbitrary number of bits. For instance, 63 and 64 differs in 6 bits, and hence are at a *Hamming* distance of 6 in the Boolean cube. A *Gray* code by definition has the property that successive integers differ in precisely one bit. The most frequently used Gray code for the embedding of arrays in Boolean cubes is a *binary-reflected* Gray code [15,28,35]. This Gray code is periodic. The code preserves adjacency for any loop (periodic one-dimensional lattice) of even length, and for loops of odd length one edge in the loop is mapped into a path of length two [15]. For the embedding of multi-dimensional arrays, each axis may be encoded by the *binary-reflected* Gray code. The embedding of an $N_1 \times N_2 \times \ldots \times N_d$ array requires $\sum_{i=1}^{d} \lceil \log_2 N_i \rceil$ bits. The *expansion*, i.e., the ratio between the consumed address space and the actual array size, is $2^{\sum_{i=1}^{d} \lceil \log_2 N_i \rceil} / \Pi_{i=1}^{d}$, which may be as high as $\sim 2^d$ [7,10]. The expansion can be reduced by allowing some successive array indices to be encoded at a Hamming distance of two. The *dilation* is the maximum Hamming distance between any pair of adjacent array indices. Every two-dimensional array can be embedded with minimum expansion and dilation 2 [3]. Minimum expansion dilation 2 embeddings for a large class of two-dimensional arrays are given in [10], which also provides a technique for reducing the expansion of higher dimensional arrays. Minimal expansion dilation 7 embeddings are possible for all three dimensional arrays [4]. Embeddings with dilation 2 for many three dimensional arrays are given in [9].

The lattice emulation by a binary-reflected Gray code embedding is part of the standard programming environment on the Connection Machine system. In CM-Fortran, array axes are by default encoded in a binary-reflected Gray code for the off-chip segment of the address field. In the other Connection Machine languages, the Gray code encoding is invoked by configuring the Connection Machine as a lattice of the appropriate number of dimensions. The benefit of the lattice emulation feature is twofold: the virtual processors are assigned to physical processors such that the communication requirements are minimized, and lattice organized computations are often easier to express by virtue of programming constructs corresponding directly to the operations in the problem domain.

# 5  Scientific Libraries

A library of basic scientific routines on data parallel computers must be capable of handling two forms of concurrency: concurrent execution of multiple, independent problems, and concurrent execution of a single problem. In a single instruction stream architecture, such as the Connection Machine, the independent problems must require the same instructions for a good load balance. For instance, multiple matrix multiplications, multiple factorizations, or multiple FFT's can be performed concurrently. The need for multiple operations of the same type occurs frequently in the solution of partial differential equations on data parallel architectures, as illustrated in the next section. In this section, we give two specific examples of concurrency in a single operation by briefly describing how matrix multiplication and the FFT are performed in the library routines available on the Connection Machine.

## 5.1  Matrix Multiplication

The matrices to be multiplied are in general distributed across several, but not necessarily all, processors. In many cases, several different matrix products will be formed concurrently in disjoint sets of processors. Within each set, some data motion is required to compute a matrix product. Typically, each processor will have several matrix elements of each operand assigned to it. A suitable local matrix multiplication algorithm is required in addition to a global algorithm that implements the appropriate data motion given the allocation of the input and output matrices.

### 5.1.1 The basic algorithm

The current Connection Machine library routine for matrix multiplication is based on mesh emulation. The data motion for multiplying two matrices on a mesh can be partitioned into two phases; Alignment and Multiplication. The purpose of the alignment is to make sure that the range of "inner" indices for one matrix is a subset of the range of "inner" indices of the other matrix [15,21,20]. All processors can concurrently perform a multiplication and an addition. The data motion during the multiplication phase preserves this property. The essence of the data motion is best illustrated for the multiplication of two $P \times P$ matrices on a $P \times P$ mesh of processors [2].

**forall** $i, j \in \{0, 1, \ldots, P-1\} \times \{0, 1, \ldots, P-1\}$ **do**

**Alignment:**

$$a(i,j) \leftarrow a(i, (i+j) \bmod P)$$
$$b(i,j) \leftarrow b((i+j) \bmod P, j)$$

**Muliplication:**

$$c(i,j) \leftarrow c(i,j) + a(i,j) \times b(i,j)$$
**for** $k := 1$ **to** $P-1$
$$a(i,j) \leftarrow a(i, (j+1) \bmod P)$$
$$b(i,j) \leftarrow b((i+1) \bmod P, j)$$
$$c(i,j) \leftarrow c(i,j) + a(i,j) \times b(i,j)$$
**endfor** $k$
**endforall** $i, j$

By a binary-reflected Gray code encoding of array indices, the data motion in the above algorithm only requires nearest neighbor communication. All array indices are extended to the nearest greater power of two. For the multiplication of matrices of arbitrary shapes on any size Boolean cube configured multiprocessor, it is necessary to generalize the algorithm to the following cases:

- The set of processors are configured with $N_0$ processors along axis zero, and $N_1$ processors along axis one, $N_0 \neq N_1$.

- A submatrix per processor instead of a single element

- Arbitrary $P$, $Q$, and $R$

- Parallelization of the loop on the "inner" index.

The first generalization is necessary for several reasons. One reason is that the number of processors $N$, to which the matrices are allocated, may not be a square. Another reason is that for small matrices and a large number of processors, the operands for a matrix product may only extend over a subset of processors, even if only a single element of each operand is assigned to each processor. In this case, as well as in the case where each operand has several elements assigned to each processor, the optimum configuration of the physical processors is an array of the same shape as that of the product matrix $C$ [21]. The need for the second and third generalization is apparent. The last generalization is motivated by the fact that the number of processors $N$ may be significantly greater than the number of elements $PR$ of the product matrix $C$. For $N \geq PR$, all three loops in a Fortran 77 code for matrix multiplication by the standard algorithm may be parallelized. The third axis, i.e., the axis for the "inner index" can be instantiated partially, or totally, in space. With all three axes instantiated in space, the operands are assigned to orthogonal planes. For instance, the matrix $A$ can be assigned to the plane defined by axes zero and two, the matrix $B$ to the plane defined by axes one and two, and the matrix $C$ to the plane defined by axes zero and one. The matrix $A$ is copied along axis one, and the matrix $B$ along axis zero. The matrix $C$ is obtained by reduction along axis two.

### 5.1.2 Arbitrary Matrix Shapes and Sizes

We assume that a submatrix of each operand is stored in each processor to which any matrix element is assigned, and that the submatrices in different processors are of the same shape and size. We first define a matrix multiplication algorithm for a two-dimensional mesh of processors. The third axis, i.e., the axis of the inner index, is entirely instantiated in time. The set of processors are assumed to be configured with $N_0$ processors along axis zero and $N_1$ processors along axis one. If $N_0 \neq N_1$, then the range of the inner index for the two operands is clearly different in every processor. The alignment must assure that the inner index range for one operand is a subset of the inner index range of the other operand. This property must be maintained during the multiplication. In the case of the Connection Machine implementation, the ratio $\frac{N_0}{N_1} = 2^s$ for some integer $s$. For a small matrix and sufficiently many processors, the third axis can also be parallelized, and the set of processors are configured as an array with three axes.

Configuring the Connection Machine processors with the same number of processors along each axis of a two-dimensional array, and using Gray code encoding of the axes, allow the alignment and data motion during multiplication to be based entirely on processor addresses [18]. The data motion during the multiplication phase implements an *all-to-all broadcasting* [25] within rows for the matrix $A$ and columns for the matrix $B$: $C \leftarrow C + A \times B$. The broadcasting is accomplished by cyclic rotation. Memory requirements are conserved which is an important property for the multiplication of matrices having large submatrices allocated to each processor. The minimum number of rotation steps along axis zero is $N_0 - 1$ and along axis one $N_1 - 1$. The algorithm implemented on the Connection Machine moves a complete submatrix assigned to a processor when communication is needed. No local data motion is required. If $N_0 > N_1$, then $\frac{N_0}{N_1}$ rotation steps are performed along axis zero for every rotation step along axis one. Only a fraction of the local submatrix of the matrix subject to the fewest rotation steps is used in a local matrix-matrix multiplication for each rotation step along the longest processor axis. But, for the matrix subject to the largest number of rotation steps, the entire submatrix is used for each rotation step. Figure 2 illustrates the data motion of the matrices $A$ and $B$ for $N_0 = 4$ and $N_1 = 8$.

The alignment along the longest axis is performed as if the processor array was square with the number of processors along an axis equal to the number of processors along the shortest axis. The alignment along the shortest axis is performed as if the processor array was square with the number of processors along an axis equal to the number of processors along the longest axis. With this alignment, the multiplication can be accomplished by the minimum number of rotation steps along each axis [18].

With the third axis entirely instantiated in time, the matrix product requires $2\frac{P}{N_0}\frac{R}{N_1}Q$ arithmetic operations in sequence. The arithmetic time is proportional to the number of matrix elements per physical processor of the matrix $C$, and the length of the "third" axis, $Q$. The communication time for the multiplication phase is proportional to $\frac{PQ}{N_0}$ for the matrix $A$ and to $\frac{QR}{N_1}$ for the matrix $B$. The alignment phase carried out as shifts along the axes of the mesh requires approximately the same amount of time as the multiplication phase if the cyclic shifts only can be performed in one direction at a time. For the Connection Machine implementation, the router is used for the alignment, and the time for large values of $N_0$ and $N_1$ is considerably less than the time required by a sequence of cyclic shifts. However, the time for alignment on a Boolean cube network can be further improved by optimizing the cyclic shifts [15].

The algorithm presented above is correct for all array shapes and sizes, and matrix shapes and sizes. However, the processor utilization can be improved when $Q$ is small compared to $P$ and/or $R$, and to $N_0$ and $N_1$. The matrix $C$ is computed in-place. Only the set of processors to which $C$ is allocated participate in the arithmetic operations. If $Q < R$, then only $Q$ columns of $C$ are computed concurrently. The set of $Q$ columns is a function of the step of the algorithm such that at the end of the algorithm all columns of $C$ are computed. Similarly, if $Q < P$, only $Q$ rows are computed concurrently in any step.

By replicating $A$ $\min(\frac{R}{Q}, \frac{N_1}{Q})$ times along axis one, and matrix $B$ along axis zero $\min(\frac{P}{Q}, \frac{N_0}{Q})$ times, all processors to which the matrix $C$ is allocated are used.

The optimum aspect ratio of the physical processor array is the same as the aspect ratio of the product matrix $C$. If there is only one matrix element of each operand per processor, then the optimum aspect ratio of the physical machine is 1, since it is desirable to minimize the maximum axis length: $N_0 = N_1 = \sqrt{N}$.

## ALIGN

**A**

| | | | | | | | |
|---|---|---|---|---|---|---|---|
| 00 | 01 | 02 | 03 | 04 | 05 | 06 | 07 |
| 10 | 11 | 12 | 13 | 14 | 15 | 16 | 17 |
| 22 | 23 | 24 | 25 | 26 | 27 | 20 | 21 |
| 32 | 33 | 34 | 35 | 36 | 37 | 30 | 31 |
| 44 | 45 | 46 | 47 | 40 | 41 | 42 | 43 |
| 54 | 55 | 56 | 57 | 50 | 51 | 52 | 53 |
| 66 | 67 | 60 | 61 | 62 | 63 | 64 | 65 |
| 76 | 77 | 70 | 71 | 72 | 73 | 74 | 75 |

**B**

| | | | | | | | |
|---|---|---|---|---|---|---|---|
| 00 | 11 | 22 | 33 | 44 | 55 | 66 | 77 |
| 10 | 21 | 32 | 43 | 54 | 65 | 76 | 07 |
| 20 | 31 | 42 | 53 | 64 | 75 | 06 | 17 |
| 30 | 41 | 52 | 63 | 74 | 05 | 16 | 27 |
| 40 | 51 | 62 | 73 | 04 | 15 | 26 | 37 |
| 50 | 61 | 72 | 03 | 14 | 25 | 36 | 47 |
| 60 | 71 | 02 | 13 | 24 | 35 | 46 | 57 |
| 70 | 01 | 12 | 23 | 34 | 45 | 56 | 67 |

## ROTATE

**←A**

| | | | | | | | |
|---|---|---|---|---|---|---|---|
| 02 | 03 | 04 | 05 | 06 | 07 | 00 | 01 |
| 12 | 13 | 14 | 15 | 16 | 17 | 10 | 11 |
| 24 | 25 | 26 | 27 | 20 | 21 | 22 | 23 |
| 34 | 35 | 36 | 37 | 30 | 31 | 32 | 33 |
| 46 | 47 | 40 | 41 | 42 | 43 | 44 | 45 |
| 56 | 57 | 50 | 51 | 52 | 53 | 54 | 55 |
| 60 | 61 | 62 | 63 | 64 | 65 | 66 | 67 |
| 70 | 71 | 72 | 73 | 74 | 75 | 76 | 77 |

**B**

| | | | | | | | |
|---|---|---|---|---|---|---|---|
| 00 | 11 | 22 | 33 | 44 | 55 | 66 | 77 |
| 10 | 21 | 32 | 43 | 54 | 65 | 76 | 07 |
| 20 | 31 | 42 | 53 | 64 | 75 | 06 | 17 |
| 30 | 41 | 52 | 63 | 74 | 05 | 16 | 27 |
| 40 | 51 | 62 | 73 | 04 | 15 | 26 | 37 |
| 50 | 61 | 72 | 03 | 14 | 25 | 36 | 47 |
| 60 | 71 | 02 | 13 | 24 | 35 | 46 | 57 |
| 70 | 01 | 12 | 23 | 34 | 45 | 56 | 67 |

## ROTATE

**←A**

| | | | | | | | |
|---|---|---|---|---|---|---|---|
| 01 | 02 | 03 | 04 | 05 | 06 | 07 | 00 |
| 11 | 12 | 13 | 14 | 15 | 16 | 17 | 10 |
| 23 | 24 | 25 | 26 | 27 | 20 | 21 | 22 |
| 33 | 34 | 35 | 36 | 37 | 30 | 31 | 32 |
| 45 | 46 | 47 | 40 | 41 | 42 | 43 | 44 |
| 55 | 56 | 57 | 50 | 51 | 52 | 53 | 54 |
| 67 | 60 | 61 | 62 | 63 | 64 | 65 | 66 |
| 77 | 70 | 71 | 72 | 73 | 74 | 75 | 76 |

**↑B**

| | | | | | | | |
|---|---|---|---|---|---|---|---|
| 20 | 31 | 42 | 53 | 64 | 75 | 06 | 17 |
| 30 | 41 | 52 | 63 | 74 | 05 | 16 | 27 |
| 40 | 51 | 62 | 73 | 04 | 15 | 26 | 37 |
| 50 | 61 | 72 | 03 | 14 | 25 | 36 | 47 |
| 60 | 71 | 02 | 13 | 24 | 35 | 46 | 57 |
| 70 | 01 | 12 | 23 | 34 | 45 | 56 | 67 |
| 00 | 11 | 22 | 33 | 44 | 55 | 66 | 77 |
| 10 | 21 | 32 | 43 | 54 | 65 | 76 | 07 |

Figure 2. Matrix multiplication on a 4 × 8 array.

| $P \times Q \times R$ | Gflops/s | $P \times Q \times R$ | Gflops/s | $P \times Q \times R$ | Gflops/s |
|---|---|---|---|---|---|
| $64 \times 4 \times 1$ | 7.84 | $64 \times 4 \times 8$ | 9.20 | $64 \times 4 \times 32$ | 9.38 |
| $64 \times 8 \times 1$ | 11.98 | $64 \times 8 \times 8$ | 13.53 | $64 \times 8 \times 32$ | 13.71 |
| $64 \times 16 \times 1$ | 16.35 | $64 \times 16 \times 8$ | 17.74 | $64 \times 16 \times 32$ | 17.88 |
| $64 \times 32 \times 1$ | 17.35 | $64 \times 32 \times 8$ | 18.02 | $64 \times 32 \times 32$ | 18.11 |
| $64 \times 64 \times 1$ | 19.08 | $64 \times 64 \times 8$ | 19.55 | $64 \times 64 \times 32$ | 19.61 |
| $64 \times 256 \times 1$ | 20.38 | $64 \times 256 \times 8$ | 20.51 | $64 \times 256 \times 32$ | 20.52 |

Table 3. Performance data for the local matrix kernels.

### 5.1.3 Parallelizing the third axis

The set of processors participating in the multiplication using the extended algorithm is approximately $\min(P, N_0) \times \min(R, N_1)$. If $P < N_0$, and/or $R < N_1$, only a subset of size $\sim PR$ of all $N$ processors are used. If $N >> PR$, then substantially improved processor utilization can be achieved by also instantiating the third axis at least partially in space. To generate two instances in space, we partition the matrix $A$ as $(A_0 A_1)$ and $B$ as $\left(\frac{B_0}{B_1}\right)$, where $A_0$ and $A_1$ are $P \times \frac{Q}{2}$ matrices and $B_0$ and $B_1$ are $\frac{Q}{2} \times R$ matrices. Then, the products $A_0 B_0$ and $A_1 B_1$ are computed on disjoint sets of $PR$ processors. Each set is configured as a mesh. By employing the algorithm above for each mesh, each product only requires approximately $\frac{Q}{2}$ steps. Denote the two meshes 0 and 1. Mesh 0 contains twice as many copies of $A_0$ and $B_0$ as before, but no copies of $A_1$ and $B_1$. Mesh 1 contains twice as many copies of $A_1$ and $B_1$ as in the case of a single mesh, and no copies of $A_0$ and $B_0$.

We introduce a third processor axis for the enumeration of the meshes of size $\sim PR$ each. The product $A \times B$ is obtained through a plus-reduction along the third axis, the axis of the inner index $Q$. The number of planes in the third dimension is $\min(Q, \frac{N_0}{P} \times \frac{N_1}{R}) = 2^{\min(q, n-p-r)}$ and the length of the "parallelized" inner axis is $\hat{Q} = 2^{\min(0, q+p+r-n)}$.

### 5.1.4 CM implementation issues

The local matrix multiplication kernel makes use of matrix-vector kernels. These kernels read a segment of a vector into the registers of the floating-point unit, then use it for a matrix-vector multiplication. The timings for a few matrix shapes are given in Table 3. The standard allocation scheme for arrays implies that, in general, the configuration of the physical machine is different for arrays of different shapes. However, for the algorithms described above, the operands need to be allocated assuming the same physical machine. The reconfiguration of the set of physical processors to a common shape is performed as part of the alignment. The alignment is performed by the Connection Machine router. Timings are given in Table 4.

| $P \times Q \times R$ | 8k | 16k | 32k | 64k |
|---|---|---|---|---|
| $128 \times 1024 \times 1024$ | 117 | 196 | | |
| $256 \times 1024 \times 1024$ | 190 | 321 | 582 | 968 |
| $512 \times 1024 \times 1024$ | 281 | 496 | 825 | 1468 |
| $256 \times 256 \times 256$ | 107 | 210 | 316 | 488 |
| $512 \times 512 \times 512$ | 199 | 362 | 558 | 1062 |
| $1024 \times 1024 \times 1024$ | 357 | 664 | 1045 | 1936 |
| $2048 \times 2048 \times 2048$ | | | 1829 | 3463 |
| $4096 \times 4096 \times 4096$ | | | | 5814 |

Table 4. Performance in Mflops/s of the non-local matrix multiplication.

## 5.2   The Fast Fourier Transform (FFT)

The network interconnecting processor chips in the Connection Machine forms a 12-dimensional Boolean cube. In a Boolean cube of $N = 2^n$ nodes, $n$ bits are required for the encoding of the node addresses. Every node $u = (u_{n-1}u_{n-2}\ldots u_m \ldots u_0)$ is connected to nodes $v = (u_{n-1}u_{n-2}\ldots \overline{u}_m \ldots u_0)$, $\forall m \in [0, n-1]$.

A radix-2 butterfly network for $P$ inputs and outputs has $P(p+1)$ nodes. Let the node addresses of the butterfly network be $(y_{p-1}y_{p-2}\ldots y_0|z_{t-1}z_{t-2}\ldots z_0)$, where $t = \lceil \log_2(p+1)\rceil$. The butterfly network is obtained by connecting node $(y|z)$ to the nodes $(y \oplus 2^{p-1-z}|z+1)$ and $(y|z+1)$, $z \in [0, p-1]$, where $\oplus$ denotes the bit-wise exclusive-or operation. For the computation of the radix-2 FFT the last $t$ bits can be interpreted as time. The network utilization defined as the fraction of the total number of nodes that are active at any given time is $\frac{1}{t}$. During step $z$, the communication is between ranks $z$ and $z+1$. Complex multiplications are made in rank $z$ for decimation-in-time FFT and rank $z+1$ for decimation-in-frequency FFT.

By identifying all nodes with the same $y$ value and different $z$ values, node $y$ becomes connected to nodes $y \oplus 2^z$, $\forall z \in [0, p-1]$, which defines a Boolean $p$-cube. All nodes participate in every step in computing an FFT on $P$ elements on a $p$-cube. In step $z$, all processors communicate in dimension $z$. Only $\frac{1}{p}th$ of the total communications bandwidth of the $p$-cube is used. The full arithmetic power, instead of only half, can be used by splitting the butterfly computations between the pair of nodes storing the data. Each node performs 5 real arithmetic operations. This splitting of butterfly computations was implemented on the Connection Machine model CM-1. The parallel arithmetic complexity for computing an FFT on $P = 2^p$ complex elements on a Boolean $n$-cube becomes $5\lceil \frac{P}{N}\rceil \log_2 P$ real arithmetic operations, ignoring lower order terms. The speed-up of the arithmetic time is $\min(P, N)$ The communication complexity is $3\lceil \frac{P}{N}\rceil \log_2 N$ element exchanges in sequence.

In computing an FFT on $P$ complex elements distributed evenly over $N = N < P$ processors, there are $\frac{P}{N}$ elements per processor. If the cyclic assignment is used, then the first $p-n$ ranks of butterfly computations are local to a processor. The last $n$ ranks require inter-processor communication. For consecutive assignment the first $n$ steps require inter-processor communication, and the last $p - n$ steps are local to a processor. If the data is allocated in a bit-reversed order, then the order of the inter-processor communication and the local reference phases are reversed.

The embedding defined above is the binary encoding of array indices. Every index is directly identified by an address in the address space. For arrays embedded by a binary-reflected Gray code array, elements that differ by a power of two greater than zero are at a distance of two, i.e., $Hamming(G(i), G(i + 2^j)) = 2, j \neq 0$ [16]. Even though the elements to be used in a butterfly computation are at a Hamming distance of two, it is still possible to perform an FFT with $\min(p, n)$ nearest neighbor communications [19]. The current Connection Machine implementation assumes that the array axes are encoded in binary encoding. If the data is encoded by Gray code, then an explicit reordering to binary order is performed before the FFT computation. The Connection Machine router is currently used for this reordering. An optimum reordering is given in [15].

If there is only one element per processor, then every element is either involved in a computation or a communication. With multiple elements per processor, the communication efficiency can be increased from $\frac{1}{\min(p,n)}$ to $\frac{\min(p,n)}{n}$, which for $p > n$ is one. The increased communication efficiency is achieved by communicating concurrently in as many dimensions as possible.

## 5.2.1 Maximizing the communication efficiency

The radix-2 FFT implemented on the Connection Machine makes use of pipelining to achieve a high utilization of the communication (and computation) resources. For details of the implementation, and alternate implementations see [26,27]. High radix FFT's are discussed in [13]. With $N$ processors performing $\frac{N}{2}$ butterfly computations concurrently, $\frac{P}{N}$ butterfly computations must be performed sequentially in each stage. In each of the first $n$ butterfly stages, the lowest order $p-n$ bits are identical for the pair of data elements in a butterfly computation. The first $n$ butterfly stages can be viewed as consisting of $\frac{P}{N}$ independent FFT's, each of size $N$ with one complex data element per processor. This property was used in [14] for devising sorting algorithms on Boolean cubes. The independent FFT's can be pipelined. Every FFT performs communication in processor dimensions $n - 1, n - 2, \ldots, 0$. Each FFT is delayed by one communication with respect to the preceding one. After the $n$ butterfly stages with inter-processor communication, the remaining $p - n$ stages are entirely local. The high order $n$ bits identify $N$ different FFT's of size $\frac{P}{N}$ each.

The number of complex data element transfers in sequence for the pipelined FFT is $n + \frac{P}{N} - 1$. The communication efficiency, measured as (the sum of the communication resources used over time)/((total number of available communication resources)*(time)),

| Axis length | Time msec | Mflops /s |
|---|---|---|
| 32 | 1.126 | 1455 |
| 64 | 2.184 | 1800 |
| 128 | 4.130 | 2222 |
| 256 | 8.326 | 2519 |
| 512 | 17.446 | 2705 |
| 1024 | 38.452 | 2727 |
| 2048 | 78.796 | 2928 |
| 4096 | 167.645 | 3002 |
| 8192 | 355.822 | 3065 |

Table 5. Performance for 2048 concurrent local radix-2 DIF FFT.

| FFT | Time (msec) | | | | Mflops/s | | | |
|---|---|---|---|---|---|---|---|---|
| | 2k | 4k | 8k | 16k | 2k | 4k | 8k | 16k |
| 128×128 | 34.1 | 16.7 | 13.6 | 7.4 | 1075 | 1101 | 673 | 621 |
| 512×512 | 574 | 292 | 136 | 72 | 1315 | 1291 | 1390 | 1308 |
| 2048×2048 | | | 2213 | 1313 | | | 1668 | 1405 |
| 32×32× 32 | 99.5 | 50.4 | 24.2 | 13.7 | 791 | 780 | 811 | 720 |
| 64×64× 64 | 548 | 411 | 198 | 108 | 1378 | 919 | 956 | 875 |
| 128×128×128 | | | 1611 | 885 | | | 1093 | 995 |

Table 6. Performance for some two- and three-dimensional radix-2 DIF FFT.

for the stages requiring communication is $\frac{\frac{P}{N}}{n+\frac{P}{N}-1}$, $p \geq n$. The efficiency is approximately one for $\frac{P}{N} \gg n$.

Table 5 gives the performance for a collection of local, radix-2 FFT as a function of size for a Connection Machine system model CM–2. The same data are plotted in Figures 3 and 4. The performance data is for single precision floating point data. Both decimation-in-time and decimation-in-frequency FFT are implemented. Some sample timings for two- and three-dimensional radix-2 FFT are given in Table 5.2.1.

A higher radix FFT yields a better performance by a better utilization of the memory bandwidth, and a better load balance for the inter-processor communication stages [26,27]. Figure 5 illustrates the difference in performance for the FFT computations local to a processor for a mix of radix-4 and radix-8 kernels. The local maxima are for data sets of size $8^s$ for some $s$.

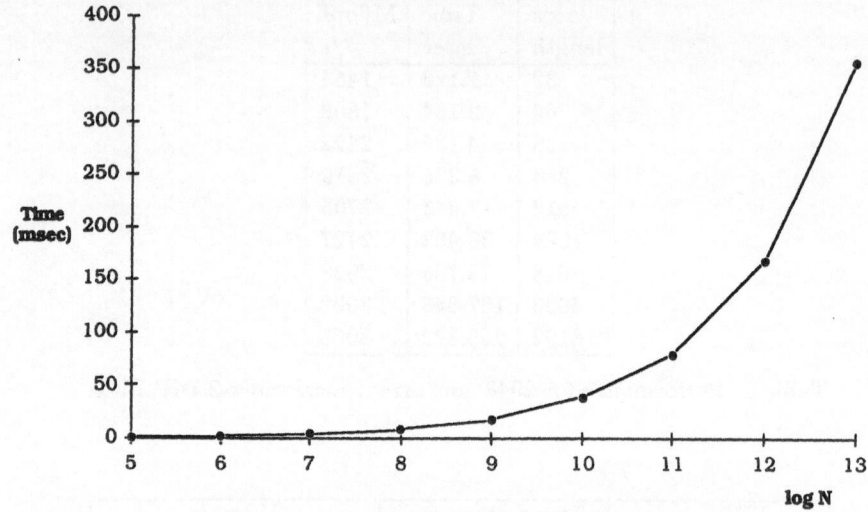

Figure 3. The execution time for local radix-2 FFT.

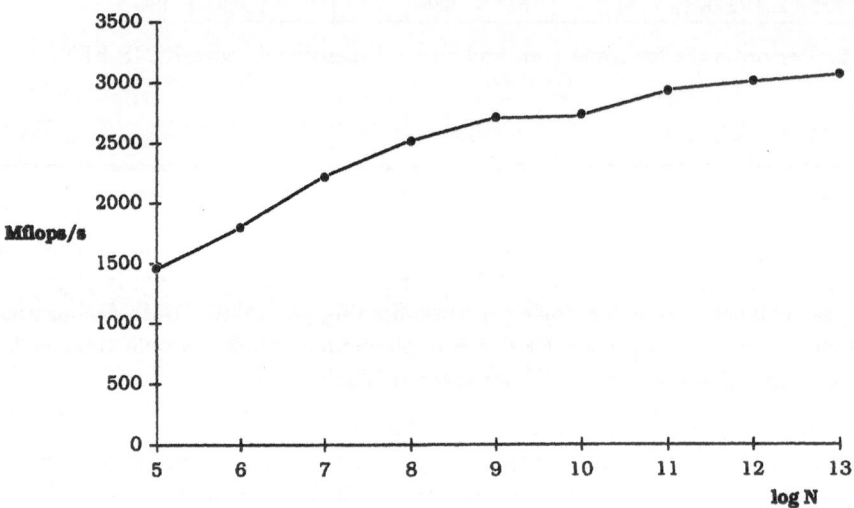

Figure 4. The floating-point rate for 2048 local radix-2 FFT's.

251

Comparison of Radix-2, -4, and -8 Kernels

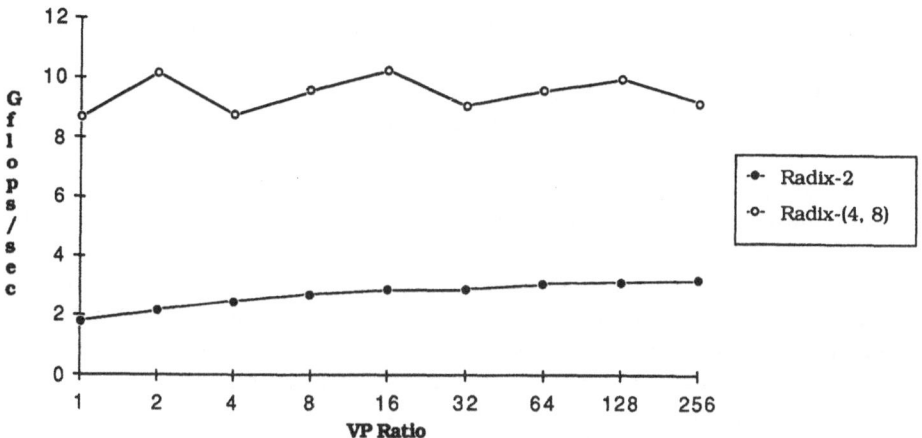

Figure 5. Performance of local radix-2 and radix-4/8 FFT computations.

# 6 Data Parallel Applications

In this section we present two applications. The first is the solution of the compressible Navier-Stokes equations, and the second the computation of the forward propagation of acoustic waves in the ocean. Both problems are formulated in three spatial dimensions. The purpose with these applications is to illustrate how problems are formulated for a data parallel computer and some of the functions that are needed. The examples are also intended to make it obvious that multiple concurrent instances of a computation often are both necessary and occur naturally. Both examples are based on finite difference techniques. An explicit technique is used for the Navier-Stokes problem, and an implicit technique for the underwater acoustics problem.

## 6.1 A compressible Navier-Stokes flow solver

The Navier-Stokes equation describes the balance of mass, linear momentum and energy, and models the turbulent phenomena that occur in viscous flow. In three dimensions, the equations in conservative form are

$$\frac{\partial \mathbf{q}}{\partial \tau} = \frac{\partial \mathbf{F} + \mathbf{F}_\nu}{\partial \xi} + \frac{\partial \mathbf{G} + \mathbf{G}_\nu}{\partial \eta} + \frac{\partial \mathbf{H} + \mathbf{H}_\nu}{\partial \zeta}, \tag{1}$$

where the variable vector $\mathbf{q}(\xi, \eta, \zeta, \tau)$ has five components: one for density, three for the linear momentum in the three coordinate directions $x, y$ and $z$, and one component for the total energy. The coordinates of the physical domain are $x, y$ and $z$, whereas $\xi, \eta$ and

$\zeta$ are coordinates in the computational domain. $\mathbf{F}, \mathbf{G}$ and $\mathbf{H}$ are the flux vectors and $\mathbf{F}_\nu, \mathbf{G}_\nu$ and $\mathbf{H}_\nu$ are the viscous flux vectors.

For regular computational domains, the solution can be approximated by discretizing the domain by a three-dimensional grid. The grid may be stretched in order to get a good resolution of the boundary layers without an unnecessarily large number of grid points in the interior. A stretched grid is topologically equivalent to a regular grid, and any efficient embedding of such grids can be used advantageously. In [32,33], an explicit finite difference method is used. To stabilize the numeric method, artificial viscosity is introduced through a fourth order derivative. Centered difference stencils are used. The approximation of the derivatives of the flux vectors is second order accurate. There are two difference stencils being used in each lattice point, and the stencils vary for interior points, points on or close to a surface, edge, and corner. Given the directional dependence there are one interior stencil, six face stencils, 12 edge stencils, and eight corner stencils of second order accuracy. The number of different types of stencils increases with the order of approximation. A central difference stencil with $2N + 1$ points in each of three dimensions gives rise to a total of $8N^3 + 12N^2 + 6N + 1$ stencils because of the boundaries. For the derivates, $N = 1$ in our case, and for the artificial viscosity, $N = 2$, and the total number of stencils are 27 and 125, respectively. All these stencils are subgraphs of the stencil in the interior, and can be represented by a set of vectors [32]. A three step Runge-Kutta method is used for the integration.

For the computations, the Connection Machine was configured as a regular 3D-grid. There are approximately 170 variables per virtual processor (grid point). The maximum virtual processor ratio with $8k$ bytes per physical processor is 8. The subgrid for each processor is a $2 \times 2 \times 2$ grid. By using difference stencils on or close to the boundaries that are subgraphs of the stencils in the interior, the stencils can be implemented as EOSHIFT operations in Fortran 8X. With periodic boundary conditions, CSHIFT should be used. The measured bandwidth for nearest neighbor communication in this grid was on the average about 2.5 Gbytes/s for a Connection Machine with 64k processors. The execution time as a function of the virtual processor ratio and the machine size are given in Table 7. The aspect ratio of any pair of dimensions of the physical domain was either one or two, and the size ranging from $16 \times 16 \times 32$ to $64 \times 64 \times 64$.

From Table 7, it is clear that the execution time per time step is independent of the machine size, as expected. The execution time as a function of the virtual processor ratio is shown in Figure 6. The processor utilization increases by a factor of 2.75 as the virtual processor ratio increases from 1 to 8. The work increases by a factor of 8, but the execution time only by a factor of 2.9. Figure 7 shows the floating-point rate at a virtual processor ratio of 8. With this virtual processor ratio, grids with up to 524,288 points were simulated.

In the Navier-Stokes code, the operations in each virtual processor consist of stencil computations applied to vectors of length five. Each floating-point processor performs a three dimensional convolution (on vector arguments) within the physical subdomain

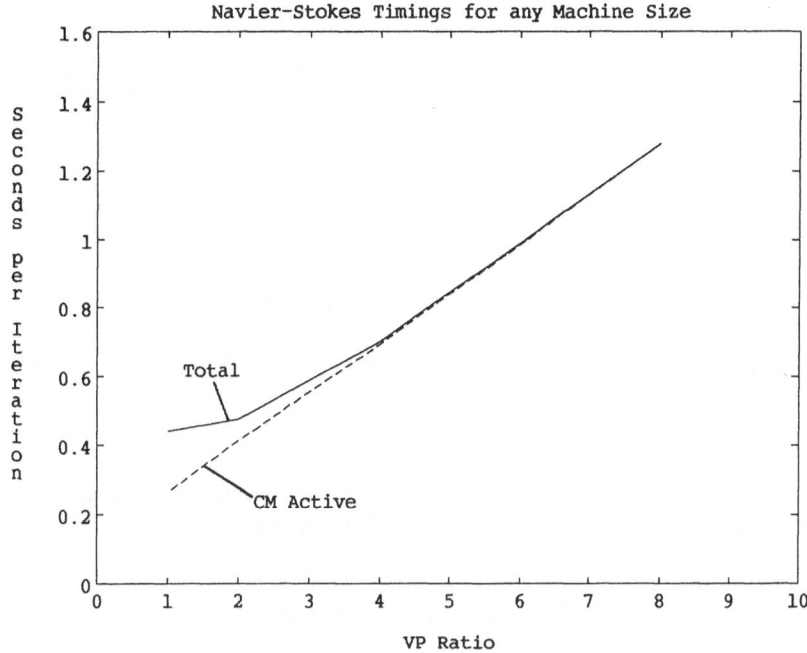

Figure 6. The Execution Time for a Single Time Step.

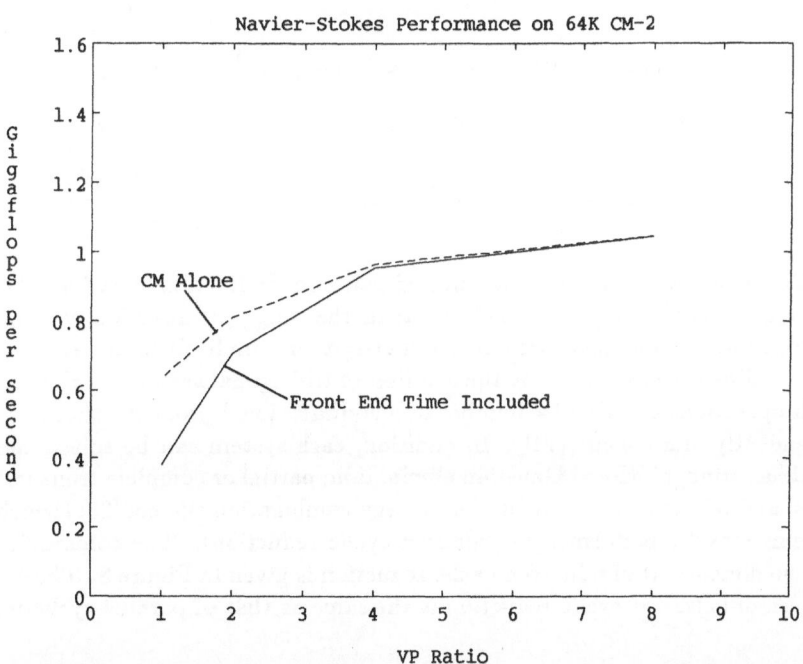

Figure 7. The Execution Speed for the Compressible Navier-Stokes Solver.

| virtual | Machine Size | | | | | |
| processor | 8k | | 16k | | 32k | |
| ratio | CM-time | Total time | CM-time | Total time | CM-time | Total time |
| 1 | 2.61 | 4.43 | 2.61 | 4.42 | 2.61 | 4.42 |
| 2 | 4.12 | 4.74 | 4.12 | 4.74 | 4.12 | 4.74 |
| 4 | 7.07 | 7.08 | 7.04 | 7.05 | 7.01 | 7.02 |
| 8 | 12.70 | 12.70 | 12.83 | 12.83 | 12.83 | 12.83 |

Table 7. Execution Time for Different Virtual Processor Ratios.

mapped into the memories of the processors served by a floating-point unit. In addition, the computation of the flux vectors requires matrix-vector multiplication and matrix-matrix multiplication on small matrices and vectors in each virtual processor. Optimized routines for these operations were not available at the time this code was implemented and evaluated with respect to performance. Incorporating optimized routines is expected to increase the performance by a factor of about three.

## 6.2 Acoustic field computation by an Alternating Direction Method

The forward propagation of acoustic waves by the so called Wide Angle Wave Equation [30] implies the solution of an equation of the form

$$(1 + \frac{1}{4}(1 - \delta)X)(1 - \frac{1}{4}Y)u(r + \Delta r) = (1 + \frac{1}{4}(1 + \delta)X)(1 + \frac{1}{4}Y)u(r) \qquad (2)$$

where $k_0$ is a reference wave number, and $n(r, \theta, z) = k(r, \theta, z)/k_0$ $\delta = ik_0\Delta r$, and

$$X = \frac{1}{k_0^2}\frac{\partial^2}{\partial z^2} + (n^2(r, \theta, z) - 1), \quad \text{and} \quad Y = \frac{1}{k_0^2 r^2}\frac{\partial^2}{\partial \theta^2}$$

This equation is a parabolic approximation of Helmholtz equation. The solution to the equation above can be marched out in the range $(r)$ direction with an Alternating Direction Method [36,29]. Tridiagonal matrix-vector multiplications are performed in the $\theta$ and $z$ directions, followed by the solution of tridiagonal systems in the same directions. Both operations consist of a number of one-dimensional problems that can be solved independently and concurrently. In addition, each system can be solved concurrently by substructuring, pipelined Gaussian elimination, partial or complete transposition of equations, and odd-even cyclic reduction, or any combination thereof [23] (which for multiple systems may be performed as balanced cyclic reduction). The communication pattern (in one dimension) of odd-even cyclic reduction is given in Figure 8. The communication pattern of balanced cyclic reduction is the same as that of parallel cyclic reduction [11].

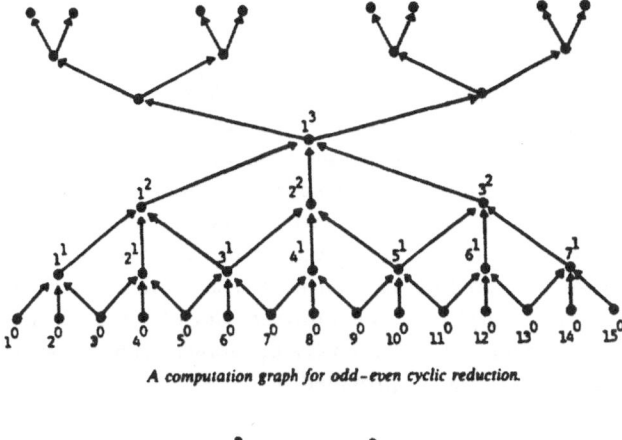

A computation graph for odd-even cyclic reduction.

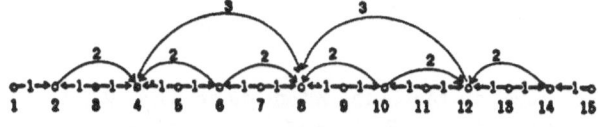

A storage conserving computation graph for cyclic reduction.

Figure 8. The communication topology for odd-even cyclic reduction.

The communication is defined by the grid and the difference stencil for the matrix-vector multiplication, but for the solution of the tridiagonal systems of equations, the communication depends on the selected algorithm. For pipelined Gaussian elimination, communication in the form of a Hamiltonian path is required. For equation transposition, the communication is equivalent to *all-to-all personalized communication* (or *all-to-some some-to-all* personal communication), which can be performed through butterfly network communication [22,25]. For balanced cyclic reduction, communication is required in the form of a data manipulator network. The communication requirements for odd-even cyclic reduction is a subtree of the data manipulator graph with the root at the top center and the leaf nodes being all nodes at the bottom level.

In the Connection Machine implementation, the processors are configured as a two dimensional grid. The tridiagonal systems are solved by substructured elimination followed by odd-even cyclic reduction for the reduced system of equations. The performance for the substructuring phase is about 1 Gflops/s without using any optimized library routines. The reduction phase in the current implementation uses a straightforward implementation of odd-even cyclic reduction. By using balanced cyclic reduction instead, a higher processor utilization and better performance can be achieved [24]. Note, that with the lattice emulation there is no need to perform a transposition of the data when the computations switch from one axis in the physical domain to the other. Communication time and storage accesses are the same for both directions.

# 7  Summary

Supercomputers with a performance in the Tflops/s range are becoming technically and economically feasible to build in state-of-the-art technologies. Such computers will have thousands to tens of thousands of processing units interconnected by a bounded degree network. The most critical resources with respect to performance are the communication and memory subsystems. The efficient utilization of these resources is imperative to high performance. We have presented a few examples of how data allocation, data motion between processors, and algorithms can be chosen such that good utilization of data parallel architectures is accomplished. Performance data for matrix multiplication, Fast Fourier Transforms, a compressible Navier-Stokes solver, and an underwater acoustics code on the Connection Machine are provided.

## Acknowledgement

The access to a Connection Machine system model CM –2 with 64k processors provided by the Advanced Computing Facility of the Los Alamos National Laboratories is gratefully acknowledged. The assistance provided by Ralph Brickner of the Advanced Computing Facility in carrying out the measurements was invaluable.

# References

[1] *Lisp release notes. Thinking Machines Corp., 1987.

[2] L.E. Cannon. *A Cellular Computer to Implement the Kalman Filter Algorithm.* PhD thesis, Montana State Univ., 1969.

[3] M.Y. Chan. *Dilation-2 Embeddings of Grids into Hypercubes.* Technical Report UT-DCS 1-88, Computer Science Dept., University of Texas at Dallas, 1988.

[4] M.Y. Chan. *Embeddings of 3-Dimensional Grids into Optimal Hypercubes.* Technical Report, Computer Science Dept., University of Texas at Dallas, 1988. To appear in the Proceedings of the Fourth Conference on Hypercubes, Concurrent Computers, and Applications, March, 1989.

[5] Monty M. Denneau, Peter H. Hochschild, and Gideon Shichman. The switching network of the TF-1 parallel supercomputer. *Supercomputing Magazine*, 2(4):7–10, 1988.

[6] W. Morven Gentleman. Some complexity results for matrix computations on parallel processors. *J. ACM*, 25(1):112–115, January 1978.

[7] I. Havel and J. Móravek. B-valuations of graphs. *Czech. Math. J.*, 22:338–351, 1972.

[8] W. Daniel Hillis. *The Connection Machine*. MIT Press, Cambridge, MA, 1985.

[9] Ching-Tien Ho and S. Lennart Johnsson. *Embedding Meshes in Boolean cubes by Graph Decomposition*. Technical Report YALEU/DCS/RR-689, Department of Computer Science, Yale University, March 1989.

[10] Ching-Tien Ho and S. Lennart Johnsson. On the embedding of arbitrary meshes in Boolean cubes with expansion two dilation two. In *1987 International Conf. on Parallel Processing*, pages 188–191, IEEE Computer Society, 1987.

[11] Roger W. Hockney and C.R. Jesshope. *Parallel Computers*. Adam Hilger, 1981.

[12] J.W. Hong and H.T. Kung. I/O complexity: the red-blue pebble game. In *Proc. of the 13th ACM Symposium on the Theory of Computation*, pages 326–333, ACM, 1981.

[13] Michel Jacquemin and S. Lennart Johnsson. *Radix-4 and radix-8 FFT on the Connection Machine*. Technical Report , Thinking Machines Corp., 1989. in Preparation.

[14] S. Lennart Johnsson. Combining parallel and sequential sorting on a Boolean n-cube. In *1984 International Conference on Parallel Processing*, pages 444–448, IEEE Computer Society, 1984.

[15] S. Lennart Johnsson. Communication efficient basic linear algebra computations on hypercube architectures. *J. Parallel Distributed Comput.*, 4(2):133–172, April 1987.

[16] S. Lennart Johnsson. *Odd-even cyclic reduction on ensemble architectures and the solution tridiagonal systems of equations*. Technical Report YALE/DCS/RR-339, Dept. of Computer Science, Yale University, October 1984.

[17] S. Lennart Johnsson. *Optimal Communication in Distributed and Shared Memory Models of Computation on Network Architectures*, page . Morgan Kaufman, 1989.

[18] S. Lennart Johnsson, Tim Harris, and Kapil Mathur. *Matrix Multiplication on the Connection Machine*. Technical Report , Thinking Machines Corp., 1989. in Preparation.

[19] S. Lennart Johnsson and Ching-Tien Ho. *Emulating Butterfly Networks on Gray Code Encoded Data in Boolean Cubes*. Technical Report , Department of Computer Science, Yale University, 1989. in Preparation.

[20] S. Lennart Johnsson and Ching-Tien Ho. Expressing Boolean cube matrix algorithms in shared memory primitives. In *The Third Hypercube Conference*, pages 1599–1609, ACM, 1988.

[21] S. Lennart Johnsson and Ching-Tien Ho. Matrix multiplication on Boolean cubes using generic communication primitives. In *Parallel Processing and Medium Scale Multiprocessors*, pages 108–156, SIAM, 1989. (Presented at the ARMY workshop on Medium Scale Parallel Processing, Stanford University, January 1986).

[22] S. Lennart Johnsson and Ching-Tien Ho. Matrix transposition on Boolean n-cube configured ensemble architectures. *SIAM J. Matrix Anal. Appl.*, 9(3):419–454, July 1988.

[23] S. Lennart Johnsson and Ching-Tien Ho. *Multiple tridiagonal systems, the alternating direction method, and Boolean cube configured multiprocessors.* Technical Report YALEU/DCS/RR-532, Dept. of Computer Science, Yale University, New Haven, CT, June 1987.

[24] S. Lennart Johnsson and Ching-Tien Ho. *Optimizing Tridiagonal Solvers for Alternating Direction Methods on Boolean Cube Multiprocessors.* Technical Report YALEU/DCS/RR-679, Department of Computer Science, Yale University, January 1989.

[25] S. Lennart Johnsson and Ching-Tien Ho. Spanning graphs for optimum broadcasting and personalized communication in hypercubes. *IEEE Trans. Computers*, ():, September 1989.

[26] S. Lennart Johnsson, Ching-Tien Ho, Michel Jacquemin, and Alan Ruttenberg. Computing fast Fourier transforms on Boolean cubes and related networks. In *Advanced Algorithms and Architectures for Signal Processing II*, pages 223–231, Society of Photo-Optical Instrumentation Engineers, 1987.

[27] S. Lennart Johnsson, Robert L. Krawitz, Douglas MacDonald, and Roger Frye. *Radix-2 FFT on the Connection Machine.* Technical Report , Thinking Machines Corp., 1989. in Preparation.

[28] S. Lennart Johnsson and Peggy Li. *Solutionset for AMA/CS 146.* Technical Report 5085:DF:83, California Institute of Technology, May 1983.

[29] S. Lennart Johnsson, Yousef Saad, and Martin H. Schultz. Alternating direction methods on multiprocessors. *SIAM J. Sci. Statist. Comput.*, 8(5):686–700, 1987.

[30] Ding Lee, Yousef Saad, and Martin H. Schultz. *An efficient method for solving the three-dimensional wide angle wave equation.* Technical Report YALEU/DCS/RR-463, Department of Computer Science, Yale University, October 1986.

[31] Michael Metcalf and John Reid. *Fortran 8X Explained.* Oxford Scientific Publications, 1987.

[32] Pelle Olsson and S. Lennart Johnsson. *A Dataparallel Implementation of Explicit Methods for the Three-dimensional Compressible Navier-Stokes Equations.* Technical Report CS-89/4, Thinking Machines Corp., February 1989.

[33] Pelle Olsson and S. Lennart Johnsson. *A Study of Dissipation Operators for the Eulers Equations and Three-Dimensional Channel Flow.* Technical Report CS-89/3, Thinking Machines Corp., February 1989.

[34] Seymor Parter. The use of linear graphs in Gaussian elimination. *SIAM Review*, 3(2):119–130, 1961.

[35] E M. Reingold, J Nievergelt, and N Deo. *Combinatorial Algorithms*. Prentice-Hall, Englewood Cliffs. NJ, 1977.

[36] R. Richtmyer and K.W. Morton. *Difference Methods for Initial-Value Problems*. Wiley-Interscience, 1967.

[37] Howard J. Siegel. *Interconnection Networks for Large Scale Parallel Processing*. Lexington Books, 1985.

# Solving the Shallow Water Equations on the Cray X-MP/48 and the Connection Machine 2

PAUL N. SWARZTRAUBER[1,2] and RICHARD K. SATO[1]

[1] National Center for Atmospheric Research, 1850 Table Mesa Drive, Boulder, Colorado 80307, USA, which is sponsored by the National Science Foundation.
[2] The manuscript was completed while this author was visiting the Research Institute for Advanced Computer Science, NASA Ames Research Center, Moffett Field, CA 94035.

## Abstract

The shallow water equations in Cartesian coordinates and two dimensions are solved on the Connection Machine 2 (CM-2) using both the spectral and finite difference methods. A description of these implementations is presented together with a brief discussion of the CM-2 as it relates to these specific computations. The finite difference code was written both in C* and *LISP and the spectral code was written in *LISP. The performance of the codes is compared with a FORTRAN version that was optimized for the Cray X-MP/48.

**1. Introduction.** The Connection Machine 1 (CM-1) is a single instruction multiple data (SIMD) computer with 64K one-bit processors [3] that has been of significant interest to the computer science community. However its performance is below that of other supercomputers and consequently it has not generated significant interest among institutions that require state-of-the-art performance. However, a departure from the "one-bit" philosophy of the CM-1 resulted in the CM-2 with a reported peak performance capability that is superior to most supercomputers and has therefore generated a broad base of interest in the supercomputing community.

Topics in Atmospheric and Oceanic Sciences
© Springer-Verlag Berlin Heidelberg 1990

Although the fully configured CM-2 is still commonly referred to as a 64K one-bit processor machine, its increased performance is due to the addition of 2K Weitek 32-bit floating point processors. Each Weitek is rated at 16 Mflops and therefore the peak rate of the CM-2 is presented as 32 billion floating point operations per second (32 Gflops). It therefore became necessary for the scientific computing community in general and the atmospheric science community in particular to examine the performance of the CM-2 on their particular type of computations. As we will show, the performance of the CM-2 on the shallow water equations is significantly less than 32 Gflops but it is nevertheless in a range that continues to make it interesting to scientific supercomputer users.

In a previous paper [8] we presented early results of the finite difference model which will be updated in this paper together with new results for a spectral model. The finite difference code is highly vectorizable and parallelizable and runs at near peak rates on most computers. Consequently it provides a optimistic view of the performance that is possible for problems in the atmospheric sciences.

The shallow water equations constitute a greatly simplified weather prediction code [7] that has already been used to benchmark a number of machines [4]. Their solution is a relatively small but necessary step in the process of determining the applicability of the SIMD architecture to climate and weather prediction. Ultimately these tests must be conducted on the sphere where harmonic transforms are computed for the spectral method.

A compressible model should also be implemented. An explicit model with local communication should be straightforward compared to a implicit model which requires global communication. A list of problems that can be used to evaluate the suitability of massively parallel processing (MPP) is given in Table 1 below.

The entries in Table 1 can be combined to provide a sequence of models and experiments that could assist in determining the suitability of massively parallel computing to the problems in the atmospheric and related sciences. For example, we chose first to solve the shallow water equations in Cartesian coordinates using finite differences on the Connection Machine. Our next step was to implement the spectral method for the shallow

Table 1

PARALLEL GEODYNAMICS PROJECTS

| Equations | Geometries | Methods |
|-----------|------------|---------|
| Shallow Water | Cartesian | Finite Difference |
| Incompressible | Spherical | Spectral |
| Compressible | Irregular | Semi-Lagrangian |

water equations in Cartesian coordinates. We will present the results of these experiments in the sections that follow; however, we begin with a brief description of the Connection Machine.

**2. The Connection Machine 2.** Consider now several components of the CM-2 and concepts that facilitate the implementation of scientific computations.

HOST

There is only one copy of the program which resides on the host or front end which is physically distinct from the CM-2 but controls its execution. A CM-2 can support up to four front end systems that must currently be either a DEC VAX 8000 series, Symbolics 3600 system or a Sun 4/280. Programs are developed, stored, compiled, loaded, and executed on the front end. Instructions to parallel variables are passed over an interface to the microcontroller where they are broadcast for execution by all processors simultaneously. Program steps that do not involve operations on parallel variables are executed on the front end. Hence the CM-2 is essentially an extension to the front end which provides computational power for operations on parallel variables. Note that any serial portion of the code must therefore run at the speed of the host.

The CM-2 does not support multi-programming in the traditional sense. That is, several jobs cannot share the same processors but rather a job must complete execution before another job can be allowed access to the processors. However, a CM-2 can be configured as several smaller subsystems and each of these subsystems can be simultaneously running independent jobs each from a separate front end machine.

## PROCESSORS

The CM-2 is a massively parallel SIMD machine. A fully configured CM-2 has 64K (65,536) single-bit processors that are packaged 16 per chip. For every two chips (32 processors), there is a Weitek 3132 floating point chip. The CM-2 operates at an 8.0 MHz clock speed and the Weitek processors can produce an add and a multiply per cycle for a peak computational rate of 16 million floating point operations per second (Mflops). The 2048 Weitek processors therefore combine for a total peak floating point computational rate of 32 billion floating point operations per second (Gflops). In practice the performance is considerably below this figure because of the communication between the CM chips and the Weitek 3132. The actual performance figures are presented in the sections that follow.

## MEMORY

Each processor has 8K (8192) bytes of memory for a total of 512 megabytes of memory for a 64K processor system. Each processor can access data from its memory at a rate of 5 megabits per second. The total memory bandwidth is therefore over 300 gigabits per second. A processor can only operate on data that is in its own memory but the processors are logically interconnected so that data can be transferred between processors.

## COMMUNICATIONS

The 4096 CM-2 chips are connected in a 12-dimensional hypercube network. Communication is bit-wise rather than packet-wise with a maximum of $5 \times 10^{10}$ bits transmitted per second (aggregate). However, like peak computational rates, the peak communication rates are not achieved in practice, particularly by the router. General communication between processors is handled by a router that can transmit messages of any length. The throughput of the router depends on the length of the messages and on the access patterns with typical bandwidths from .1 to 1 billion bits per second.

A second, faster mechanism for communications between chips is a programmable NEWS grid that can be configured to support multi-dimensional configurations. The NEWS grid allows processors to transmit data between processors according to a regular rectangular pattern. For example, in a 2-dimensional grid, each processor could simultaneously execute the same instruction to fetch data from its neighbor to the north, east, west or south. Use of the NEWS grid eliminates the overhead of the router. This feature will also be available in FORTRAN which is currently in beta test phase.

## VIRTUAL PROCESSORS

The CM-2 is a data parallel computer. For variables that are declared to be parallel, each processor will have a single element of that variable in its memory. If this number exceeds the number of physical processors then virtual processors are created. When an operation is performed on a parallel variable, each processor performs the operation on its element. For example, if A, B, and C are declared as parallel variables, then each processor will have a single element from each of the three arrays and a parallel operation such as C = A + B will be executed by each processor using its A, B, and C variables.

Through the concept of virtual processors, the CM-2 can support applications that require more processors than are physically available. For example, parallel variables with 64K elements will have one value associated with each processor. If, however, parallel variables have 256K elements, then four virtual processors are created for each physical processor at the time the CM-2 is initialized for the application.

There are several advantages to an SIMD architecture:

a)  It is not necessary to synchronize the processors which simplifies code development, debugging and fault isolation.

b)  The issue of whether or not to parallelize is nonexistent. The nagging issue that persists throughout the multiprocessing community is whether or not to encourage users to multitask or to simply run separate job streams.

c)  The host or front end contains the only copy of the code. Multiple copies of a weather code would require a substantial memory resource since these codes can occupy several Mbytes of memory. Indeed the code would occupy all of the memory if the CM-2 were configured as a multiple instruction multiple data (MIMD) machine that required multiple copies of the code. This aspect of an SIMD computer highlights the fundamental design philosophy of the CM; namely, that the electronics should be dedicated to items such as data and arithmetic where it is used frequently rather than memory for multiple copies of a code where its use is relatively infrequent. Memory for address space can also be minimal since arrays rather than their individual elements are referenced.

There are also some disadvantages to SIMD.

a)  The scalar part of any code will perform at the speed of the host. This simply emphasizes what is already known from Amdahl's rule, namely that scalar code must be kept at a minimum for efficient multiprocessing. For certain computations such as the physical parameterizations in atmospheric models, this may require a creative reformulation of the computations.

b)  Computations on smaller arrays take the same time as computations on larger arrays. The computing time depends only on the length of the formula and is independent of the size or dimensions of the array. If the formulas for the boundary and interior of a domain have about the same arithmetic operations per point then they will require about the same amount of computing time.

c) Nested gridpoint conditionals are expensive since all possible branches must be computed. The appropriate alternative is then selected via a mask. However this increase in computation is bounded once the model is fixed unlike the computations as a function of grid size.

**3. The finite difference model of the shallow water equations.** The benchmark code for the CM-2 is a simple atmospheric dynamics model based on the shallow water equations [7] which represent a primitive but useful model of the dynamics of the atmosphere. Its performance provides a rough estimate of the performance of more complex atmospheric models on various computer systems. This estimate is generally of the high side since state-of-the-art models would have a higher percentage of scalar computations that would run on the host or front end system.

The shallow water equations consist of the following system of three time dependent, partial differential equations. Two additional equations define relationships for the vorticity and for the height field.

$$\frac{\partial u}{\partial t} - \zeta v + \frac{\partial H}{\partial x} = 0$$

$$\frac{\partial v}{\partial t} + \zeta u + \frac{\partial H}{\partial y} = 0$$

$$\frac{\partial P}{\partial t} + \frac{\partial}{\partial x}(Pu) + \frac{\partial}{\partial y}(Pv) = 0$$

where $u$ and $v$ are velocities in the $x$ and $y$ direction respectively; $P$ is pressure; $H = P + \frac{1}{2}(u^2 + v^2)$ is the height field and $\zeta$ is the vorticity given by

$$\zeta = \frac{\partial u}{\partial x} - \frac{\partial v}{\partial y}.$$

This form of the shallow water equations was used by Sadourney [7] because it yielded a conservative finite difference approximation. Using these equations and his numerical formulation, a solution was obtained on a rectangular domain in Cartesian coordinates with periodic boundary conditions in both directions. The leapfrog time differencing scheme was used and second order centered finite difference approximations were used for spatial derivatives.

The model is in-memory with variables declared as two-dimensional arrays that are the size of the grid plus 1 (an extra row and column are defined for storage of periodic boundary values). For a grid of 256×256 points, the memory requirement on the Cray X-MP/48 is about one million words. Secondary storage is not used and hence there was no requirement for I/O to external devices. The computationally intensive portions of the code consist of three doubly-nested DO-loops in which various quantities are computed over the two-dimensional grid. Each of the loops is evaluated in a separate subroutine and the computations in the loops account for more than 90% of the floating point operations in the code. The inner loops are easily vectorized and for systems like the CM-2 which support parallel execution on multiple processors, the computations that correspond to both the inner and outer loops can be executed in parallel.

The FORTRAN code appeared in [4] where it was used to benchmark several supercomputers other than the CM-2 for meteorological modeling. The FORTRAN code was converted to *LISP and on TMC's System V the LISP it ran at 1.4 Gflops. This figure was extrapolated from a speed of .18 Gflops on the NCAR 8K system. Both a C* and *LISP version were coded for the CM-2 and these codes appear in [8]. The benchmark results are presented in Table 2. All cases were run on an 8K processor system using 32-bit precision floating point arithmetic. Performance figures on the CM are usually reported for a full system but as extrapolations from figures that are obtained from a smaller system. This is justified by the observation that the additional processors simply add additional flops if the size of the grid is increased without an increase in communication or the scalar part of the computation.

The first three rows of Table 2 give the performance achieved by an unoptimized C* version of the model. Calls to a lower level language (PARIS) subroutine library were

| | | | | | Table 2 | | |
|---|---|---|---|---|---|---|---|

Mflops for Shallow Water Code on Thinking Machine CM-2

| Case | Language | CM Size | Model Grid | VP ratio | Mflops (measured) | Mflops (for full 64K sys) |
|------|----------|---------|------------|----------|-------------------|----------------------------|
| A | Unopt C* | 8192 | $64 \times 128$ | 1 | 77 | 616 |
| B | Unopt C* | 8192 | $64 \times 4096$ | 32 | 102 | 816 |
| C | Unopt C* | 8192 | $256 \times 256$ | 8 | 102 | 816 |
| D | Opt C* | 8192 | $64 \times 128$ | 1 | 112 | 896 |
| E | Opt C* | 8192 | $64 \times 4096$ | 32 | 164 | 1312 |
| F | Opt C* | 8192 | $256 \times 256$ | 8 | 183 | 1464 |
| G | *LISP | 8192 | $512 \times 512$ | 32 | 215 | 1720 |

required to access data from neighboring processors. For example, the line

CM_get_from_west_always(&pW, &p, FLEN)

fetches the value of the variable "p" from the processor to the west and stores it in the local variable "pW". The grid size was specified as a power of two in order to efficiently map it onto the the CM-2 and to obtain the highest computational rates. For Case A, a model grid of 64×128 was specified to correspond to the 8K processors of the CM-2 and hence there was a physical processor for each point of the model grid giving a VP ratio of 1.

For Case B, the model grid was specified as 64×4096 which has 32 times more points than the number of physical processors giving a VP ratio of 32. Case C was a run with a 256×256 model grid. Again there are more points than the total number of physical processors, and int this configuration, each physical processor represents a 4×2 array of virtual processors for a VP ratio of eight. For both Case B and Case C, the performance achieved was 110 Mflops on the 8K system which corresponds to 880 Mflops on a full 64K system. Although the performance improved when the VP ratios increased from one to eight and from one to 32, performance did not improve when the VP ratio increased from eight to 32. This is evidently because the communications overhead is less efficient for the 64x4096 configuration than for the 256×256 configuration.

**Optimized C* Version**

The optimized C* version and the unoptimized C* version are compared in the next three entries of Table 2. In the optimized version, additional subroutines were added to the lower-level subroutines library and invoked in the C* program. These additional routines allowed more efficient use of the processors by overlapping communications with computations. For example, if a value was fetched from a neighboring processor and was to be added to another variable a "fetch and add" instruction was issued rather than a "fetch" instruction followed by an "add" instruction. The line

CM_get_from_west_with_f_add_always (&cu, &cu, FLOAT)

fetches the value of "cu" from the west neighbor and adds it to the local value of "cu". The optimized C* version resulted in Mflops rate improvements of 35%, 49%, and 66% over the unoptimized version for the three cases that were run with a top rate of 1464 Mflops for the 256×256 grid.

**\*LISP Version**

The \*LISP version resulted in the highest performance rates although the corresponding model configuration 512×512 was not run for the optimized C* version. The \*LISP version ran at 215 Mflops on the 8K processor system. Assuming a linear scale up in performance, a 64K system would execute at a rate of 1.7 Gflops if a VP ratio of 32 is maintained. The *LISP compiler is considered to be more highly optimized than the C* compiler.

## Cray X-MP Version

The Cray X-MP version is written in FORTRAN which is highly vectorized and a listing of the code is provided in [8]. The 256×256 grid requires one million words of memory and executes at 148 Mflops on a one processor Cray X-MP. The 512×512 case requires 3.9 million words of memory and also executes at 148 Mflops. A microtasked version runs at 560 Mflops on the full Cray X-MP/48. The Cray X-MP floating point operations are performed with 64-bit precision compared with 32-bit precision on the CM-2.

## Observations

As with all computational models of multidimensional phenomena, the benchmark includes loops of lower dimension. For example, periodic boundary conditions are implemented in the FORTRAN version by one dimensional loops in which the left boundary is copied to the right as well as the top to the bottom. These loops are usually implemented on a serial or vector computer without much consideration since it is known that their contribution to the computing time is an order of magnitude below that of the two dimensional loops. However, when implemented in this manner on the Connection Machine, the total computing time (for the *LISP version) was increased by a factor of 2.5 to 4 depending on the VP ratio. The largest increase corresponded to the highest VP ratios for which the least amount of interprocessor communications is required and hence the highest computational rates are achieved. Although the boundary computations are performed on a subset of the grid, they require the same amount of time as computations on the interior when the amount of arithmetic per point is the same. In addition, the communications that were explicitly specified by the code are not directly supported by the NEWS communication.

These reduced rates were not representative of the rates that could be obtained via alternate algorithmic and coding efforts and therefore the periodic boundary conditions were implemented using the periodic communications that are available using NEWS communication. This was achieved by simply branching around the code that explicitly implemented the boundary conditions and increasing the number of grid points to the nearest power of two.

This modification in the computation is representative of the efforts that will be necessary to implement existing codes on a parallel processor. The periodic boundary conditions could be implemented using more efficient communications methods such as the scan instruction or perhaps in microcode using the hypercube communications. These efforts are seen as typical for any significant change in computer architecture and not unlike the efforts that were required to adapt codes to early vector architectures.

**4. The spectral model of the finite difference equations.** A natural next step in the progression of models toward weather and climate prediction on the CM-2 is one in which the spectral method is used to solve the shallow water equations. The techniques that are used to ensure stability of the computations for the spectral method are different from those used for the finite difference method and therefore the conservative form of the equations that were used above in Section 3. for the finite difference method can be replaced by the more traditional form of the shallow water equations.

$$\frac{\partial u}{\partial t} + u \frac{\partial u}{\partial x} + v \frac{\partial u}{\partial y} - fv = -g \frac{\partial h}{\partial x}$$

$$\frac{\partial v}{\partial t} + u \frac{\partial v}{\partial x} + v \frac{\partial v}{\partial y} + fu = -g \frac{\partial h}{\partial y}$$

$$\frac{\partial h}{\partial t} + u \frac{\partial h}{\partial x} + v \frac{\partial h}{\partial y} + h \left( \frac{\partial u}{\partial x} + \frac{\partial v}{\partial y} \right) = 0$$

where $g = 9.8 \ m/sec^2$, $f = 10^{-4}/sec$, and $\bar{h} \approx 1 \times 10^4$.

These equations are solved subject to periodic boundary conditions using the computational approach that is outlined below.

1.  The spectral representations of $u$, $v$, and $h$ are computed using a multiple FFT program developed by TMC [5]

2.  The upper third of the coefficients are truncated to significantly reduce (but not completely eliminate) aliasing. This approach was also used in [1].

3.  The spatial derivatives $u_x$, $u_y$, $v_x$, $v_y$, $h_x$, and $h_y$ in physical space are computed by formal differentiation of the spectral representations computed in 1. followed by the inverse FFT.

4.  The time derivatives $u_t$, $v_t$, and $h_t$ are computed by substituting the spatial derivatives into the right hand side of the shallow water equations given above.

5.  The solution is advanced to the next time level using "leap frog" time differencing.

This approach is in fact called the pseudo-spectral method which is a significant variant of the spectral method. A comparison of three different methods for solving differential equations is given in [1]. The steps 1 through 5 were implemented in FORTRAN/CAL on the Cray X-MP/48 and in *LISP by Oliver McBryan on the CM-2. The codes were optimized for the particular machine rather than line for line translations from one machine to the other. The results are given below in Table 3.

### Observations

a)  Most of the computing time is in subroutine FFT991 which is a fast Fourier transform (FFT) written in Cray machine language by Clive Temperton at ECMWF. Because of the difficulty in converting this code to a multitasked version, the use of four processors was made possible by separating the rectangle into four subdomains and creating four separate tasks which called FFT991 on these subdomains. There is some overhead in this operation since the length of the vectors is divided by four.

b)  The time for an ordered transform on the CM-2 was prohibitive because of the cost of the global communication required for bit reversal. An ordered transform on the CM-2 takes two to three times as long as an unordered transform [6]. However,

| Resolution | Mflops | | | Time/Grid point | | |
|---|---|---|---|---|---|---|
| | X-MP (1 proc) | X-MP (4 proc) | CM2 (65K proc) | X-MP (1 proc) | X-MP (4 proc) | CM2 (65K proc) |
| 256 × 256 | 125 | 397 | 601 | 2.41 | .75 | .77 |
| 512 × 512 | 129 | 476 | 806 | 2.59 | .70 | .64 |
| 2048 × 2048 | - | - | 1162 | - | - | .53 |

**Table 3**

**Spectral Shallow Water Equations Model**

**Cray X-MP/48 and Thinking Machine CM-2**

NOTE: The X-MP/48 time is for 64 bit precision and
the CM-2 time is for 32 bit precision.

since the order of the FFT in spectral space is not particularly relevant to the solution which is presented in physical space, it is not necessary to order the FFT. The spectral representations of the derivatives were computed by multiplying the bit reversed coefficients by wave numbers that were also in bit reversed order. The inverse FFT then accepts the bit reversed coefficients and provides the derivatives on the grid which is ordered in physical space.

c)  The spectral code makes extensive use of the FFT and indeed the performance on the shallow water equations is about the same as the performance on the FFT. The vectorization and parallelization of the FFT is nontrivial [9], [10], particularly when compared with the finite difference method. Nevertheless methods have been developed over the years that enable the FFT to perform at near peak rates and FFT codes have been repeatedly groomed for optimum performance on the Cray X-MP. This fact should be kept in mind when comparing the performance of the computers since the FFT on the CM-2 is both a new code and a new implementation of the FFT. It was written by the staff at Thinking Machines Corporation [5] under the supervision of Lennart Johnsson and was provided to NCAR on a beta test basis.

d)  Tempertons FFT on the Cray X-MP/48 is for real sequences and requires about half the time required for a complex transform. On the other hand the TMC FFT is for complex sequences. This apparent disadvantage was for all practical purposes over-come by combining two real variables into a single complex variable which could then be transformed concurrently [6]. Although the complex transform would have to be postprocessed to recover the real transforms, it was not necessary for the reasons already mentioned in b) above. It was simply sufficient to multiply each coefficient by its corresponding wave number followed by an inverse transform which yielded the derivatives of two distinct variables as the real and imaginary parts respectively. This technique works best if the number of variables is even which unfortunately is not the case here were the variables are $u$, $v$, , and $P$. However it is possible to extract two different derivatives, say with respect to both $x$ and $y$ with one complex transform [6].

e)  CM FORTRAN is currently under beta test at several institutions and is expected to make the CM-2 more attractive to the scientific computing community.

**5. Conclusions.** A status report has been provided on a continuing project; namely, to determine the suitability of massively parallel processing to the computational needs of the atmospheric science community. As of this report, two data points have been obtained. The first is 1.5 Gflops for an elementary weather model in Cartesian coordinates using the finite difference method and the second is about 1 Gflop for the same model using the spectral method. These results are encouraging and certainly do not preclude further investigation into the suitability of massively parallel processing and in particular the suitability of SIMD architectures. Optimum codes for the CM-2 were compared with codes that were optimized for the Cray X-MP/48 rather than a comparison of Gflops for the two machines on the same code.

The Gflops that were obtained on the CM-2 were not obtained on the first runs and indeed more than a modest effort was required by individuals with considerable knowledge about parallel computing and the machine itself. This effort must be weighed against any increase in flops/dollars. A substantial increase in performance is necessary to justify the effort that is required to convert a major production code to a new architecture. This study is a beginning and much work remains to be done before a definitive statement can be made regarding the future relationship of SIMD and computing in the atmospheric and related geosceinces. In particular more data points should be obtained for more complex problems, perhaps selected from Table 1 in section 1.

## ACKNOWLEDGEMENTS

We are grateful to the staff at Thinking Machines Corporation for their support of this project and in particular we wish to thank Dr. Lennart Johnsson and his staff for their counsel and for providing the CM FFT under beta test. We would also like to thank Dr. John Richards (TMC) for developing the C* version of the finite difference model and Dr. Oliver McBryan at the University of Colorado for his extensive support of this project including the development of *LISP codes for both the spectral and finite difference methods.

### Key words and phrases:

Parallel computing
Shallow water equations
Weather prediction

# REFERENCES

[1]   G.L. Browning, J.J. Hack and P.N. Swarztrauber, *A comparison of three numerical methods for solving differential equations on the sphere*, The Monthly Weather Review, 117(1989), pp. 1058-1075.

[2]   *Connection Machine Model CM-2 Technical Summary*, Thinking Machines Technical Report HA87-4, April, 1987.

[3]   D. Hillis, *The Connection Machine*, MIT Press, Cambridge, MA., 1985.

[4]   G.-R. Hoffman, P. N. Swarztrauber, and R. A. Sweet, *Aspects of using multiprocessors for meteorological modeling*, in: Multiprocessing in Meteorological Models, G.-R. Hoffmann and D. F. Snelling, eds., Springer-Verlag, New York, 1988.

[5]   L. Johnsson, R. Krawitz, and R. Frye, *Computing radix-2 FFT on the Connection Machine*, Technical Report, Thinking Machines Corp., Cambridge, MA, 02142, 1989.

[6]   O.A. McBryan, *Connection Machine application performance*, Rpt. CU-CS-434-89, Department of Computer Science, University of Colorado at Boulder, Boulder, Colorado, 80309, 1989.

[7]   R. Sadourney, *The Dynamics of finite-difference models of the shallow-water equations*, Journal of Atmospheric Sciences, 32, (1975), p680-689.

[8]   R.K. Sato, and P.N. Swarztrauber, *Benchmarking the Connection Machine*, Proceedings of the IEEE Supercomputer Conference, Orlando Florida, November 14-18, 1988, pp. 304-309.

[9]   P.N. Swarztrauber, *Vectorizing the FFT's*, In: Parallel Computations, G. Rodrigue, ed., Academic Press, New York, 1982.

[10]  P.N. Swarztrauber, *Multiprocessor FFTs, Parallel Computing*. 5 (1987), pp. 197-210.

# Some Computational Fluid Dynamics Applications on the Intel iPSC/2

R. M. CHAMBERLAIN[1] and G. CHESSHIRE[2]

[1] Intel Scientific Computers, Pipers Way, Swindon, SN3 1RJ, Wiltshire, UK
[2] Intel Scientific Computers, Beaverton, OR 97006, USA

ABSTRACT

Several fluid-dynamics codes have been ported to the Intel iPSC/2. In particular, Nekton*
is a package for modelling the incompressible Navier-Stokes equations. Nekton is based
on spectral elements and incorporates a conjugate-gradient solver. The elements are
distributed over the nodes of the iPSC/2 and the solver uses a microcoded matrix-matrix
product on each node to obtain performance of 10 MFlops per node. Passage* is a package
for modelling the steady Navier-Stokes equations for internal flows such as ducts and
turbines. It uses a block-structured finite-difference approach to domain decomposition
which lends itself naturally to parallel implementation. FLO87* is a compressible
Navier-Stokes code using finite-volume multigrid methods for fast convergence on steady
flows. Its logically cubical computational domain is divided into blocks to be distributed
over the iPSC. Each of these codes has a straight-forward parallel decomposition because of
the regularity of its data structures and dependencies.

---

\*     Passage is a registered trademark of Technalysis Inc.

Nekton is a registered trademark of Nektonics Inc.

FLO87 is a registered trademark of Intelligent Aerodynamics.

Topics in Atmospheric and Oceanic Sciences
© Springer-Verlag Berlin Heidelberg 1990

1.    INTRODUCTION

Accurate and fast computer simulation of fluid flows is an important part of the design and manufacture of new products in many areas. Typically the speed and accuracy of such simulations have to be compromised by the user because of the limitations in computer power, resources and memory. As serial computers reach fundamental limits in performance, many engineers are turning to parallel computers to provide them with the computing necessary for their design work.

The iPSC/2 family of concurrent supercomputers give the design engineer the computer power and memory for fast and accurate simulations. Three examples are given in this paper. The iPSC/2 contains a number of processors or "nodes", each with their own local memory. Calculations are carried out on data in a processor's local memory and data can be transferred between processors' local memory by message passing. There is no shared memory or shared resources to limit the scalability of the system.

Each processing node contains an Intel 80386 and 80387. Both scalar and vector accelerators are available. The Weitek 1167 is available as an accelerator for scalar portions of the code and a vector processor speeds up vector parts of programs. The peak performance of the vector board is 6.6 MFlops (64-bit precision) and 20 MFlops (32-bit precision). The memory at each node can be chosen to be 1,4,8 or 16 MBytes.

The communication between processors uses proprietary Direct Connect hardware. The communication is implemented so that the time for passing a message between any pair of nodes is essentially the same (whether they are directly linked or the message must go through intermediate nodes). As far as the programmer is concerned, it is only necessary to specify the message buffer and its destination; the rest is handled by the operating system and the hardware. Further, because the communication time is the same between all pairs of nodes, it is not necessary for the user to worry about the topology of the processor interconnection and fitting his problem into this topology.

Concurrent Workbench includes features, which ease the use of programming such a machine. There is an automatic vectoriser VAST2 for the vector board and full parallel debugger to find bugs in both the calculations and communications. The front end for the iPSC/2 is an Intel 301, but software is available to run the iPSC/2 from Sun Workstations. In this way, the engineer can perform his I/O directly on the Sun, using all the familiar graphics for pre and post-processing.

The programming languages for the iPSC/2 are Fortran and C with additional library subroutines to handle the communications. Typically, and in all the examples in this paper, there are existing sequential codes in Fortran. Much of this code, which handles the input and output of the problem, is unchanged and runs on the engineer's workstation or the Intel 301. It is only the central numerical kernel which is parallelised. Usually there is a domain decomposition, so that each processor does all the calculations for a sub-region of the domain. For solving partial differential equations, values of the domain are usually only immediately dependent on other local values and so it is only necessary to communicate boundary values between processors. Thus substantial parts of the numerical kernel are left unchanged and only when boundary values of the subdomain are accessed is communication needed.

## 2.   NEKTON

The problem is to model incompressible fluid flow in complex two and three dimensional geometries. Heat transfer can also be simulated in the model. Examples includes radiator design, computer system cooling and glass forming. The governing equations are :

$$\frac{\partial u}{\partial t} \quad + \quad u.\nabla u \quad = \quad -\nabla p + \frac{1}{Re} \nabla^2 u + f$$

$$\nabla.u \quad = \quad 0 \qquad\qquad (1)$$

$$\frac{\partial T}{\partial t} \quad + \quad u.\nabla T = \quad \alpha \nabla^2 T$$

where u is the velocity, p is the pressure, T is the temperature, Re is the Reynolds number and f is the external force.

A spectral element method is used, which combines a high order finite element method with spectral techniques. The finite element method allows the geometric flexibility to model complex domains and the spectral techniques ensure rapid convergence. The domain is divided into relatively few, large spectral elements on which the variables are approximated by N-th order tensor-product polynomial expansions. Convergence to the exact solution is achieved by increasing the order of the approximation, rather than by increasing the number of elements.

The system (1) is discretised and the resulting systems of equations are solved by conjugate gradient algorithms. (Multigrid would be an alternative). Currently conjugate gradients with a diagonal preconditioner are used. For an Nth order element, there are $N^3$ nodal points. All the nodal points in an element are coupled in the discrete Laplacian matrix, A (say), so A contains dense $N^3 \times N^3$ submatrices corresponding to each element. In the conjugate gradient step, a matrix-vector multiply Ax is needed to calculate the residual. Ostensibly this requires $O(N^6)$ multiplies per element but fortunately the tensor product structure of the matrix allows the product to be computed as 6N matrix - matrix products of NxN matrices for a total of $6N^4$ multiplies. The rest of the conjugate gradient step requires some vector operations of $O(N^3)$.

For a parallel implementation of the spectral elements method, elements are assigned to processors. The matrix-vector product described above can all be done in parallel without communication. There are two sets of communications required. Firstly some global inner products are needed. These can be done using library subroutines. The second and major set of communication is the exchange of nodal points on the boundary between processors holding adjacent elements. This can be done in parallel and the ratio of computation to communication is proportional to $N^2$. For these high order approximations, communication is not dominant and the algorithm is well suited to the iPSC/2.

This spectral element approach has been implemented on the iPSC/2 with vector processors, using 32-bit precision arithmetic. The standard vector library calls run at 3-4 megaflops per processor, but a custom microcode matrix multiply was written, which runs at 8-15 megaflops per processor. An example problem with 64 tenth order elements takes about 0.107 seconds per conjugate gradient step. This corresponds to a total system performance of 100 megaflops. For comparison, a VAX 8700 runs at under 1 Megaflop and a single processor Cray-XMP achieves 75 MFlops on the same problem.

## 3.  PASSAGE

For modelling internal flows through complex geometries (eg. turbines, pumps, etc.) the steady Navier Stokes equations can be used. PASSAGE (see Ecer et al (1987)) is a package for simulating steady flows through stationary or rotating three dimensional geometries using either the steady Navier-Stokes equations or approximations (Euler equations or incompressible Euler or Navier-Stokes equations).

The equations are solved by means of the Clebsch transformation

$$u = \nabla \varnothing + S \nabla \eta + I \nabla \lambda + t$$

where $\emptyset$, S, $\eta$, I and $\lambda$ are the Clebsch variables and t is due to the viscous strees. The advantage of the Clebsch variables is that approximations to the Navier Stokes equations can be solved simply by omitting unnecessary terms.

Passage is based on a block structured finite element approach. The domain is divided into three-dimensional blocks and on each side of a block a boundary condition for inflow, outflow or solid wall is applied. The interfaces between blocks are handled by specifying inflow boundary conditions on one side of the interface and outflow on the other side. In each block, local iterations are applied to the discretisations to converge to the solution.

For a typical application, different approximations to the Navier Stokes equations are appropriate in different regions, (eg. the Euler equations or the potential flow equations may be sufficient in the interior of the flow). In this case, Passage is first used with the potential flow equations to approximate the flow over the whole domain and to identify regions where more accurate approximations are required. The simulation is then repeated with the appropriate levels of approximation for each region, which allows complex flows to be modelled efficiently.

On the iPSC/2 the blocks are distributed among the processors and the only communication between the nodes is for the block interface conditions. By adjusting the size of the blocks it is possible to achieve good load balance between processors. In this way nearly perfect perfect speed-up has been demonstrated. Typically the memory requirements for PASSAGE are high and on conventional sequential supercomputers, much of the time is spent paging through memory. By having the memory distributed amongst the processsors, this paging is eliminated and the performance of PASSAGE on a 16 processor iPSC/2 with scalar accelerators can exceed that of a conventional supercomputer.

## 4. FLO87

FLO87 uses explicit finite-volume methods to find steady solutions to the compressible Euler equations and Navier-Stokes equations for transonic aerodynamic flow in three dimensions. FLO87 was developed by Jameson and further details are available in Jameson(1983) and Jameson(1987). The governing equations are

$$\frac{\partial \mathbf{u}}{\partial t} + \frac{\partial \mathbf{F_1}}{\partial x} + \frac{\partial \mathbf{F_2}}{\partial y} + \frac{\partial \mathbf{F_3}}{\partial z} = 0$$

where

$$\mathbf{u} = \begin{pmatrix} \rho \\ \rho u \\ \rho v \\ \rho w \\ \rho E \end{pmatrix}, \quad \mathbf{F_1} = \begin{pmatrix} \rho u \\ \rho u^2 + p \\ \rho uv \\ \rho uw \\ (\rho E + p)u \end{pmatrix}, \quad \mathbf{F_2} = \begin{pmatrix} \rho u \\ \rho uv \\ \rho v^2 + p \\ \rho vw \\ (\rho E + p)v \end{pmatrix}, \quad \mathbf{F_3} = \begin{pmatrix} \rho u \\ \rho uw \\ \rho vw \\ \rho w^2 + p \\ (\rho E + p)w \end{pmatrix}$$

and

$$E = \frac{1}{2}(u^2 + v^2 + w^2) + \frac{p}{(\gamma - 1)\rho}.$$

A typical problem for FLO87 is transonic flow over a wing-body configuration. The computational domain is a cubic grid wrapped around the wing in a "C" shape. One side of the grid forms the surface of the wing and folds onto itself behind the trailing edge of the wing and beyond the wing tip. An adjacent side of the grid forms the body to which the wing is attached. The other four sides of the grid are inflow and outflow boundaries.

The steady solution is found by discretising the problem in space using the finite-volume method and discretising in time using a five-step Runge-Kutta method. Starting from an arbitrary initial condition, the simulation runs for as many time steps as it takes to reach a steady solution. Since the goal of the computation is to compute an accurate steady solution as quickly as possible and not to compute the details of the evolution of the flow from the initial state, this particular Runge-Kutta method is chosen for stability with large time steps rather than for accuracy. For the same reason, FLO87 allows the time step to vary over the grid, so that it is locally as large as stability conditions permit. With this change, the steps are no longer real time steps, but iterations towards a steady solution of the Euler equations.

A sequence of steady solutions is found for a given problem on successively finer grids; the solution to each provides the initial state for the next. Since the number of iterations needed to reach a steady solution from a given initial state increases as the grid is refined (the amount of work per iteration also increases), it makes sense to use as coarse a grid as possible. Although the accuracy of the solution on the coarse grid is not sufficient as a final result, it provides an initial state which is already close to the steady solution on a finer grid. Using this initial state, fewer iterations are needed on the finer grid to reach a steady solution. This procedure may be continued on still finer grids until finally the solution is as accurate as desired.

The distinguishing feature of transonic flows is the presence of shock waves - discontinuities in one or more of the solution components. In order to make the iterations stable in the presence of discontinuities, FLO87 includes both second-order (Laplacean) and fourth-order dissipation terms. The second-order dissipation provides stability around shocks, while the fourth-order dissipation ensures smooth transition between shocks.

Inclusion of these dissipation terms turns the problem of finding a steady solution into an elliptic system of equations. This makes it amenable to solution by the multigrid method, which FLO87 uses to accelerate convergence to a steady solution. Thus, while for an outer iteration it solves problems on successively finer grids, at the same time it uses an inner multigrid iteration over the current grid and coarser grids. The innermost iteration within the multigrid method still consists of a modified time step as described above.

The computational domain for FLO87, a rectangular three-dimensional grid, is decomposed on the iPSC/2 into a grid of identical blocks or sub-grids, one per processor, with one plane overlap for each face of a sub-grid. Since the number of processors is a power of two, the number of sub-grids in each directions is in general also a power of two. On a 64-processor machine, for example, this leaves open the possibility that the domain may be divided in one, two or three directions (that is, 64 x 1 x 1, 8 x 8 x 1 or 4 x 4 x 4). The constraint that each sub-grid should be identical, even for the problem on the coarsest grid and at the coarsest multigrid level, requires only that the number of cells on the coarsest grid should be a multiple of the number of processors in that direction.

Within this constraint, the optimal number of sub-grids in each direction is a trade-off between the vector length for efficient vector operations, the cost of computing boundary conditions, message latency and message bandwidth. Efficient vector operations favour dividing the grid in the two directions with the least number of cells, so that the third direction on each sub-grid has the longest possible vector length. An expensive boundary condition on one face of the grid suggests dividing that face up among as many processors as possible. The effect of message latency can be reduced by dividing the grid up into slabs to lower the number of messages. However the total message traffic can be minimised by dividing the grid into cubes. Finding the optimal way to divide up the grid is best done experimentally.

The only communication needed between processors is for transfer of the solution between adjacent faces of sub-grids and for global commutative operations. For example, at the beginning of each iteration of the finite-volume method, it is necessary for each processor to have the latest values of the solution components in all the cells adjacent to each cell that processor is to iterate on. This requires that each processor should send the solution on each face of its sub-grid to the processor that owns the neighbouring sub-grid. This is sufficient also for computing the second-dissipation terms; the fourth-order dissipation terms require communication of an additional face of results in each direction. The multigrid cycle does not require any further communication, because of the constraint that the coarsest grid should be divided equally among the processors. If we relax this constraint, inter-level transfers in the multigrid cycle require further communications. The only communication required, other than between sub-grids, is for some global commutative operations. For example, the global maximum of the current error is used as a check for convergence.

Due to the high ratio of computation to communication and the regularity of the domain, this algorithm obtains good load balance and high parallel efficiency on the iPSC/2. On a 64-processor iPSC/2 machine with scalar accelerators and with 4 Megabytes of memory per processor, a large problem runs with approximately 70 percent efficiency, approximately three times faster than a Convex C2.

## 5. CONCLUSIONS

This paper has described three complete fluids dynamics applications running on the iPSC/2. The conversion of these applications to the iPSC/2 was straightforward and supercomputer performance has been achieved by these codes. This demonstrates that distributed memory multiprocessors are very suitable for solving large fluid dynamics applications. In the future, larger, more accurate and faster simulations will be possible on these machines than on the fastest sequential supercomputers.

References

Ecer, A., Akay, H.U., and Sheu, W.H., 1986 Variational Finite Element Formulation for Viscous Compressible Flows, Symposium on Numerical Methods for Compressible Flows. ASME Winter Meeting, Anaheim, California, December 7-12.

Jameson, A., 1983 The evolution of computational methods in aerodynamics. Transactions of the ASME, Journal of Applied Mechanics, Vol. 50, pp 1052-1070.

Jameson, A., 1987 Successes and challenges in computational aerodynamics, AIAA 87-1184-VPCP.

# Parallel Implementation of Advection Calculations in Pseudospectral Atmospheric Hydrodynamical Models

Zaphiris D. Christidis

IBM Research, T. J. Watson Research Center, P. O. Box 704, Yorktown Heights, N. Y. 12598, USA

## ABSTRACT

A domain decomposition algorithm was used for the parallel solution of the time dependent Advection-diffusion or Conservation of Pollutant Mass equations, by employing Fourier pseudospectral methods. The core of the calculation was based on the application of the two-dimensional (2-D) FFT for each time step. Parallelization was achieved by the concurrent application of the 1-D FFT by $NP$ processors. The computational domain was partitioned in $NP^2$ rectangular blocks, and each dependent variable was discretized and placed in a global memory. Thus each processor had a fast access to any sequence of data blocks for reconstructing columns or rows for the FFT applications.

The performance of the algorithm has been studied on shared memory systems, using EPEX, which is an environment for parallel execution on IBM System/370, IBM RT/PC and the Research Parallel Processing Prototype (RP3). Results have been obtained by solving a two-dimensional, time dependent hyperbolic equation. It was found that the parallel efficiency of the method is satisfactory for large size problems.

Topics in Atmospheric and Oceanic Sciences
© Springer-Verlag Berlin Heidelberg 1990

# 1. INTRODUCTION

Over the past years a number of sophisticated numerical methods have been developed for the solution of the hydrodynamic equations. The object of many of these methods has been to minimize numerical errors in the approximations for spatial derivatives. The nature of these errors is numerical dispersion, distortion, and formation of numerical instabilities. A recent and widely favored method with high accuracy and efficiency is the *pseudospectral* method (Fox and Orszag [3], Orszag [3]). This method consists of expanding the unknown dependent variables of an equation in a series of global, orthogonal and complete set of functions (basis functions), and demanding that the equation be exactly satisfied at a set of points in the space domain (grid or collocation points). The best choice of basis set depends on the particular geometry of each individual problem. Nevertheless, the Fourier or the Chebyshev functions are very popular and commonly used due to their efficiency in applications where Fast Fourier Transforms are employed. Thus for example in most problems where gradients of the dependent variables need to be evaluated, it is quite common to perform FFTs, evaluate the gradients analytically in the spectral domain, and then with a inverse FFT , return to the space or grid domain having calculated the gradients with high order of accuracy.

The object of this paper is to demonstrate an approach for the implementation of pseudospectral methods on experimental multiprocessor computer systems, within IBM. In the next section, an overview of all the parallel computer systems at IBM research will be given. Section 3 describes the pseudospectral method, while in section 4, the basic strategy for its parallel implementation is presented. Section 5 discusses the initial conditions for a test problem, section 6 contains results of parallel runs in different systems, and finally section 7 concludes this paper by summarizing the parallel run results.

# 2. PARALLEL PROCESSING AT IBM

At IBM, there are various scientific research groups, focusing on parallel computations as a modular approach to supercomputing. These groups are involved in an ongoing effort towards the development of new parallel systems, with special emphasis on hardware and software design. This effort lead to the development of several experimental parallel systems. Two of them will be examined next.

## 2.1 Advanced Computing Environment (ACE)

The ACE is an experimental multiprocessor workstation, also developed at the IBM T. J. Watson Research Center. It consists of two major sub-systems: an IBM Model 6150-125 RT workstation, and an ACE multiprocessor workstation, which consists of up to 8 processor modules and up to 80 MB of shared memory. Each ACE processor module consists of the the same 32-bit RISC microprocessor, and memory management unit as in the RT, with a 20 MHz Motorola 68881 floating-point co-processor, and 8MB of local memory. Inter-processor communication is achieved through a bus at a peak rate of 80 Mbytes/second. The IBM RT acts as a host providing the user interface and all I/O services for the ACE, which appears physically as a co-processor attached to the RT, sharing the same file systems residing on hard disks on the RT. Currently Ace runs the Mach operating system, while parallel program development is done in FORTRAN and C.

## 2.2 Loosely Coupled Array of Processors (lCAP)

At IBM Kingston, a scientific applications research group under IBM Fellow Enrico Clementi, has developed two experimental systems composed of IBM-3081/3084 as front end processors, coupled to ten FPS-164/264 attached array processors from Floating Point Systems, Inc. These parallel systems have a peak performance of about 110/550 MFLOPS. To achieve faster communication (IO's) between attached processors (APs), five locally shared bulk memories, and a globally shared memory of bandwidth 44 Mbytes/second have been added to each of the two systems in a ring configuration. In addition to the bulk memories, FPS-made busses, capable of connecting all ten APs together have been installed. The FPS bus allows the APs to transfer data between processors at the rate of 22 Mbytes/second. Programming development is done in Fortran and Assembler. Recently the FPS-164/264 have been replaced with a cluster of IBM-3090/400 with Vector, and the Parallel Fortran Prototype in combination with the old lCAP software are used for parallel programming.

## 3. THE PSEUDOSPECTRAL METHOD

In order to apply the pseudospectral method, the following two-dimensional hyperbolic equation will be used as a model equation.

$$\frac{\partial G}{\partial t} = -\bar{u}\frac{\partial G}{\partial x} - \bar{v}\frac{\partial G}{\partial y} \tag{1}$$

The function $G = G(x, y, t)$ is the unknown dependent variable, $x$, $y$ and $t$ are the independent space and time variables, while the quantities $\bar{u}$ and $\bar{v}$ are the mean $x$ and $y$ components of the current or wind vector $\bar{V}$. The above equation is also known as the advection or transport equation, as it describes the transport of the conservative quantity $G$ by the mean flow $\bar{V}$. Thus an advection operator can be defined as

$$\Lambda \equiv -\bar{u}\frac{\partial}{\partial x} - \bar{v}\frac{\partial}{\partial y}. \tag{2}$$

By assuming that the space and time domain is discretized according to,

$$\begin{aligned}
x_i &= x_0 + (i-1)\Delta x & i &= 1, 2, \cdots, NX \\
y_j &= y_0 + (j-1)\Delta y & j &= 1, 2, \cdots, NY \\
t_n &= t_0 + n\Delta t & n &= 0, 1, \cdots, NT - 1
\end{aligned} \tag{3}$$

two sets of polynomials can be defined in a complex form as follows:

$$\Psi_{l,k}(x,t) = \sum_{m=-NX/2}^{NX/2} \psi_m(t)e^{imxd_x} \tag{4}$$

$$\Psi_{k,m}(y,t) = \sum_{l=-NY/2}^{NY/2} \psi_l(t)e^{ilyd_y} \tag{5}$$

where $d_x = 2\pi/(NX\Delta x)$, $d_y = 2\pi/(NY\Delta y)$, $\exp(i\varphi) = \cos(\varphi) + i\sin(\varphi)$. Each polynomial $\Psi_{l,k}(x,t)$ in relation (4) is a trigonometric interpolant along a grid line parallel to the $x$ axis, while each polynomial $\Psi_{k,m}(y,t)$ in relation (5) is a trigonometric interpolant along a grid line parallel to $y$ axis. Therefore the derivatives of the polynomials of (4) could be used to obtain approximations to the values $\partial G/\partial x$ of the sample equation (1) at the points of the grid $\mathcal{N} = x_i \times y_j$. Similarly, the derivatives of the polynomials of (5) could be used to obtain approximations to the values $\partial G/\partial y$ at the points of the grid $\mathcal{N}$. This kind of discretization of the first order space derivatives is called a pseudospectral Fourier discretization (Zlatev et. al. [3]).

A host of explicit numerical schemes for advancing the solution in time can be applied. In this study, the following three-level time differencing scheme was used to approximate the time derivative in Eq. (1).

$$\frac{\partial G}{\partial t} \approx \frac{G^{n+1} - G^{n-1}}{2\Delta t} \tag{6}$$

The above scheme is known as the "Leapfrog" or mid-point rule (Haltiner and Martin [3]). The mid-point rule is second order accurate in $t$, $(O[(\Delta t)^2])$, and conditionally stable when applied to Eq. (1). The stability criterion is quite similar to the Courant-Friedrichs-Lewy (CFL) condition for computational stability (Christidis [3]). Furthermore the requirement for computer storage in its application is minimal, as only two time levels of the independent variable $G$ needed to be stored for the performance of one time step.

To illustrate how the pseudospectral method can be applied, without any loss of generality, only the $x$ component of the operator $\Lambda$ is considered. Thus, equation (1) is discretized only at the grid points $x_i$ given by the first of Eq. (3). Then the discrete set of values of $G$ is replaced by a continuous function given by the finite Fourier series,

$$G(x,t) = \sum_{m=-NX/2}^{NX/2} g_m(t)e^{imxd_x} \tag{7}$$

while the Fourier coefficients $g_m$ are calculated by means of the Fourier Transform ($FT$) defined as,

$$g_m(t) = \frac{1}{NX}\sum_{x_i} G(x_i,t)e^{-imx_id_x} \tag{8}$$

The space derivatives can be evaluated at the grid points $x_i$ by using (7) and (8),

$$\frac{\partial G(x_i, t)}{\partial x} = \sum_{m=-NX/2}^{NX/2} imd_x g_m(t)e^{imx_i d_x}.$$
(9)

The procedure described by relation (7) is the Inverse Fourier Transform. The relation (9) can also be classified as a part of the inverse Fourier Transform, since it tran the spectral variables $\{g_m\}$ to grid point variables $\{G(x_i, t)\}$.

If the full two dimensional form of operator $\Lambda$ is used, the $y$ gradients of $G$ can be calculated in the same manner as in (9), e.g. by using the trigonometric interpolants defined by (5). All other calculations must be performed on the grid points only.

If two-dimensional (2-D) Fourier transforms are performed, then the procedure requires $(NX^2 \times NY^2)$ numerical operations. By the use of the one-dimensional (1-D) FFT the number of simple operations is reduced to $O[NX \times NY \times \log_2(NX \times NY)]$ provided that $NX$ and $NY$ are powers of 2.

## 3.1 Numerical Algorithm

A general algorithm will be given describing the procedure followed for solving Eq. (1) two dimensions.

**Step A.** For each grid line parallel to the $x$ axis, execute the following operations:

1. *Perform a forward FFT using the values of $G(x, y)^n$ on the $x$ grid line.*

2. *Calculate approximations to the Fourier coefficients $\partial \tilde{G}/\partial x$, of $\partial G(x, y)^n/\partial x$ on the $x$ grid line.*

3. *Perform an inverse FFT on $\partial \tilde{G}/\partial x$ on the $x$ grid line. Thus $\partial G(x, y)^n/\partial x$ is obtained.*

4. *Form the product $F_x^n \equiv -\bar{u}(x, y)^n \partial G(x, y)^n/\partial x$ on the same grid line.*

**Step B.** For each grid line parallel to the $y$ axis, execute the following operations:

1. *Perform a forward FFT using the values of $G(x, y)^n$ on the $y$ grid line.*

2. *Calculate approximations to the Fourier coefficients $\partial\tilde{G}/\partial y$ of $\partial G(x,y)^n/\partial y$ on the y grid line.*

3. *Perform an inverse FFT on $\partial\tilde{G}/\partial y$ on the y grid line. Thus $\partial G(x,y)^n/\partial y$ is obtained.*

4. *Form the product $F_y^n \equiv -\bar{v}(x,y)^n \partial G(x,y)^n/\partial y$ on the same grid line.*

**Step C.** Finally $G(x,y)^{n+1}$ is obtained at the points of the grid mesh $\mathcal{N}$, by:

1. *Adding the quantities $F_x^n$ and $F_y^n$.*

2. *Using the values of $G(x,y)$ at the $n-1$ time step.*

3. *Applying the Leapfrog time-integration algorithm to calculate approximations of $G(x,y)$ for the $n+1$ time step.*

In general, Fourier pseudospectral methods are very efficient and highly accurate. The only drawback is that the application of the FFT implies periodic boundary conditions in space.

In the following section, the parallel implementation of the problem as described by Eq. (1) is discussed in more detail.

## 4. PARALLEL IMPLEMENTATION

The discrete Fourier transform of two dimensional data structures by the FFT technique was considered for parallel implementation as a series of one dimensional FFTs which is applied in succession to each dimension of the data structure. Parallelism was achieved not by a decomposition of the FFT algorithm, but through the segmentation of the data as evenly as possible among processors, followed by the concurrent application of 1-D FFTs on arrays resident in each processor.

Pseudospectral methods have global interpolation properties. In order to calculate gradients of functions, it is necessary to have simultaneous information about all the variables defined on grid lines. The application of the 2-D FFT makes use of this global information, by requiring data exchange through the global memory between attached processors.

Once the gradients have been calculated, time marching algorithms may be evenly partitioned among independent processors without undue concern for the size of the grid. Considering for simplicity, a two dimensional data structure, distributed in columns

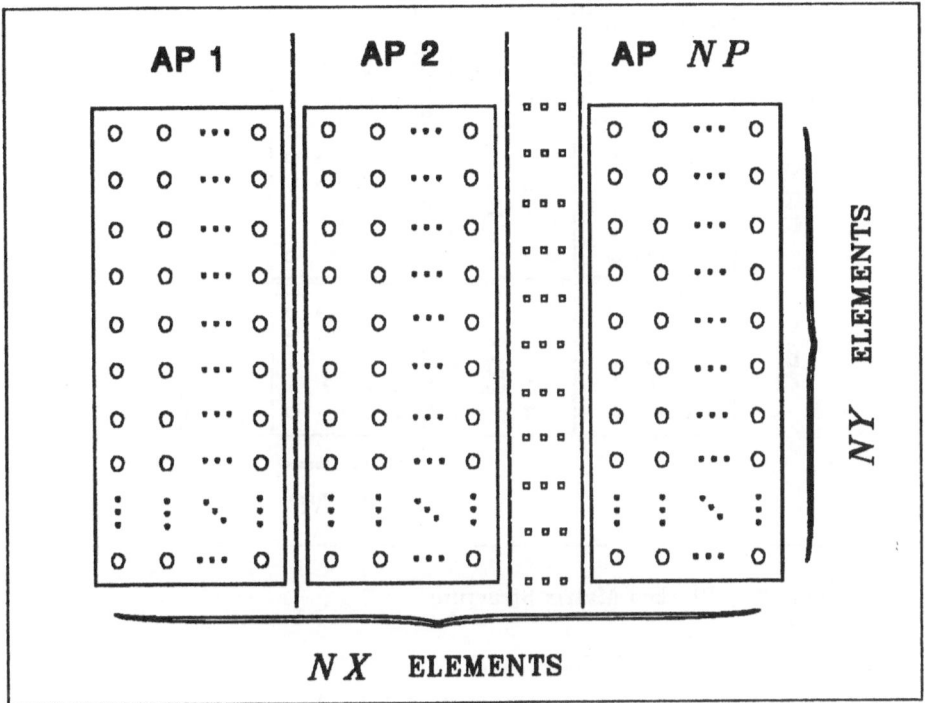

Figure 1.   Distribution of Matrix Elements Among Processors

between the processors (Fig. 1). Functionally the execution of a 2-D FFT on a matrix $\overline{\overline{G}}$ of size $NX \times NY$ can be viewed as the result of performing $NX$ series of one 1-D FFTs of length $NY$ (column-wise), followed by $NY$ series of 1-D FFTs of length $NX$ (row-wise). It should be noted that the above procedure is structurally equivalent to alternating direction algorithms.

On a parallel machine consisting of $NP$ processors, the number of columns $NX$, of the matrix $\overline{\overline{G}}$, must be distributed evenly among the $NP$ processors so each processor has a fraction, $NCOL = NX/NP$, in its local memory. Then parallel execution can be initiated with each processor performing a total of $NCOL$ 1-D FFTs of length $NY$. Thereafter to complete the process, each processor is required to execute 1-D FFTs on equal segments ($NROW = NY/NP$) of the rows of the intermediate result matrix. This requires an effective transpose of the intermediate results. As a final step, another exchange of rows and columns is performed, in order to regain the original distribution of data. For purposes of illustration a sample matrix of block size $3 \times 3$ is shown with the desired segmentation among 3 processors in Fig 2.

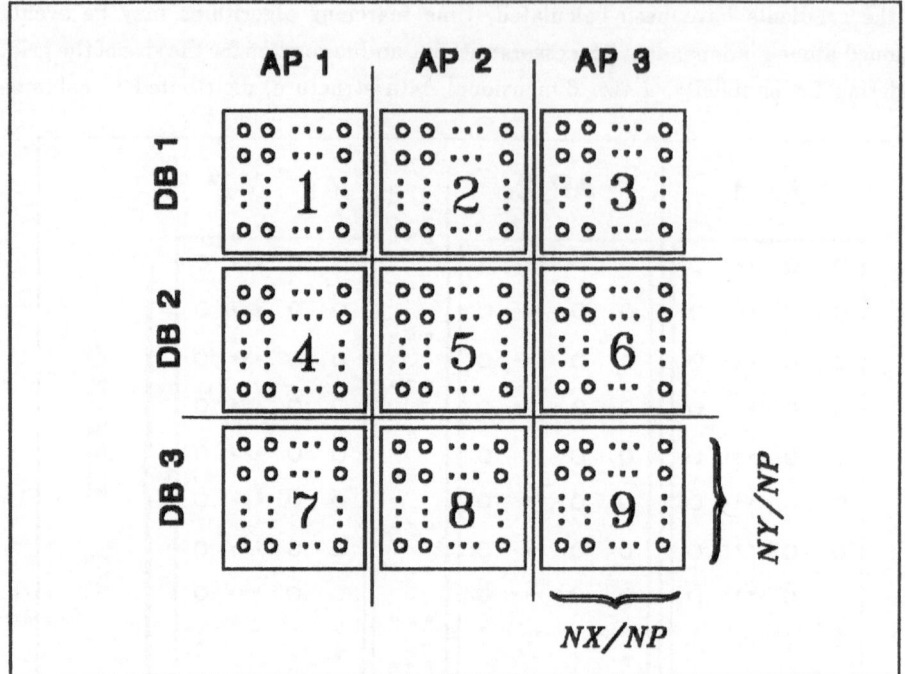

Figure 2.   Blocked Matrix Structure

## 4.1 The "Matrix Slice" Method

The basic idea in the development of this method, is to eliminate any extra procedures required for data sorting and transfers, by arranging a priori, the order in which the matrix elements are stored. Again as before, the original matrix is distributed column-wise between APs, and it is assumed without loss of generality that the matrix is square. However, the columns are sliced in a number of blocks equal to the number of APs employed. Consider the 3 processor example where each such block is order $n$; data is arranged contiguously in the order $(1, 1), (1, 2), \cdots (1, n), (2, 1), \cdots (2, n), \cdots (n, n)$. In the labelling of Fig. 2, blocks #1, #4 and #7 are organized contiguously in this manner for fast access by processor number one. Similarly processors two and three have an identical data structure with blocks #2, 5, 8 and #3, 6, 9 respectively.

The procedure then followed in the effective transpose, consists of mapping these data blocks onto one global memory in order to reconstruct the full matrix accessible by all contributing APs. Thus the sequence of each processor acquiring row data, consists of performing one block-column write, followed by $NP - 1$ or $NP$, block-row reads for a square or rectangular matrix respectively. Synchronization is achieved by allocating $NP$ words of common memory as inter-processor mailboxes which receive updates on processor writes or reads. A synchronization check is performed by each processor prior to each read, to ensure that before processing the the $NX$ direction the $NY$ direction is processed. Thus each AP before a read, updates its status and checks the status of the other APs. If at least one of them is not ready, it rechecks the whole "mailbox" until everybody is ready. Once everybody is ready, the AP executes a column-wise or row-wise read.

The data undergoing the IO's are "scrambled" in the sense that the contiguous memory locations are occupied by non-contiguous elements. Nevertheless the data is "unscrambled" while the actual FFT is performed with minimal computational cost. Specifically, originally contiguous elements are restored by being "picked" from the scrambled matrix via stride parameters calculated initially. So in a parallel code, the original matrix size is given and a number of parameters are calculated once and for all. These parameters are global (common) and local for all APs, and include the number of initial columns/rows per AP, and vector pointers containing stride parameters for the choice of the right elements when performing the column/row part of the FFT. The method is quite general and it is described in more detail in Christidis *et. al.* [3].

## 5. TEST PROBLEM

In order to illustrate the parallel implementation, of the Fourier pseudospectral method, equation (1) was solved by partitioning in data blocks the initial condition $G(x, y, 0)$ and the flow components $\bar{u}(x, y)$, $\bar{v}(x, y)$ among three APs (Fig. 2).

Periodic boundary conditions in $x$ and $y$ were implemented and the local $G$ gradients were calculated in each AP via the concurrent use of the 1-D FFT by discretizing Eq. (1) with respect to the space variables $x$ and $y$. In turn each AP partition marched forward in time applying the explicit mid-point rule and exchanging the same amount of data per time step. The data exchange was synchronized to assure proper parallel execution of the FFTs and related numerical operations between the APs.

The initial condition chosen, was a conical distribution defined as,

$$G(x, y, t = 0) = \begin{cases} G_0(1 - \bar{r}/r) & \text{as } \bar{r} < r \\ 0 & \text{as } \bar{r} \geq r \end{cases} \tag{10}$$

where $\bar{r} = [(x - x_0)^2 + (y - y_0)^2]^{1/2}$ is a position vector and $r$ is the base radius of the cone with a maximum height $G_0$ with center located at $x_0, y_0$. The flow field was chosen to be of constant angular velocity, e.g.,

$$\begin{aligned} u &= -y \\ v &= x. \end{aligned} \tag{11}$$

Figure 3 shows a vector plot of the flow field, while Fig. 4 shows a perspective

VELOCITY VECTOR FIELD

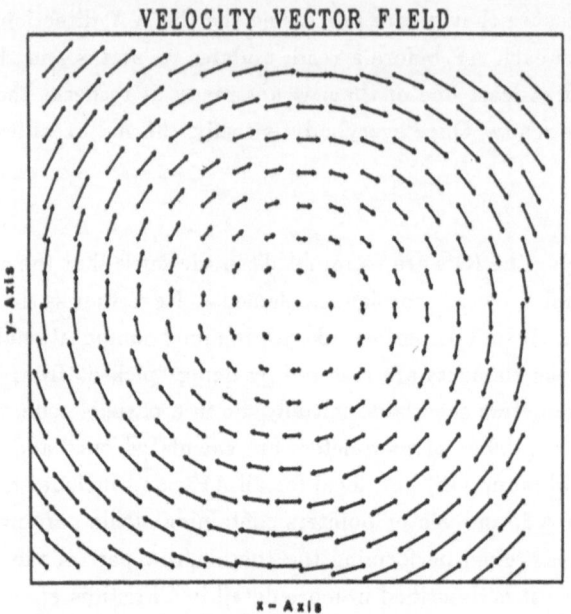

Figure 3.   Vector Flow Field V

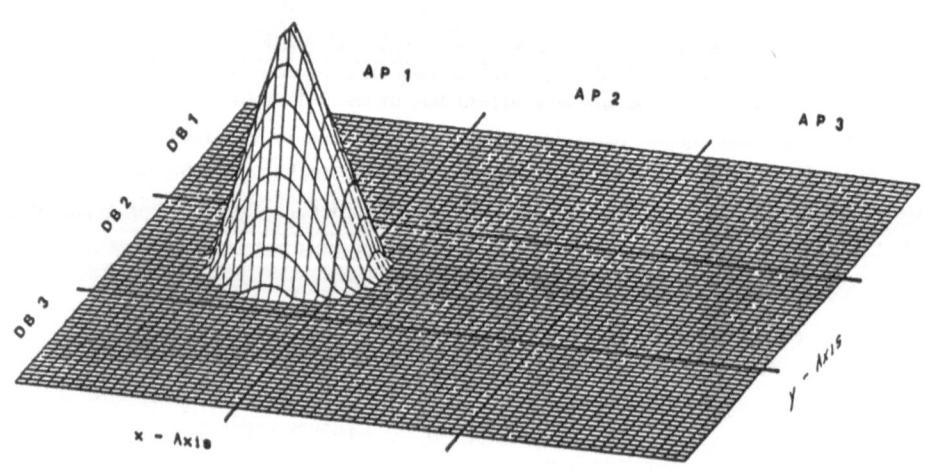

Figure 4.   Perspective Plot of the Initial Conical Distribution

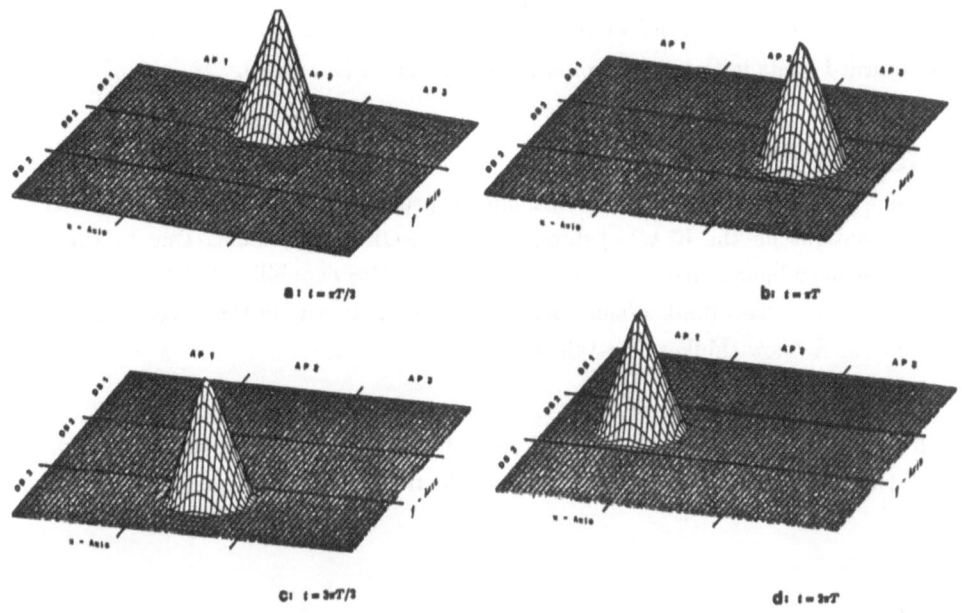

Figure 5. Calculated Conical Distribution at: a) $t = \pi T/2$, b) $t = \pi T$, c) $t = 3\pi T/2$ and d) $t = 2\pi T$.

plot of the initial condition, as distributed in data blocks among the three processors. The above conditions constitute the well-known problem of the rotation of a conical distribution in a uniform velocity field, which is widely used as a test case in numerical analysis of differencing schemes (Chock [3], Zlatev [3]).

A full rotation is performed for any time period $T = 2\pi$. The solution to Eq. (1) is constant along the characteristics $x + iy = (x_0 + iy_0)e^{it}$ which yields $G(x, y, 0) = G(x, y, 2\pi)$ (Gottlieb and Orszag [3]).

In discretizing Eq. (1), the number of grid points in $x$ and $y$ was taken to be $NX = NY = 64$, while the number of time steps $NT$ for one full rotation was set to 1400.

As it is shown in Fig. 4, a large portion of the conical distribution is in processor #1 and in the data block #2 of the same processor. Figure 5a-5d displays the solution of Eq. (1) for the time periods $t = \pi T/2$, $\pi T$, $3\pi T/2$ and $2\pi T$ respectively. For $t = \pi T/2$ the calculated conical distribution is found in processor # 2 and the first data block, while for $t = \pi T$ the distribution has moved into processor #3 in the second data block. In turn, the same conical distribution has travelled through processor #3 and at $t = 3\pi T/2$

has returned to processor #2 at the third data block. Finally the conical distribution has returned to its initial position after a full rotation ($t = 2\pi T$).

## 6. RESULTS

Equation (1) was solved for one through 10 processors, and for different horizontal grid resolutions on the lCAP system, and the ACE workstation. Due to the small shared memory bandwidth as compared to the FLOPS capability of lCAP, the system is classified as a "fast-thinker/slow-talker", in conjunction with the ACE which can be classified as a "slow-thinker/fast-talker".

Performance curves on the lCAP and ACE for the the parallel solution of Eq. (1) employing the "Matrix Slice" algorithm are illustrated in Figs. 6 and 7. The speedup

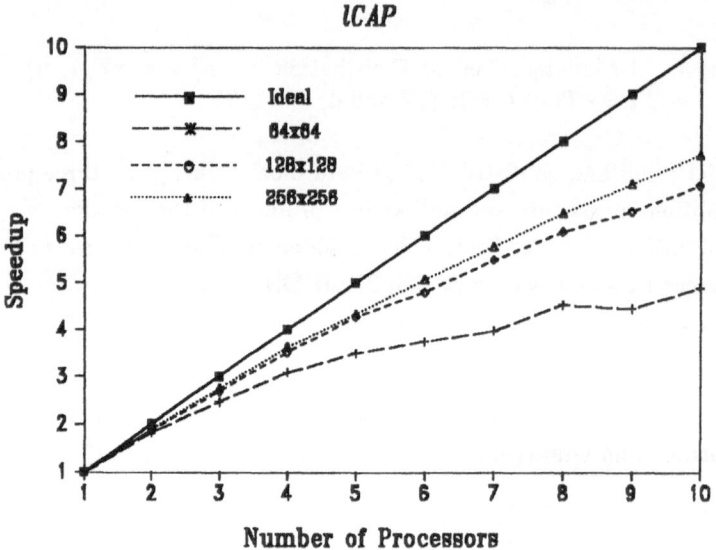

Figure 6.    lCAP Speedup Curves for Different Problem Sizes

$S$ in a parallel run can be defined as:

$$S = \frac{T_1}{T_n} \tag{12}$$

In the ideal situation in which the run is completely parallelizable, the computation is perfectly divisible among the processors. Thus, $T_n = T_1/n$, and from Eq. (12), it is obtained $S = n$. This behavior is represented by the straight solid line in Fig. 6. In the actual parallel run, the obtained speedup rates deviated from the ideal as the number of APs increased. This was mainly attributed to the,

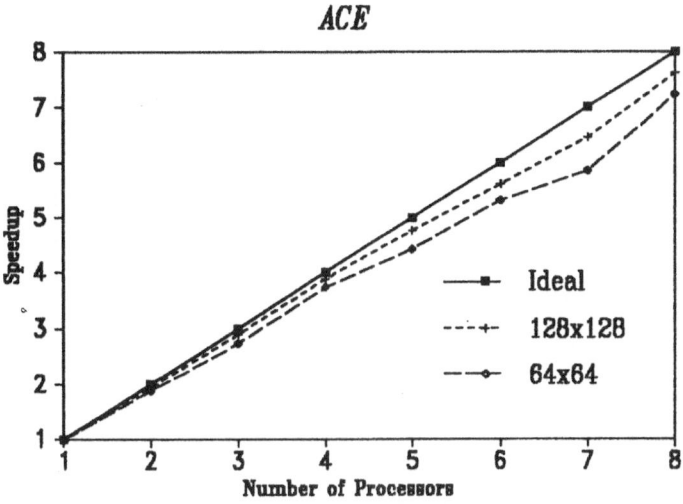

Figure 7. ACE Speedup Curves for Different Problem Sizes

1. necessary inter-processor communication for data swapping,

2. uneven data load balance among the processors and

3. operating system overhead.

In the absence of sequential portions in the computer code and if $T_d$ is the total elapsed time due to the fixed overhead in communication and memory latency, a theoretical speedup curve can be defined as,

$$S = \frac{T_1}{T_n + T_d}. \tag{13}$$

In the lCAP system $T_d$ is of the same order of magnitude as $T_n$ for small matrix sizes ($64 \times 64$), mainly due to the communications overhead, and the relatively small amount of numerical calculations. Load balancing is also important in the case of 2, 4 or 8 APs where each processor has the same amount of square data blocks. On the ACE workstation, inter-processor communication is very efficient as compared with the time spend for floating point numerical calculations. This is evident, as speedups of 7.2 and 7.6 are experienced with 8 processors for $64 \times 64$ and $128 \times 128$ problem sizes.

Figure 8 shows the ratio of processor intercommunication time $T_d$, over total parallel execution time $T_n + T_d$ as a function of the number of APs and problem sizes on lCAP.

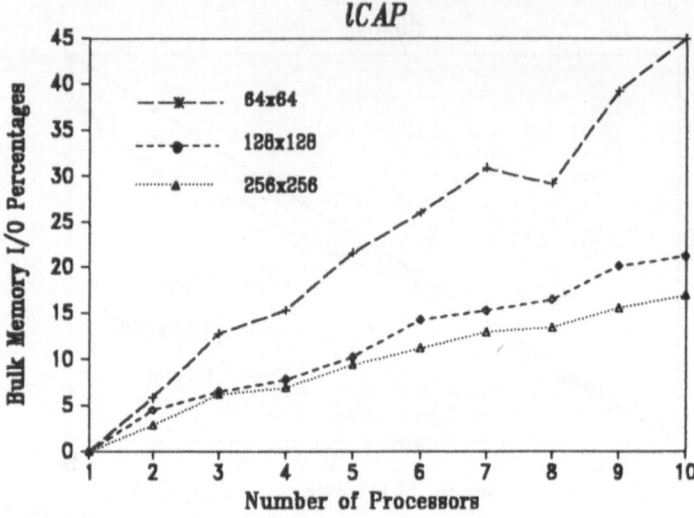

Figure 8.   ICAP Processor Intercommunication Ratios

It can be seen that the communication increases with the number of APs, but for large problem sizes remains under 20 percent of the total parallel run.

An important factor in a parallel run is the actual time spent for processor synchronization. If the amount of work in each processor is different, all processors will not reach the synchronization barrier at the same time, and therefore some must wait for others in order to exchange data. Hence, the quantity $T_d$ can be considered as the sum of the actual data transfer time $T_c$ and the processor synchronization time $T_w$. In figure 9 the ratio of processor synchronization time $T_w$, over total parallel execution time $T_n + T_d$ is plotted as a function of the number of APs and problem sizes.

It is seen that the synchronization time increases with increasing number of APs, as more processors compete to perform bulk memory IO's through the memory channel bottlenecks and also due to unequal distribution of work among processors. On the other hand, when the load is well balanced, the amount of synchronization is minimal.

For the $256 \times 256$ case on ICAP, a speedup of 7.7 was realized with 10 processors; the total time involved in data transmission and rearrangement was approximately a factor of 6 less than that employed in calculation. Less rewarding results were obtained for smaller problem sizes due to the fixed overhead in inter-processor communication.

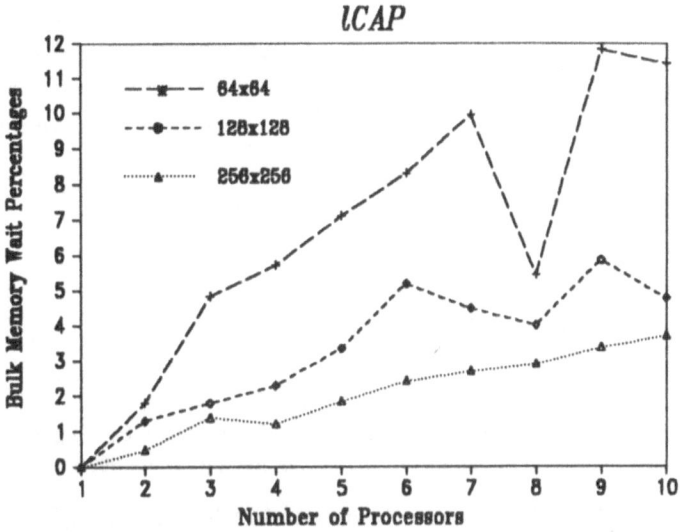

Figure 9.    ICAP Processor Synchronization Ratios

## 7. DISCUSSION AND CONCLUSIONS

A domain decomposition method was used in order to perform a numerical simulation of a rotating conical section on the ICAP system. The equation describing the physical problem was the hyperbolic transport equation, and it was solved using Fourier pseudospectral methods in space, with an explicit finite differencing scheme in time. Because of the necessity of frequent data transfer among the different processors the efficiency is strongly dependent upon the strategies and hardware chosen for communications. In this study the "Matrix Slice" method in combination with the shared bulk memories was used in order to perform efficiently the Fast Fourier Transforms, necessary for the space discretization in the pseudospectral method. The reasons for the use of this particular scheme were that

- it was very easy to code,
- the data transfers are done in place, because the auxiliary storage for sending/receiving is a part of the common bulk memory, and not of the local memory,
- it can work with as much data as can be held in bulk memory,
- there is no complicated algorithm for data transfer among processors to work against the highly efficient FFT.

Here it was demonstrated that the "Matrix-Slice" method leads to an efficient algorithm for parallel implementation of pseudospectral methods. It can handle very large rectangular matrices, and is intended to be a user-transparent routine that can be called from a Fortran program. The data processing part is general and FFT routines may be replaced with ease with any ADI solvers.

# REFERENCES

D. P. Chock, 1985: A Comparison of Numerical Methods for Solving the Advection Diffusion Equation-II. *Atmospheric Environment*, 19, 571-586.

Z. D. Christidis, 1986: Hydrodynamic Mesoscale Modeling of Atmospheric Transport and Pollutant Deposition in the Vicinity of a Lake. Ph.D. Thesis, Atmospheric and Oceanic Science Department, University of Michigan, Ann Arbor.

Christidis, Z. D., V. Sonnad, and D. Logan, 1987: Parallel Implementation of a 2-D Fast Fourier Transform on a Loosely Coupled Array of Processors. IBM Kingston Technical Report, KGN-68.

E. Clementi, D. Logan, A. Capotondi, Z. Christidis, and V. Sonnad, 1986: Solving Engineering Problems In Parallel With a Loosely-Coupled Array of Processors. Computers in Mechanical Engineering.

E. Clementi, and D. Logan, 1987: Parallel Processing with a Loosely Coupled Array of Processors System. IBM Kingston Technical Report, KGN-43.

D. G. Fox, and S. A. Orszag, 1973: Pseudospectral approximation to two-dimensional turbulence. *J. Comput. Phys.*, 11, 612-619.

D. Gottlieb, and S. A. Orszag, 1977: *Numerical Analysis of Spectral Methods: Theory and Applications*. SIAM, Philadelphia, PA, 200 pp.

G. J. Haltiner, and R. T. Williams, 1979: *Numerical Prediction and Dynamic Meteorology.*, Wiley, 477 pp.

S. A. Orszag, 1972: Comparison of pseudospectral and spectral approximation.*Stud. Appl . Math.*, 51, 253-259.

Z. Zlatev, R. Berkowicz and L. P. Prahm, 1983: Stability Restrictions on Time Stepsize for Numerical Integration of First-Order Partial Differential Equations. *J. Comput. Phys.*, 51, 1-27.

Z. Zlatev, R. Berkowicz and L. P. Prahm, 1983: Three-dimensional Advection-Diffusion Modeling for Regional Scale. *Atmospheric Environment*, 17, 491-499.

# Asymptotic Parallelism of Weather Models

Tuomo Kauranne[1]

European Centre for Medium-Range Weather Forecasts, Shinfield Park, Reading,
Berkshire RG2 9AX, U.K.

## 1. Introduction

We investigate a number of commonly used discretization techniques and parallel computer
architectures for their potential to asymptotically produce an accurate representation of the
dependent variables in operational weather models. This can be formulated as a heuristic
optimization problem: *"Which combination of a numerical solution technique and a parallel
interconnection topology will, subject to a uniform time constraint, asymptotically produce the
most accurate discrete solution to the primitive equations, when the number of processors (hence
the "cost") is allowed to approach infinity?"*

Solutions to this problem will give us guidelines to the choice of both the algorithm and the
computer architecture to be used in implementing operational weather models on massively
parallel computers with thousands of processors. The analysis is analogous to the introduction
of asymptotic computational complexity of numerical algorithms in the 1960's, as opposed to
mere empirical test runs and CPU timings common at the time.

## 2. Limitations of asymptotic analysis

An assumption taken for granted here is that the number crunching capacity is the most valuable
and also the most expensive property of a computer aimed at scientific applications. Auxiliary
features like the communication subsystem should be built to guarantee maximum utilization
of the computing power. Thus most applications should be computation bound when implemented on a parallel computer.

When solving the primitive equations we are dealing with a nonlinear system of time-dependent
partial differential equations. What will be said below qualitatively applies to most problems
of this type. The physical scale analysis underlying the validity of the primitive equations, or
some augmented system, however, is different for each such system.

The analysis also hinges upon a number of characteristic constants for parallel computers, like
the number of processors, processor speed, communication startup time and communication
bandwidth. These may vary in several orders of magnitude from one parallel computer to another,
but are (incorrectly) treated in asymptotic analysis as negligible, all the same (as usual,
asymptotic analysis here means "up to a constant multiplier").

It is these two factors, the meteorological scale analysis and the characteristic constants of the
parallel computer at hand, that finally determine the quantitative applicability of the results
presented to any particular machine and algorithm.

---

**1** This work has been supported by TEKES, The Finnish Technology Development Centre,
under the program Finsoft III: Parallel Algorithms.

We have not questioned here the usefulness of increased spatial accuracy, which is the principal purpose of running weather models on a massively parallel computer. This seems to be a very complicated issue. Numerical implementations of the primitive equations cannot describe all physical phenomena relevant to more accurate local weather forecasts. This is why we need parametrization. Improving the resolution implies, however, that processes which will now be explicitly resolved will have to be removed from parametrization schemes, in order to retain the energetical consistency of the system of equations. In particular, convection, gravity wave drag and turbulence modelling would have to be reconsidered, whereas radiation, precipitation and surface processes could probably remain intact. This may be a considerable additional theoretical and programming effort. Violating consistency on the other hand is likely to spoil any gains in forecast accuracy resulting from increased resolution.

The inadequacy of present parametrization schemes and defects in the quality and density of observations produce an error that will eventually make further improvements in spatial resolution futile, as far as improvements in forecast accuracy are concerned. However, there seems to be neither convincing evidence nor broad agreement on whether these limits have already been violated with the model resolutions at present in use. Besides, with advances in meteorological theory, improvements in the quality and scope of satellite observations and adoption of more sophisticated model initialization techniques, these limits are constantly extended.

Some other limitations inherent in the primitive equations, such as the assumptions of hydrostatic balance and the incompressibility and inviscidity of air, will also produce a model error which will eventually dominate the forecast error. Their removal, although probably tedious from the programming point of view, is theoretically simpler, since all of these assumptions have been consciously adopted during a systematic simplification process starting from the Navier-Stokes equations and finishing in the primitive equations. This process can be backtracked to produce a new, more complete set of equations for atmospheric motion.

Anyway, as it is apparently going to take several years before the first operational weather models are implemented on a massively parallel computer, it seems reasonable to investigate the parallelizability of weather models, considering also large increases in spatial resolution. It is hoped that such investigations will, at the very least, remove one potential obstacle from the way to more accurate weather forecasts: that of inadequate computer capacity.

We restrict ourselves here to MIMD type parallel computers. SIMD computers have the benefits of being more easily programmable and, being simpler, more cost-effective. There are some drawbacks, however, which may offset these benefits. Since the primitive equations are non-linear, it is plausible that in the future adaptive numerical techniques and more complicated implicitly treated parts with locally varying coefficients will play an increasingly important role in their numerical treatment.

Adaptive techniques are rather difficult to implement on a SIMD computer which requires the problem to be far more rigidly parallelizable than when implemented on a MIMD parallel computer. The same applies to irregular geometries which, of course, are not a great problem in global atmospheric models as yet. However, a more accurate treatment of various discontinuities in the atmospheric flow may well result in the use of tracking techniques to locate the discontinuities. On such an occasion, a SIMD computer is rather unforgiving and easily suffers a severe loss of efficiency. The importance of more implicit time-stepping schemes will be discussed later.

Locally, however, the very concept of a differential equation is closely related to SIMD-type parallelism. Solutions to differential equations with smooth coefficients and data are typically locally smooth. They are then amenable to uniform local approximation. Uniform treatment is exactly what SIMD parallelism is all about.

Translated into a recommendation to computer architectures, the above remarks seem to indicate that there is an optimally efficient architecture which combines a global MIMD-structure with local SIMD-processing. Such a computer could take advantage of both the large-grain MIMD parallelism and the small-grain SIMD parallelism inherent in all physical models based on differential equations. Indeed, many of the more recent parallel machines take advantage of this balance, usually in the form of separate vector units attached to each independent processor.

It seems at the moment that even though new and improved automatic parallelizing compilers appear on the market constantly, a truly efficient parallelizing compiler for a massively parallel computer is very difficult to write. The basic reason for this is the excessive generality of Fortran (or most other programming languages) as a formal language. As has been widely recognized, parallelism is in no way naturally expressed in the constructs of the present standard Fortran. On the contrary, parallelism easily gets very delicately intervowen into program statements, requiring very sophisticated reasoning of the compiler to be detected. This makes the compiler difficult to write and expensive to run.

On the other hand languages like Occam, intended to make parallelism in programs as explicit as possible, do not seem to make programming, for instance, novel algorithms for simulation of the atmosphere, any easier. That is, they do not naturally express notions relevant to mathematical modelling of physical phenomena. Consequently, both techniques available to write a parallel weather model seem to lead to rather involved computer programming.

In the case of operational weather models, considerable performance benefits from massively parallel computers would probably justify rewriting the model to suit the architecture. Similar applications exist elsewhere, too. Such applications are often important models of physical phenomena. In most cases, the underlying physics is parallelizable at the molecular level. This is reflected in the mathematical structure of the models which are based on differential operators. Differential operators are local which means that the result of the operation of such an operator at any physical location only depends on the values of the forcing function in an arbitrarily small neighbourhood of that location. The boundary conditions complicate matters somewhat, but functional analysis tells us that in a local linearization, the effects of the boundary conditions can be separated from the effects of the principal operator.

It is on this physical and mathematical level that the parallelism of an atmospheric model is "natural". Programming languages are capable of describing immensely more general structures than physical models. Hence this natural parallelism is very easily lost when a model is translated into a computer program. Preserving the parallelism calls for very disciplined programming.

In the investigations below we limit ourselves strictly to the viewpoint of preserving the natural parallelism of physical models. The conclusions drawn are not relevant when thinking of such a model as a fixed computer program. Their significance lies in an analysis of the possibility of combining locality of operation, hence parallelizability, and the relevant physics. Programming is seen as a secondary activity that should serve to preserve the parallelism facilitated by the physics. In no case can programming contribute parallelism that would not be inherent in the physics - this would imply that the model is inconsistent. In this sense our viewpoint is very much "special purpose", aimed at gaining optimal benefit from parallel computers for a certain specific application: an operational weather model.

In the following analysis the focus is on the effects of parallelism alone. Hence we assume that the individual processors in our massively parallel computer have a certain fixed computational power. Since the model has to run within a fixed time slot, say five hours, this implies that there is a fixed upper limit to the number of floating point operations any individual processor can perform. When going to more massively parallel computers, the gain has to come from the parallelism alone. This isolation of the effects of parallelism on the numerical modelling of the atmosphere is our principal purpose.

## 3. Basic numerical features of weather models

The asymptotic serial computational complexity of a single forecast run of a weather model based on the primitive equations is determined by

$$C h^{-1}\left(n_v (h^{-2})^g (-\log h) + n_v^2 h^{-2}\right) \tag{1}$$

where

$h$ = spatial grid length

$n_v$ = number of vertical layers

$g$ = complexity exponent of the elliptic solver used for solving the implicitly treated part

$C$ = a generic constant

The complexity estimate (1) assumes that the linearized discrete equation is either inverted explicitly in the vertical, as in semi-implicit spectral models, or its vertical eigenmodes are calculated explicitly once and for all. This latter strategy decouples the primitive equations in the vertical and is normally adopted in semi-implicit grid point models. The inverse and eigenmode matrices thus obtained are full $n_v$ by $n_v$ matrices that remain constant throughout the computational domain. The last term in (1) corresponds to the complexity of multiplying the forcing terms by one or the other.

As long as $n_v$ remains small compared to $h^{-1}$, the direct inversion or eigenmode calculation is an efficient procedure. If the vertical grid spacing is refined with h to retain the present aspect ratio of computational cells, however, there will be a point after which the last term in (1) will become dominant. Then it will become more efficient to solve the vertical part of the equations explicitly each time. This will, in particular, be the case with spectral models as the vertical part in this case is a banded equation with a bandwidth depending only on the degree of approximation in the vertical. Band solvers will reduce the last term in (1) to $n_v h^2$. In the case of grid point methods, it seems necessary to resort to a fully three-dimensional solution algorithm. This will asymptotically result in the same complexity $O(n_v h^2)$ as with spectral methods, if an optimal elliptic solver is used, but the constant of proportionality in (1) will be larger than when band solvers can be used. Some optimal elliptic solvers will be briefly described below.

The estimate (1) also assumes that an explicit or a semi-implicit time-stepping scheme is used. The factor (- log h) in (1) is essential in the case of conventional semi-implicit methods, but can be dropped if an explicit method or an optimally efficient elliptic solver is used. In both cases, a form of the Courant-Friedrichs-Lewy stability condition has to be respected. This is reflected in the very first term $C h^{-1}$ in (1). In the case of an explicit method, the speed that determines the CFL limit is the speed of sound, since primitive equations do not filter the Lamb wave. In the case of semi-implicit methods, including spectral methods, the critical speed depends on the degree of implicity.

Since in a semi-implicit method the horizontal part of the divergence equation is always treated implicitly, the sound speed limit can be violated. If diffusion and the terms responsible for gravity waves are also treated implicitly, the limiting velocity is the speed of advection. Semi-implicit methods have consequently a threefold advantage over explicit methods in the longest time-step facilitated by stability, because fastest atmospheric wind speeds are roughly three times slower than sound. If part of the advection term is treated implicitly, as in the case of semi-Lagrangian techniques, a further relaxation in the CFL limit by a constant factor is achieved. The limiting speed is now the maximum rate of change of wind speeds. An analogous reasoning applies to the gravity wave terms if only the linearized part is treated implicitly, as is customary. However, the asymptotic complexity still remains the same, as expressed by (1). In fact, asymptotically, explicit methods will eventually beat any semi-implicit methods when the grid length h is allowed to approach zero. This follows from the factor (- log h) in the complexity estimate (1) for semi-implicit methods. It will eventually offset any constant-factor benefits resulting from a less stringent CFL condition, as long as some CFL criterion still has to be respected.

Only a nonlinear, fully implicit (in the sense that all dynamic terms are treated implicitly) time-stepping scheme incorporating a non-aliasing techique can totally ignore the CFL condition, as far as stability is concerned, which drops the factor $h^{-1}$ from (1). These have been too expensive to use so far, but optimal elliptic solvers may well change the situation. In all semi-implicit and implicit methods, n, two-dimensional or one three-dimensional, at least partially elliptic, system has to be solved each time-step. This is done implicitly in spectral methods by diagonalizing the system in the horizontal with explicit latitudinal Fourier and meridional Legendre transforms. In this case, the implicitly treated part is a Helmholtz equation. Normally, solving elliptic equations increases the complexity exponent g from one, which is the minimum, to some fractional number between one and three, depending on the type of elliptic solver used. In the case of spectral methods based on spherical harmonics $g = 3/2$, because of the complexity of the discrete Legendre transform.

Implicit treatment of a set of terms is equivalent to assuming a balance condition induced by these terms to hold instantaneously. This amounts to filtering any waves resulting from a physical delay in restoring the balance. Because the primitive equations are fundamentally hyperbolic, rather than parabolic, there are certain waves we definitely want to resolve. Consequently, we do not want to violate the Nyquist limit induced by these waves by using an excessively long time-step. If the time-step would be longer than half the period of these waves, they would alias into waves with a longer period. There is thus a meteorological upper limit to the length of the time-step, possibly of the order of a couple of hours.

Advances in numerical analysis during the last two decades have led to the discovery of several techniques to solve elliptic equations with optimal complexity, i. e. retaining $g = 1$. Some examples of such techniques are the multigrid method [Hack1] and some domain decomposition methods [Wid], which at the same time belong to the class of preconditioned conjugate gradient methods.

In the multigrid method, an elliptic equation is solved on a sequence of grids, in a defect correction fashion. The idea is that because conventional point-wise iterative methods, while they smooth out the high-frequency error quickly, converge slowly when the error also has low frequency components, these should be removed before applying any iteration. This can be done more effectively on a coarser grid, where the high frequency error cannot be represented at all. Hence, we alternate a coarse grid correction procedure and a local smoothing iteration on a finer grid, to arrive at a very fast solver. The idea is applied recursively and we finally have a logarithmically (in the number of degrees of freedom in the model) growing number of grids, which are traversed back and forth serially during the solution. The seriality in traversing between levels is dictated by stability. There is also a nonlinear version of the multigrid method, the so-called Full Approximation Scheme (FAS), which has the same optimal complexity in mildly nonlinear problems (i.e. problems that remain elliptic uniformly over the nonlinear parameter manifold) [Reus].

Domain decomposition methods use solutions on small subdomains as preconditioners to a global solution. Sometimes a relaxation procedure on the subdomain boundaries is also incorporated. The algorithm is naturally parallelizable by assigning the local solutions to different processors, but an optimal asymptotic complexity is only attained if this is combined with a multigrid-type hierarchy of subdomains.

On toroidal domains, also the spectral transform methods fall in the category of almost optimal elliptic solvers, since they then have the same optimal complexity as the methods introduced above, apart from the logarithmic factor. The efficiency of spectral transform methods relies upon the logarithmically growing complexity of the Fast Fourier Transform which, on a toroidal domain, can be used in all directions. The forcing function is transformed into spectral space, where applying a differential operator corresponds to multiplication of the transformed data by a polynomial. An inverse transform will subsequently restore the solution to the physical domain. Although spectral methods do not quite achieve the optimal complexity on a sphere like the atmosphere, this defect is compensated in present resolutions to some extent by superior accuracy.

When individual time steps alone are viewed, explicit methods are completely parallelizable. They do not require the gathering or dissemination of any global information and hence, on a grid-like parallel architecture, they will be able to utilize any number of processors. This is not possible with semi-implicit schemes.

The reason for this is that the parallel complexity of elliptic solvers necessarily grows at least logarithmically with the accuracy. On a compact manifold like the earth's atmosphere any strongly elliptic operator has a discrete spectrum. Consequently, the mapping from the forcing function to the solution can be represented by the spectral series

$$u(x) = \sum_{i=0}^{\infty} \lambda_i f_i v_i(x), \tag{2}$$

$$f_i = \int_{\Omega} f(x) v_i(x) dx \tag{3}$$

where

$u(x)$   = solution to the elliptic equation

$f(x)$   = forcing function

$\lambda_i$    = i th eigenvalue of the inverse of the elliptic operator

$v_i(x)$  = i th eigenfunction of the inverse of the elliptic operator

$\Omega$    = solution domain

Hence, at any point, the solution is the sum of an infinite series with quadratically decaying weights (in the case of a second order operator): the eigenvalues of the inverse operator. With increasing accuracy a rationally growing (the exact rate depending on the order of discretisation) number of terms has to be incorporated in the sum. With any fixed accuracy, the number of possible values for the sum grows linearly with the number of summands. Consequently, any program executed with finite speed and capable of separating all these different values takes a time at least logarithmically long in the number of summands. This is analogous to coding with a finite alphabet, and follows from the corresponding result for these alphabets in information theory. Elliptic solvers calculate this sum implicitly and will take a time that grows logarithmically with accuracy. This also applies to spectral methods that suffer the same penalty in the logarithmically growing parallel complexity of the Fast Fourier Transform.

This logarithmic parallel complexity, implying a corresponding parallel complexity for every single time step, means that eventually there will be a limit beyond which no further gain can be obtained from parallelism, when using semi-implicit or implicit methods. Because using faster individual processors allows an increase in parallelism, too, this is not a serious defect. In fact, the number of effectively utilizable processors grows exponentially with the speed of individual processors. As far as the total computational power in time-critical problems involving elliptic solvers is concerned, an extendible massively parallel computer is far less critically dependent on the speed of individual processors than a conventional vector supercomputer.

## 4. Requirements for parallel architectures

If we want to be able to use semi-implicit or implicit time-stepping schemes, parallel computers intended to run operational weather models will have to be able to effectively execute some of the optimal elliptic solvers described in the preceeding section. This requirement sets them a number of architectural requirements, too.

When dealing with problems of physical origin, we know that the interactions in the system to be simulated are fundamentally local. Hence, the most natural mapping between the data and the processors is one in which the computational domain is divided geometrically between processors. It is also the most communication-efficient way to pack the processors in three dimensional space. Hence, a grid interconnection topology seems a natural choice. In three dimensional simulations a three dimensional grid is required, but as long as most calculations in an atmospheric model are performed layer by layer, even a two dimensional grid will do. It may even be superior to a three dimensional grid because many of the subgrid scale processes, in particular convection, are vertical in nature, and for their calculation it is essential to store a whole column in a single processor, as long as it can fit there.

If parametrization is dealing exclusively with subgrid scale, and consequently local, phenomena, it will be trivial to parallelize as far as communication is concerned. The only remaining problem would be a possible load imbalance. This can be corrected to some extent by assigning domains of varying size to different processors. When taking the asymptotic view, load imbalance in parametrization is likely to be totally dominated by an inherent load imbalance in optimal elliptic solvers, and hence can be ignored. All in all, when a vertical storage scheme and only local parametrizations are used, it shouldn't be a problem to effectively parallelize the physics.

Elliptic equations have the property that local changes in the source function cause the solution to change at every other location, too. Hence, information must be able to traverse from any location to any other in logarithmic time, if we do not want data communications to become a bottleneck. Translated into a requirement for the parallel architecture this means that there will have to be a communication path with, at most, logarithmic (in p, the number of processors) length from any one processor to any other, the length being measured as the number of intermediate nodes on a communication link.

In explicit time-stepping the processors only need to exchange values on the grid points lying on the boundaries of the subdomains. Although with processors of fixed computational power these always constitute a fixed fraction of the grid points assigned to any individual processor, the long term implication would be that faster individual processors could afford slower communication bandwidth, in an area-to-perimeter (2 D grid topology) or a surface-to-volume (3 D grid topology) ratio. Hence, in the long run, communication bandwidth would not be a major problem.

However, it is in the nature of all known optimal elliptic solvers, whether of multigrid, domain decomposition or spectral transform type, that entire subproblems or global residuals will have to be communicated, even over long distances. The first conclusion from this is that the communication bandwidth of a successful parallel computer intended to run an operational weather model will have to be of the same order as its total computing power, hence O(p). A second conclusion is that local communication should be able to proceed in parallel on all the communication links.

A third conclusion might be that because of the globality of communication, the bandwidth will have to be even O(p log p). For example, hypercube and perfect shuffle topologies satisfy this requirement (though the latter only when a single row of processors is considered and p used to denote the number of processors in a row). As multigrid methods demonstrate, however, the volume of data that has to be communicated decays exponentially with communication distance. Hence only a logarithmic amount of data will need to be transported across the whole computer. The implication is that O(p) communication bandwidth is sufficient as long as the logarithmic maximum communication distance is fulfilled.

A remark regarding shared memory is in order. None of the fast elliptic solvers requires shared memory. All operations in multigrid and domain decomposition methods are local on each grid level. If the various levels are geometrically aligned with each other by communication links, there is no need for globally shared data, as the necessary global communication can propagate along these links. Even adaptive multigrid or domain decomposition, where we use local and global residuals to indicate a need for further grid refinement, can be implemented on a pyramidal communication topology, as demonstrated by Hoppe and Muehlenbein [HopMuh].

The FFT based methods can also be implemented without shared variables, as demonstrated e.g. by Swartztrauber [Swartz], Johnsson [John] and Chamberlain [Cham]. In this case, however, a shared memory might ease programming considerably. The communication bandwidth and even the topology required for distributed implementations of FFT is in any case the same as that provided by a banyan (perfect shuffle) memory switch.

A bus to shared memory is inherently serial asymptotically, however fast it might be. But if we should use banking and a banyan switch, memory access would only mean a logarithmically (in p) growing memory access latency and essentially no bandwidth problems. This would raise the parallel complexity of optimal solvers from $O(\log p)$ to at most $O((\log p)^2)$, if all memory is shared. As memory switches may be much faster (and much more expensive) than message passing between processors, the additional $\log p$ -factor might not show up until the number of processors has grown huge. Also an effective combination of shared and cache memories can make the factor almost invisible. Hence, because shared memory makes programming easier, it may even be beneficial, considering the numbers of processors likely to be seen in massively parallel computers in the near future.

In the preceeding paragraphs we have arrived at a number of recommendations for parallel architectures aimed at running operational weather models. In summary:

1) The interconnection topology should have a 2 D or a 3 D grid embedded in it.

2) The communication distance between any two processors should not grow faster than logarithmically with the number of processors.

3) Local communications must be able to proceed in parallel.

4) Total communication bandwidth must be of the same order as the total computing power.

5) Shared memory is not necessary. It makes the computer asymptotically slightly slower but may well be useful in modestly parallel computers.

6) Any fixed latencies that do not depend on the size of the message in distributing tasks to processors or starting up communication should be as small as possible. (The motivation to this comes from Amdahl's law, as presented in [Kau]).

Some interconnection topologies that fulfil all of these requirements are: crossbar switch based, hypercubes, perfect shuffle networks or pyramids. It is possible to embed a grid into a hypercube or a perfect shuffle network, although a good mapping is often quite nontrivial. Some others violate one or more of the requirements, for instance single bus-based computers, rings, pure grids or trees.

Some of the requirements, like 2), are controversial. Even if we can build computers with logarithmically growing connectivity, the maximum length of individual connections will necessarily grow as the cube root of the number of processors, because the processors will still have to reside in three dimensional space. Since the speed of communication between two processors in a computer with wires of constant thickness is inversely proportional to the length of the wire connecting them, requiring logarithmic connectivity implies a decrease in communication speed. Vitanyi [Vita] has analyzed this, including the case of wires of varying thickness.

Logarithmic connectivity may, however, remain somewhat irrelevant in the near future, due to developments in both processor and communication hardware and software. Hardware developments include glassfiber links with very large bandwidths. But even using conventional technology, dedicated message routing hardware with novel routing algorithms may have a drastic effect. As an example, imagine a hypercube with p = 1 048 576, i.e. $2^{20}$ processors. Every processors has 20 communication links, which is also the maximum communication distance between any two processors.

A 3 D toroidally connected grid with 1 000 000 processors, on the other hand, has a maximum communication distance of 150 (or 300, if we drop the toroidal connections, as they imply some long individual wires) between any two processors. If we now exchange the wormhole message routing strategy, which is apparently the fastest message routing strategy available to parallel computers with any topology, to the mad postman strategy, as described in [Jess], we gain a factor of eight in communication speed, if most of the processors are idle when long distance communication takes place. Because the mad postman strategy seems difficult to implement effectively on a hypercube, this reduces the effective maximum communication distance on the grid to 19, as compared to the hypercube, and there are still only six communication links per processor. On a non-toroidal grid the effective maximum communication distance is 38, but this gives the additional benefit of having local, and consequently fast, links only.

In multigrid algorithms running on a grid structured parallel computer all communication takes place between nearest active neighbours. Messages generally propagate along straight lines on the grid, crossing a maximum of 50 intermediate idle nodes on the computer described in the previous paragraph. With mad postman routing, this is reduced to the equivalent of crossing seven nodes.

## 5. Requirements for numerical techniques

In this final section we very briefly compare the benefits and drawbacks of various discretization techniques from the point of view of implementing them on a massively parallel computer. This also includes some remarks on their suitability for far higher spatial resolutions than those being used at present, as this is the principal reason to advocate massively parallel computers. Other more meteorological comparisons between some discretization techniques can be found e. g. in the discussion section of the proceedings of the workshop on Techniques for horizontal discretization in numerical weather prediction models, held at ECMWF in November 1987 [Horiz].

The biggest differences, as regards parallelization, are those between local and global discretization techniques. In spectral and pseudospectral methods the basis functions used in the Galerkin projections are global, whereas in finite difference, finite element and even in their variable order variants, like the h-p FEM [Babu], the spectral element method [KorPat] and the local spectral method [Anders], they remain local. Locality implies easy parallelizability and no need for shared variables, whereas global basis functions obviously benefit from shared memory: it is not necessary, though, even in this case, provided that the communication topology is built to provide a uniform logarithmic latency in accessing the local memories of other processors. This is the case with perfect shuffles and hypercubes.

Local methods, here collectively called grid point methods, are simple to parallelize due to their geometric coherence. Also, using optimal solvers in connection with them is fairly simple, provided the requirements of the last section are fulfilled by the parallel computer at hand. That is, we should, for instance, have a pyramidal or a hypercubic connection topology. In atmospheric wind speeds the elliptic equations used as implicitly treated parts remain strongly elliptic, which implies that multigrid type methods will attain their asymptotically optimal convergence rates. This has been recently proven even in the nonlinear FAS context by Reusken [Reus]. Strong ellipticity also implies that a quasiuniform mesh and a quasiuniform subdomain size distribution are adequate for achieving both load balance on any smoothing cycle of the multigrid algorithm and a quasiuniform error distribution after a constant number of multigrid cycles.

It is inherent, though, in multigrid type methods that the utilization of processors decreases proportionally to $1/\log p$. Every multigrid cycle requires visiting every grid level, and the parallel complexity of operations on each level is independent of the level. Only the finest levels can utilize all the processors. If we do not want the granularity to drop or use a rather fine coarsest mesh, because either may be detrimental to performance on the coarser levels, the number of processors utilized on each level decreases exponentially.

Various improved multigrid cycling strategies have been suggested e. g. by Frederickson and McBryan [FreMcB], Gannon and van Rosendale [GanRos], Chan and Tuminaro [ChaTum] and Hackbusch [Hack2]. They do keep the processors busy and improve the convergence rates, but they do not change the asymptotic complexity. The asymptotic parallel complexity of an optimal grid point method is $O(-\log h)$, if we parallelize in all three dimensions. Because parallelization in the vertical may cause communication problems in calculating convection, it may be wiser to parallelize only in the horizontal. Using $O(h^{-2})$ processors results in a $O(n_v (-\log h))$ parallel complexity. Whether we choose to use $p = O(h^{-2})$ or $p = O(n_v h^{-2})$ processors, with our assumption of fixed speed processors, $\log p$ and $-\log h$ are linearly related.

Spectral methods pose some problems in parallelization, but most of them can be overcome. Below, we describe a parallelization strategy for a pseudospectral discretization used in a global weather model using spherical harmonics as basis functions. The spectral transform is accomplished through a latitudinal Fourier transform and a longitudinal Legendre transform, layer by layer. As there is no guarantee that the strategy proposed is optimal, the conclusions based on it should be viewed as based on upper bounds to the parallel complexity of pseudospectral methods.

Thinking of massively parallel machines, as opposed to David Dent's analysis of a modestly parallel model [Dent], we cannot fit even a single latitude in the memory of an individual processor. Thus, the domain assigned to any individual processor may be a segment of a single latitude, a segment of a single longitude, or a rectangle somewhere between these two extremes. The relative computational complexity versus communication ratios of the Legendre and the Fast Fourier Transforms seem to indicate that the best rectangle is going to be fairly elongated in the latitudinal direction. In any case, corresponding locations on both the northern and the southern hemispheres should be allocated to the same processor, as they are processed sequentially in the Legendre transform phase in order to save a factor of two in computaional complexity. This is in accordance with the parallelization strategy suggested by Dent, Snelling and Carver in [Dent].

The algorithm commences with a latitudinal FFT. This can be accomplished in $O(\log p)$ time, where p is the number of processors assigned to that latitude, provided that we have at least a hypercubic connection topology between these processors. FFT is performed using the so-called binary reflected Gray code mapping, as described e.g. in Johnsson [John] and Chamberlain [Cham]. To restore the geometry, a data reorganisation has to be performed. This can be accomplished in an additional $O(\log p)$ steps, as explained in [John]. If we had had a perfect shuffle topology on each latitude, no reorganisation would have been necessary. Instead, the number of processors needed would have had to be larger by a factor $O(\log p)$.

If the data was allocated to processors on a latitudinal basis, the next step would be a transposition to longitudinal storage. This could be accomplished again in $O(\log p)$ steps using the recursive transposition algorithm [John]. This would require, however, that a full hypercube connectivity was built across the whole two dimensional grid. Hence, it seems that an optimal data allocation could be found by avoiding the transposition and experimentally searching for a latitude-longitude rectangle which produces minimum parallel execution time.

The Legendre transform, calculated by quadrature, is essentially a set of $O(n_v h^{-2})$ vector inner products, of length $h^{-1}$ each. Hence the serial complexity of all these together is $O(n_v h^{-3})$. If we only have $O(h^{-2})$ processors, the parallel complexity of the algorithm is necessarily $O(n_v h^{-1})$, which is inferior to grid point methods. If we were able to afford $O(h^{-3})$ processors (e.g. if data or model inaccuracies would preclude further increases in spatial resolution before we have exhausted our computer) we could, however, calculate the inner products in a tree-like fashion in $O(n_v \log p)$ time. We should thus end up with the same parallel complexity as with optimal grid point methods. This was achieved essentially by making use of the decay in utilization inherent in multigrid methods.

By analyzing the properties of the two discretization techniques mentioned, we can summarize the salient features of both, as far as parallelization is concerned. Spectral methods:

- A "natural" representation in global models - no boundaries.

- No excessive code rewriting - largely portable from present models.

- Optimal parallel complexity if $O(h^{-3})$ processors available, otherwise remains inferior to grid point methods.

- Suboptimal parallel complexity when moving to three dimensional implicit parts.

- Spectral (exponential) accuracy if prognostic fields are smooth (no discontinuous variables, like moisture or cloud liquid water, fronts and mountains are smeared etc.), and until limited by data or model errors.

- Necessary communication bandwidth $O(p \log p)$.

- Shared memory desirable.

- Adaptive grids difficult because tensor product basis necessary for efficiency. On the other hand, adaptive grids could prove difficult anyway, from the point of view of preserving phase speeds at grid interfaces.

- Only linear constant coefficient operators can act as implicit parts. Hence, CFL condition will have to be respected, albeit with a relaxed critical velocity limit.

The last property follows directly from the nature of harmonic analysis. By means of a spectral transform we can diagonalize a constant coefficient equation, but a variable coefficient equation, not to speak of a nonlinear one, does not admit a global spectral transform.

For optimal grid point methods the salient features are:

- Always optimal parallel complexity.

- Necessary communication bandwidth $O(p)$.

- 3 D, variable coefficient and even nonlinear implicit parts possible with the same asymptotic parallel complexity.

- Adaptive grids possible.

- Utilization decreases logarithmically.

- (Only) almost spectral accuracy possible, and even this may cause increased stability problems.

- Discontinuous and non-smooth prognostic fields not a major problem.

- Extensive code rewriting required.

- Artificial boundary conditions necessary for domain decomposition, not necessary for 'pure' multigrid.

Time accuracy has not been considered in the analysis. It would affect all semi-implicit methods similarly, and bring them closer to explicit methods in complexity. The conclusions above have been derived assuming a two-level time-stepping scheme is used, as the drive for maximum granularity tends to exhaust the memory available on a parallel computer. The effect of higher order time-marching schemes would possibly be to cut down the number of time-steps needed for accuracy, but if we have to do this at the expense of reduced spatial accuracy because of insufficient memory, we have gained nothing. Out-of-core methods have not been considered, though, and in theory they might provide some compensation, as there would be more time available for the storage and retrieval of atmospheric fields using a high-order time-stepping scheme.

## 6. References

[Anders]    Anderson, J.: A local, minimum aliasing method for use in non-linear numerical models. Submitted to Month. Weather Rev.

[Babu]      Babuska, I.: The p and h-p version of the finite element method, the state of the art. Tech. Note BN-1156, Inst. Phys. Sci. Tech., College Park, University of Maryland 1986.

[Cham]      Chamberlain, R. M.: Gray codes, Fast Fourier Transforms and hypercubes. Parallel Computing 6 (1988), pp. 225-233.

[ChaTum]    Chan, T., Tuminaro, R.: Design and implementation of parallel multigrid algorithms. Proceedings of the Third Copper Mountain Conference on Multigrid Methods. McCormick, S. (ed.), Marcel Dekker, New York 1987.

[Dent]      Dent, D.: The ECMWF model: past, present and future. Multiprocessing in meteorological models. Hoffmann, G.-R., Snelling, D. F. (eds.), Springer, Berlin 1988.

[FreMcB]    Frederickson, P., McBryan, O.: Parallel superconvergent multigrid. Proceedings of the Third Copper Mountain Conference on Multigrid Methods. McCormick, S. (ed.), Marcel Dekker, New York 1987.

[GanRos]    Gannon, D., van Rosendale, J.: On the structure of parallelism in a highly concurrent PDE solver. J. Paral. Distr. Comput. 3 (1986), pp. 106-135.

[Hack1]     Hackbusch, W.: Multigrid methods and applications. Springer, Berlin 1985.

[Hack2]     Hackbusch, W.: Talk at the Nordic Summer School on Numerical Analysis in Hanasaari, Espoo 1987.

[HopMuh]    Hoppe, H.-C., Muehlenbein, H.: Parallel adaptive full-multigrid methods on message-based multiprocessors. Parallel Comput. 3 (1986), pp. 269-287.

[Horiz]     Techniques for horizontal discretization in numerical weather prediction models. ECMWF workshop proceedings, ECMWF, Reading 1988.

[Jess]      Jesshope, C.: Operating systems and strategies for highly concurrent systems. These proceedings.

[John]      Johnsson, L.: Communication efficient linear algebra computations on hypercube architectures. J. Paral. Distr. Comput. 4 (1987), pp. 133-172.

[Kau]       Kauranne, T.: An introduction to parallel processing in meteorology. These proceedings.

[KorPat]    Korczak, K. Z., Patera, A. T.: An isoparametric spectral element method for solution of the Navier-Stokes equations in complex geometry. J. Comput. Phys. 62 (1986), pp. 361-382.

[Reus]      Reusken, A.: Convergence of the multilevel full approximation scheme including the V-cycle. Numer. Math. 53 (1988), pp. 663-686.

[Swartz]    Swartztrauber, P.: Multiprocessor FFTs. Parallel Comput. 5 (1987).

[Vita]      Vitanyi, P.: Non-sequential computation and laws of nature. Report CS-R8618, Centre for Mathematics and Computer Science, Amsterdam.

[Wid]       Widlund, O.: Optimal iterative refinement methods. TR 391, Department of Computer Science, Courant Institute of Mathematics, New York 1988.

# Meteorological Modelling on the DAP Series of Computers

C. S. VAN DEN BERGHE

Active Memory Technology Ltd., 65 Suttons Park Ave., Reading, RG6 1AZ, U.K.
(June 23, 1989)

### Abstract

The DAP is a massively parallel SIMD computer. The DAP 500 has 1024 processors and the DAP 600 4096 processors. The processors are connected to nearest neighbours through a 2-dimensional grid, to fast row and column highways to broadcast data and also to memory local to each processor. These connections make the DAP ideally suited to the types of computation found in Computational Fluid Dynamics and Meteorology. High speed connection to an external interface allows the real time visualisation of results, allowing numerical experimentation as a research and forecasting tool.

The DAP's architecture and programming style are explained with reference to a simple but typical Numerical Weather Prediction code (single layer barotropic model).

# 1    The DAP Computer

The architecture of the DAP family of computers has been discussed in numerous publications and this section is a brief overview. References [1,2,3] give more details.

The DAP family has two members, the 500 series and the 600 series. The members have the same architecture and software but differ in the scale of the implementation, the 600 series has four times as many processors as the 500 series.

## 1.1    The DAP Architecture

The DAP has four main components (as shown in Figure 1) :

- Processor element (PE) array and memory array

- Host Connection Unit

- Master Control Unit

- Fast data (I/O) channel

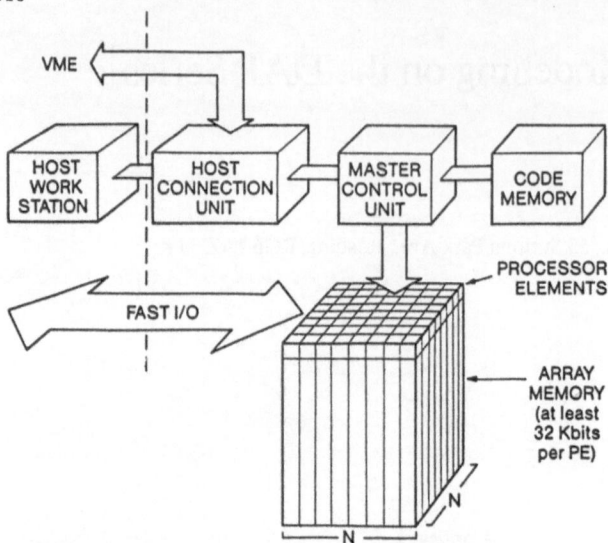

**FIG 1. DAP SYSTEM CONFIGURATION**

### 1.1.1   PE array and memory array

The single instruction multiple data architecture (SIMD, for an explanation of the classification of parallel computers see the book by Hockney and Jesshope [3]) of the DAP has 1024 (500 series) or 4096 (600 series) single bit processors arranged as a two-dimensional 32 by 32 (64 by 64) array.

Each PE is provided with connections to nearest neighbours. In addition, a bus system connects processors by rows and columns (see Figure 2). These row and column data paths provide rapid data broadcast or fetching facilities. These two forms of connection give the high level of inter-processor connectivity required by Numerical Weather Prediction and Computational Fluid Dynamics applications.

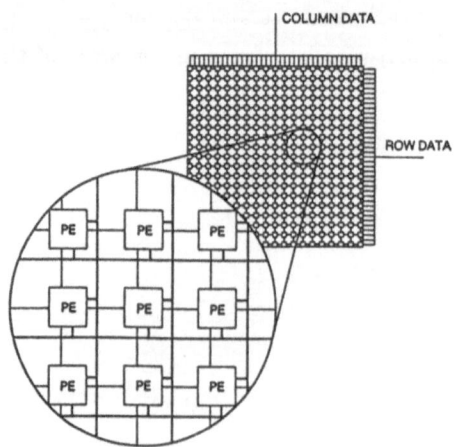

**FIGURE 2.  PE CONNECTIONS**

All PEs simultaneously execute the same instruction but an activity control register in each PE allows individual PEs or groups of PEs to be 'switched off'. The activity control bits can be generated from a predetermined pattern or from the results of a previous calculation. Activity control allows data dependent conditional processing to occur within an SIMD architecture.

The individual PEs operate in a bit serial manner, using 3 registers and logical operations between the data in these registers and the data in memory. More complex operations, such as those on floating point numbers, are coded as a sequence of operations on single bits (see below).

Each PE has a direct connection to its own memory. All PEs can transfer 1 bit to or from memory within a single machine cycle. This means that a programmer's model of the DAP memory is as a three-dimensional array of data, the PE array operating on a complete 'horizontal' plane of memory in one operation.

### 1.1.2 Host Connection Unit

The DAP is designed for connection to a host workstation which is used for program development, debugging, loading, initiating and high level control of DAP programs. The Host Connection Unit connects between the DAP and the host.

### 1.1.3 Master Control Unit

The detailed control of a DAP program is performed within the DAP by the Master Control Unit (MCU). Once a program has been initiated on the DAP the host computer can be freed for other tasks. The MCU is the source of instructions for the DAP - it takes instructions from the code memory, interprets them and controls the array of processors by broadcasting instructions to the PEs. The MCU is also a 32 bit scalar processor and performs operations on single data items (scalar operations) and well as program control (looping, jumps etc).

### 1.1.4 Fast data channel

The HCU may be used for medium speed data transfer operations such as those to or from the host workstation and filestore. However, fast data channels are provided for data intensive applications. For instance, on the DAP 510 data may be transferred to or from the array memory at up to 50 megabytes per second with the processing rate being slowed by only 4%.

Among the peripherals that are offered for connection to the fast data channel is a video output board to drive a high resolution colour display, enabling visualisation of the data used in an application while it is being processed by the DAP.

## 1.2 DAP Software

The DAP is usually programmed in an extended version of FORTRAN called FORTRAN-PLUS. The extensions to FORTRAN in FORTRAN-PLUS allow the programmer access to the features of the DAP that give high performance in application codes. FORTRAN-PLUS contains the concept of a data mode - collections of data that can be operated on with various degrees of parallelism. Matrix mode data is highly parallel and is distributed with one data item per PE : this is suitable for grid point data. Scalar mode data is operated on in the MCU, a single item at a time : this can be used for data which is global to a model. A third mode is Vector data, on which a number of PEs co-operate in calculating results for a single item in parallel with other data : this can be used to calculate boundary conditions for models. A large number of data types are supported by FORTRAN-PLUS, real data from 24 to 64 bits in steps of 8 bits, integer data from 8 to 64 bits in steps of 8 bits, logical (single bit) data and character data. The routines provided by FORTRAN-PLUS operate on the data in the most efficient form regarding both the mode of the data and its type. The language also includes expressions for shifting data from PE to PE and for a number of other common functions such as summing all the elements in an array.

An assembly level language is available. This is usually used only for high performance applications on non-standard data lengths, such as those occurring in some aspects of signal and image processing.

Floating point operations are built up from a large number of simple single bit operations (obviously provided in the system software and not the responsibility of a programmer). This means that optimisations can be made for particular operations. For instance the DAP 610 can add two sets of 32 bit floating point numbers at 40 megaflops. However, it can also find the square root of a set of 32 bit floating point numbers at 40 megaflops. The balance between these operations is very different from machines based on 32 or 64 bit calculations.

The tradeoff in the DAP software between data length and operation time means that typical control operations and conditional processing, which are usually expressed as manipulations of single bit quantities, are performed very quickly. Usually, with respect to floating point operations, the control of a program has minimal overhead on the execution time of a DAP program.

A further difference between the DAP and other distributed memory computers lies in the balance between the time taken to shift data between PEs and a typical floating point operation. The ratio between the two is typically 0.1 whereas on most other distributed memory computers it is many times greater than 1. The communications problem introduced by distributed memories and a great concern of many programs on these computers is thus alleviated on the DAP.

## 2 Barotropic vorticity equation

The Barotropic Vorticity equation is the simplest one parameter forecast equation [4] :

$$\frac{\delta}{\delta t} \nabla^2 \Psi = -V_\Psi \bullet \nabla(\nabla^2 \Psi + f) \tag{1}$$

where $\Psi$ is the streamfunction, $V$ the horizontal velocity and $f$ the Coriolis parameter.

This can be rewritten as :

$$\nabla^2 \chi + F(x, y, t) = 0 \tag{2}$$

$$F(x, y, t) = \frac{\delta \Psi}{\delta x} \frac{\delta}{\delta y} \nabla^2 \Psi - \frac{\delta \Psi}{\delta y} \frac{\delta}{\delta x} \nabla^2 \Psi + \beta \frac{\delta \Psi}{\delta x} \tag{3}$$

$$\chi = \frac{\delta \Psi}{\delta t} \tag{4}$$

These equations can be approximated using finite differences. A conservative form of Equation 3 is :

$$F_{m,n} =$$

$$\frac{1}{4d^2}((\Psi_{m+1,n+1} - \Psi_{m-1,n+1}) \nabla^2 \Psi_{m,n+1} - (\Psi_{m+1,n-1} - \Psi_{m-1,n-1}) \nabla^2 \Psi_{m,n-1}$$

$$-(\Psi_{m+1,n+1} - \Psi_{m+1,n-1}) \nabla^2 \Psi_{m+1,n} - (\Psi_{m-1,n+1} - \Psi_{m-1,n-1}) \nabla^2 \Psi_{m-1,n})$$

$$+ \frac{\beta}{2d}(\Psi_{m+1,n} - \Psi_{m-1,n}) \tag{5}$$

$$\nabla^2 \Psi_{m,n} = \frac{1}{d^2}(\Psi_{m+1,n} + \Psi_{m-1,n} + \Psi_{m,n+1} + \Psi_{m,n-1} + 4\Psi_{m,n}) \tag{6}$$

As is typical of most Numerical Weather Prediction codes, the solution of the Barotropic Vorticity equation has two major algorithmic sections :

- the calculation of values on the main finite difference grid (e.g. Equation 5)

- the solution of an elliptic equation (Equation 2)

As an example of an elliptic solver we use red-black ordered SOR [5, page 655] since this is the simplest method that raises interesting points about efficient algorithms for the DAP.

The obvious parallelism in this problem arises in the calculation of the values in Equation 5. Every grid point can be updated independently of every other and thus, in principle, the parallelism that can be exploited is of the order of the number of grid points. Every grid point needs data from itself and from it's 8 nearest neighbours. The DAP with a large number of processors and nearest neighbour connectivity is well suited to this part of the calculation. There is a high ratio between data access to neighbouring points and floating point calculations so that efficient communications are necessary.

The main computational load occurs within the elliptic solver. Red-black ordered SOR has the characteristic that on a single sweep only half the grid points are updated, the other half are updated on the following sweep. A naive implementation on the DAP would lead to only 50% efficiency since, for each sweep, half the processors are turned off (using the activity control register). However, it is possible to exploit the fact that typical simulations will use many more grid points than PEs to map the data onto the DAP so that it operates at 100% efficiency. The idea is to map the data using a so-called 'crinkled' mapping where neighbouring points are stored within the same PE memory [6]. The ordering of these points is such that a single memory plane contains all red or all black points and so these can be updated at full efficiency.

Despite the large parallelism in the problem there is a significant component of the work that is serial or involves single data items. The multiplicative factor $d$ in Equation 5 is a scalar factor which, to save space, should be stored as a scalar value and broadcast to the grid for each multiplication. The SOR procedure terminates when a convergence criterion is satisfied, typical convergence criteria are that the maximum change or the sum of changes are less than a tolerance: testing for these involves parallelism less than that of the grid.

This problem was implemented on a DAP 610. On a 128 by 128 grid with 50 SOR iterations per step, each time step took 0.006 seconds. This translates to a sustained calculation rate of 30 megaflops, a high proportion of the peak rate of 35 megaflops that the DAP 610 could achieve on this mix of operation types.

# 3 Discussion

The example discussed in the previous section shows that the DAP architecture is well matched to the types of operations typically performed in Numerical Weather Prediction calculations. These calculations are dominated by highly parallel calculations on a grid, although significant proportions of the total work must be done at a lesser degree of parallelism (e.g. scalar work). The DAP's architecture and languages match these aspects through the use of the scalar mode in FORTRAN-PLUS and the associated broadcast connections. A good choice of data mappping leads to high efficiencies on algorithms which, on first sight, are not suited to the DAP. Unlike most distributed memory architectures the overheads introduced by distributing data over many processors are small. The communications structure, nearest neighbour connection and broadcast row and column busses, match the data transfer requirements of finite difference models.

A further capability of the DAP, which is difficult to demonstrate here but which may influence the way Numerical Weather Prediction research is performed, is the data visualisation capability. As described in Section 2 the DAP can drive a video display with virtually no degradation in calculation performance. Consequently a simulation can be performed and the results viewed in *real time* and there is potential to interactively modify the parameters defining the simulation. Dynamical processes can be animated, leading to an easier and more complete understanding of the physics of the simulation.

# References

[1] D. Parkinson, D.J. Hunt, K.S. MacQueen,'The AMT DAP 500', Proceedings of the 33rd IEEE Computer Society International Conference, (1988).

[2] *DAP Series Technical Overview*, available from Active Memory Technology Ltd.

[3] R.W. Hockney and C.R. Jesshope, *Parallel Computers 2*, Adam Hilger (Bristol), (1988).

[4] J.R. Holton, *An introduction to Dynamic Meteorology*, Academic Press, (1979).

[5] W.H. Press *et al.*, *Numerical Recipes*, Cambridge University Press, (1986).

[6] S.F, Reddaway, 'Mapping Images onto Processor Array Hardware', *Parallel Architectures and Computer Vision*, ed. I. Page, Oxford University Press, pp 299-314, (1988).

# Execution of Scientific Algorithms on the Parallel Computer Parawell

H. ECKARDT

Siemens AG, Zentralabteilung Forschung und Entwicklung, Dept. ZFE F2 SYS 3,
Otto–Hahn-Ring 6, 8000 München 83, FRG

Summary:

The Parawell is a hierarchically structured parallel computer installed at Siemens Munich for investigation of parallel architectures, languages and algorithms. Some important scientific algorithms (Fourier transformation, Laplace equation, N-body problem, Mandelbrot set) have been implemented to study the performance behaviour of the Parawell. The speedup curves obtained are compared to a theoretical model. It results that the Parawell structure is well suited for many scientific applications. The total performance of the system currently used turns out to be comparable to a mainframe computer.

## 1. INTRODUCTION

The field of parallel processing is developing into an essential aspect of future computer systems. While in the 70's the continuously improving technology was the primary reason for the development of such computer systems, today there are additional arguments favouring the use of parallel computer architectures:

- High capacity demands in an expanding range of applications, both numerical and nonnumerical,

- Improved cost effectiveness by applying a large number of identical standard components instead of expensive, individual components whose technology can no longer be upgraded.

- Availability of fault tolerance and scalability of the system by using parallel concepts.

On the other hand, many problems connected with parallel processing still have to be solved in order to achieve a greater commercial break-through. Programming concepts must be found with which the inherent parallelism of an algorithm can be explicitly formulated or automatically recognized. Strategies for distributing tasks to be carried out in parallel to the individual processors available must be developed. For any particular problem the specific algorithms have to be found which are most efficient for the parallel hardware being used. A parallel computer concept can only be

Topics in Atmospheric and Oceanic Sciences
© Springer-Verlag Berlin Heidelberg 1990

successfully developed if these requirements are taken into consideration from the start.

The long range goal of the parallel processing project at the Siemens research laboratories in München-Perlach is the development of a parallel architecture concept which covers a range of commercially relevant fields of application. Towards this purpose the interdependencies between applications, algorithms, programming languages, execution models and architectural concepts must be studied. After the study of available parallel computer architectures (Klein et al. 1987) and of parallelism in the available languages (Kober 1988) practical experiments are now being run on a parallel computer which provides the testing equipment for the research areas mentioned above.

The Parawell system obtained for this purpose will be described in more detail in the following section. In the third part the programming model being at the user's disposal is described. The system performance characteristics are discussed in the fourth chapter. In the fifth part the measurement results are presented which were obtained by implementing some algorithmic kernels taken from important parallel computing applications in science and engineering.

## 2.  HARDWARE DESCRIPTION OF THE PARAWELL

The Parawell-1 parallel computer which is in use at Siemens AG in Munich since mid 1987 was developed by p1 Gmbh, a company in Munich, expanding on the ideas of the older Siemens parallel computer system SMS 201 (Kober 1977). The system is hierarchically designed and connected by busses and operates according to the master/slave principle. This means that a master node distributes tasks to its slaves. They in turn execute the tasks, which are independent of each other, and signal to the master when they have finished. The slaves themselves can also act as masters of their subordinate clusters (see Fig. 1). The existing system consists of a total of 37 processing elements.

Each master node is directly connected with its slave cluster by a 32 bit wide bus. By means of a dual port logic (DPL) it can transparently access the slave memory (see Fig. 2). A dynamically programmable address mapping unit assigns each slave memory to a virtual address range of the master. If all the slave memories are mapped to the same addresses this provides an efficient broadcast mechanism to the master. The assigning of adresses is schematically illustrated in Figure 3.

A Parawell node (Fig. 2) consists of a processor pair, Motorola 68020/68881 (12 MHz), 1-4 MByte dynamic RAM, a timer chip, an address mapping unit, a DPL and two 32-bit

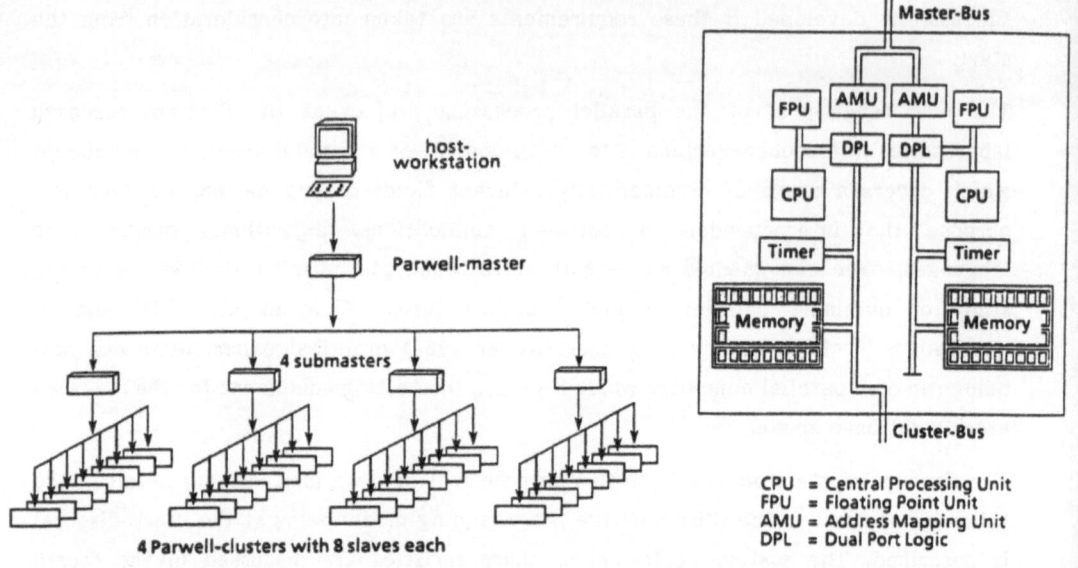

Fig. 1. Parwell-1 system structure

Fig. 2. Parwell processing element

CPU = Central Processing Unit
FPU = Floating Point Unit
AMU = Address Mapping Unit
DPL = Dual Port Logic

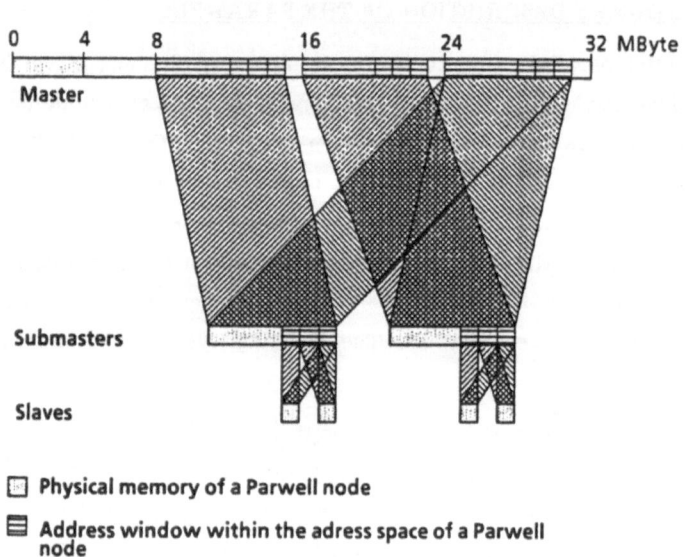

☐ Physical memory of a Parwell node

☐ Address window within the adress space of a Parwell node

Fig. 3. Address mapping in a Parwell system (3 levels)

busses which are symmetrical to each other. The input bus of a node is controlled solely by the master of the higher level. The output bus can be used to access the slaves of a cluster of the lower level. All the data exchange between two nodes must be actively executed by the common master node which has access to the memories of both partners. As a result, the communication bandwidth of the Parawell System is limited by the transfer speed of a master processor - not by the bandwidth of the bus - and can be improved by the right arrangement of clusters and increase in the number of hierarchy levels.

The Parawell master is connected by its input bus to a host work station (Siemens WS 30, identical to the Apollo DN 3000). The host handles the development of programs, system control (user-dialogue, software loading, monitoring) and execution of I/O requests. At all the nodes of the Parawell system a resident operating system core is active which is loaded when the system is initialized. It contains functions for managing processes and memory management, synchronization routines and I/O calls. They are accessed by means of a trap mechanism.

## 3.    PROGRAMMING MODELS

For the Parawell there are several user interfaces available: a hierarchical one which is adapted to the hardware architecture, a message passing interface and the parallel programming language Linda (Gelernter et al. 1984). The latter two will be focused on in Section 6. For the numerical algorithms studied, the hierarchical programming concept was used. From other studies (e.g. Detrich et al. 1988, Dongarra and Hiromoto 1984, Hiromoto 1986, Hey and Pritchard 1988) it is known that this concept permits executing numerical algorithms in parallel in a straight-forward manner. Multitasking by Cray (see for example Larson 1984 and 1988) is also based on this programming model.

User programs for the Parawell can be written in Modula-2 or Fortran and consist of two parts: a control program which runs on the master node, assigning the independent tasks to slave nodes and collecting the results, as well as a slave program which is loaded into all the slave nodes and contains the program segments to be processed in parallel. The cooperation between both programs is activated by calling the appropriate library routines linked to the programs or resident in the operating system kernel. Language extensions (except the POINTER statement used  in Fortran as described below) are therefore not required. Nevertheless introducing parallel statements to the language (for example as reported by Detrich et al. 1988) and using a precompiler could simplify the parallelization of programs.

In order to give an impression of a parallel program for the Parawell, take the example of a vector matrix multiplication of the form x=v$\underline{A}$ (vectors x and v, matrix $\underline{A}$). The corresponding sequential Fortran program reads:

```
      REAL A(20,20), X(20), V(20)
      DO 20 J=1,20
      RESULT = 0.0
      DO 10 I=1,20
10    RESULT = RESULT + V(I) * A(I,J)
      X(J) = RESULT
20    CONTINUE
```

In the parallel case, the sum over j is to be executed distributed over 20 slave nodes. The data is transferred to and from the slaves by a dynamically allocated memory area (in Fortran defined with the language extension POINTER), whose pointers to the start addresses of the data section are set by system calls. The global data are created in the broadcast area and are available to all the slaves after the master has initialised them. The local slave data (transfer area) can be accessed by the master by setting the appropriate pointer (using the routine SELECT). Data is thus communicated simply by means of assignments. The parallel Fortran program can be written in the following form:

```
C     MASTER PROGRAM
%INCLUDE 'PW_MASTER.INCL'
      POINTER /GLOBAL_PTR/ A(20,20), V(20)
      POINTER /TRANS_PTR/ ME, RESULT
      REAL X(20)
      ...
      DO 20 J=1,20
      CALL SELECT (J, JOBSTATE, TRANS_PTR)
20    ME=J
      CALL RUN_ALL ()
      CALL WAIT_ALL ()
      DO 30 J=1,20
      CALL SELECT (J, JOBSTATE, TRANS_PTR)
30    X(J) = RESULT
      ...
```

```
C      SLAVE PROGRAM
%INCLUDE 'PW_NODE.INCL'
       POINTER /GLOBAL_PTR/ A(20,20), V(20)
       POINTER /TRANS_PTR/ ME, RESULT
       RESULT = 0.0
       DO 10 I=1,20
10     RESULT = RESULT + V(I) * A(I,ME)
       RETURN
       END
```

Declarations being specific to the Parawell (including program names) can be inserted for master and slave programs as include files. The call RUN_ALL starts all the slaves and during WAIT_ALL the master waits until all the slaves have ended their program. Using SELECT, the TRANS_PTR is set at the value of the node j and the local value RESULT can be fetched back.

## 4.    SYSTEM PERFORMANCE MEASUREMENTS

In order to evaluate the capacity of a parallel computer, details on the performance of the system nodes and the communication mechanism are needed. But one can only discover how well these two operate together using sample programs as was done in Section 5.

For numerical user programs the floating point performance per single node is important. To measure the effectively uasble performance of a single processor three benchmarks were run: the Whetstone benchmark, the Linpack benchmark and the so-called Linpack kernel benchmark. The last one consists of implementing the vector instruction

$$V_3 := const. * V_1 + V_2$$

as an elementwise assignment within a loop. The results (see Table 1) depend largely on the compiler used. Since the Modula-2 compiler does not execute any global optimization, its performance values are only about half those achieved by the Fortran compiler. But even without optimization, performance of the Fortran code drops only by 20% when tested with the Whetstone benchmark. It is worth noting that the simple Linpack kernel benchmark results in nearly the same performance as the complete Linpack benchmark. Added up through all the nodes, the results for the Parawell configuration in use amount roughly to the computing power of a mainframe computer.

| | Modula-2 | Fortran |
|---|---|---|
| kilo-Whetstones/s | 419 | 910 |
| Linpack (kilo-FLOPS) | | 83 |
| Linpack kernel (kilo-FLOPS) | 55 | 86 |

Table 1. Node performance for three
benchmark programs

The maximum transfer rate of the Parawell bus is 6 MB/s. Since the data exchange must explicitly be programmed by a loop and due to delays caused by hardware (DPL), the effective transfer rate lies only between 0.5 and 1 MB/s.

## 5.    APPLICATIONS

In order to examine the speedup, the efficiency and the load distribution of the Parawell system in greater detail several algorithmic kernels from the most important application fields for parallel computers in physics and engineering were implemented. In these cases the individual tasks arising in the course of parallelizing are frequently of the same size. This permits making some general statements for the speedup.

### 5.1    General statements concerning the speedup

The most simple operating mode of the Parawell is the synchronous task processing, i.e. the master starts up all the slaves and waits until they are all finished (time $T_a$, see Fig. 4a). The subsequent communication phase (Time $T_c$ per slave) has to be executed sequentially by the master and thus requires time $k*T_c$ when k slaves are involved. The speedup (with n tasks) is obtained by

(1) $$s_k = T^1 / T^k = \{( T_a^1 + T_c^1 ) * n\} / \{(T_a^k + k*T_c^k ) * (n/k)\}.$$

The upper index of T relates to the number of slaves. Under the condition that the total communication effort is independent of k and neglecting all overheads, then:

(2) $$T_a^k = T_a^1 , \qquad T_c^k = T_c^1 .$$

Employed in eq. (1) the limit value for large k is

$$s_{k \to \infty} = T_a^1 / T_c^1 + 1 = const. ,$$

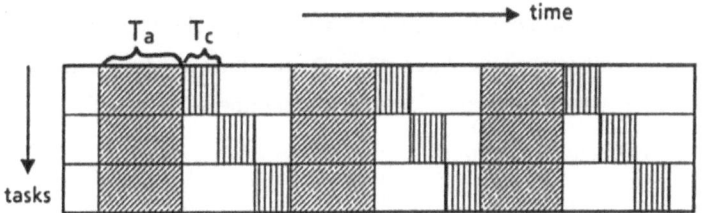

Fig. 4a. Synchronous task execution

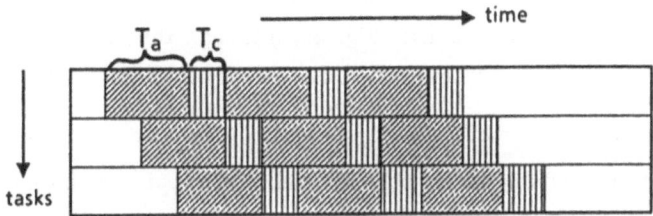

Fig. 4b. Asynchronous task execution

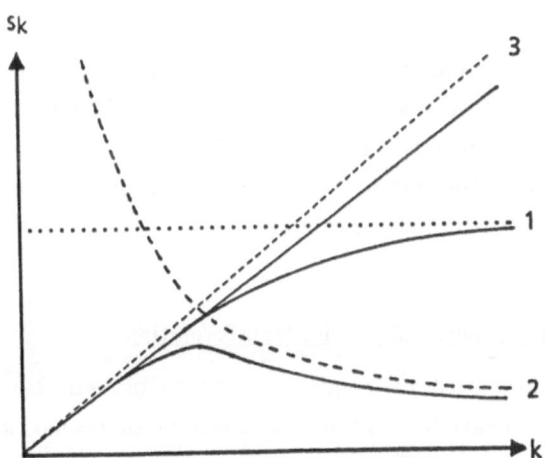

Fig. 5. Asymptotical speedup. 1: constant,
2: hyperbolical, 3: linear

i.e. a flattening out of the speedup curve to an asymptotical value (curve 1 in Fig. 5). In a more realistic situation we have to assume that an overhead is added to the communication time which is proportional to the number of slaves:

(3) $\qquad T_c{}^k = T_c{}^1 + k * \Delta T$ .

Then the speedup is weakened as much as

$$s_{k \to \infty} \sim 1 / k$$

(curve 2 in Fig. 5), i.e. the limit is zero.

Better use of the computer's resources is obtained if the communication phase is overlapped with the computing phase (Fig. 4b) although this makes the programming more difficult. The asynchronous execution mode is possible without "gaps" in time if:

(4) $\qquad T_a / T_c \geqq k - 1$ .

Neglecting the initial and end effects the speedup at this point is obtained by

$$s_k = T_1 / T_k = \{( T_a{}^1 + T_c{}^1 ) * n\} / \{(T_a{}^k + T_c{}^k ) * (n/k)\}.$$

Under close to ideal assumptions (2) the result is exactly

$$s_k = k,$$

thus, a linear speedup (curve 3 in Fig. 5). The more realistic assumption (3) however comes to

$$s_{k \to \infty} = \text{const.} \, ,$$

thus an inconvenient outcome similar to synchronous task processing (curve 1 in Fig. 5). For this reason one has to pay attention to the fact that the task size can be selected as large as to move within the area of the linear incline of Fig. 5. According to the equations above, this can be achieved by restructuring the algorithms in a way that the ratio $T_a/T_c$ is as large as possible.


## 5.2    Discrete two-dimensional Fourier transformation

The Fourier transformation is frequently used to process the data of images or measurements. This example shall be discussed in somewhat more detail since it provides an insight into the extent to which parallelizing is possible on various levels. Since we would like to use this as an instructional example we shall not be considering the fast Fourier transformation. Instead we shall deal directly with the original equations. The (complex) Fourier transform of a periodic function $F(x_{ij}) =: F_{ij}$ defined at discrete points $x_{ij} \in R^2$, i,j=0,...,n-1, is obtained by

$$\mathrm{Re}\ (G_{kl}) := 1/n^2 \sum_i \sum_j F_{ij}\ \cos\ (2\pi/n\ (ik+jl))$$

$$\mathrm{Im}\ (G_{kl}) := 1/n^2 \sum_i \sum_j F_{ij}\ \sin\ (2\pi/n\ (ik+jl))$$

with k,l=0,...,n-1. The programming of $G_{kl}$ leads in the sequential case to a four-fold loop nesting:

```
FOR l:=0 TO n-1 DO
   FOR k:=0 TO n-1 DO
      sum1:=0.0;
      sum2:=0.0;
      FOR j:=0 TO n-1 DO
         FOR i:=0 TO n-1 DO
            sum1:=sum1 + F [i,j] * Cos (...);
            sum2:=sum2 + F [i,j] * Sin (...);
         END;
      END;
      G1 [k,l]:=sum1 / FLOAT (n*n);
      G2 [k,l]:=sum2 / FLOAT (n*n);
   END;
END;
```

Obviously the part covered by the k- and l loop can be executed in parallel for all pairs k,l - or solely for all "l"s. These are the uppermost two levels of parallelism. If the i- and j loops are to be executed in parallel as well, the summation variables sum1 and sum2 have to be either exclusively accessed in each case or one has to calculate only the arithmetical expressions in parallel and to add sequentially afterwards, as done in the present case. When implementing on the Parawell the part of the above program segment to be executed in parallel must be transferred to the slave program, as demonstrated by the example in section 3. The calls for task start and communication take its place in the master program. The two variants discussed above have been implemented: the synchronous and asynchronous task execution.

The results for the speedup are shown in Fig. 6a-c. In order to prevent the computing time from increasing beyond the scale of minutes, a relatively low value for n was chosen (n=32). In the case of the greatest task size (l-parallel computation, Fig. 6a) an almost linear speedup is achieved and practically no differences between synchronous and asynchronous task distribution are discernible. This is due to the low number of tasks. When parallelism is increased (kl-parallel, Fig. 6b) the speedup is also very close to linear using asynchronous distribution, while it drops off somewhat using the

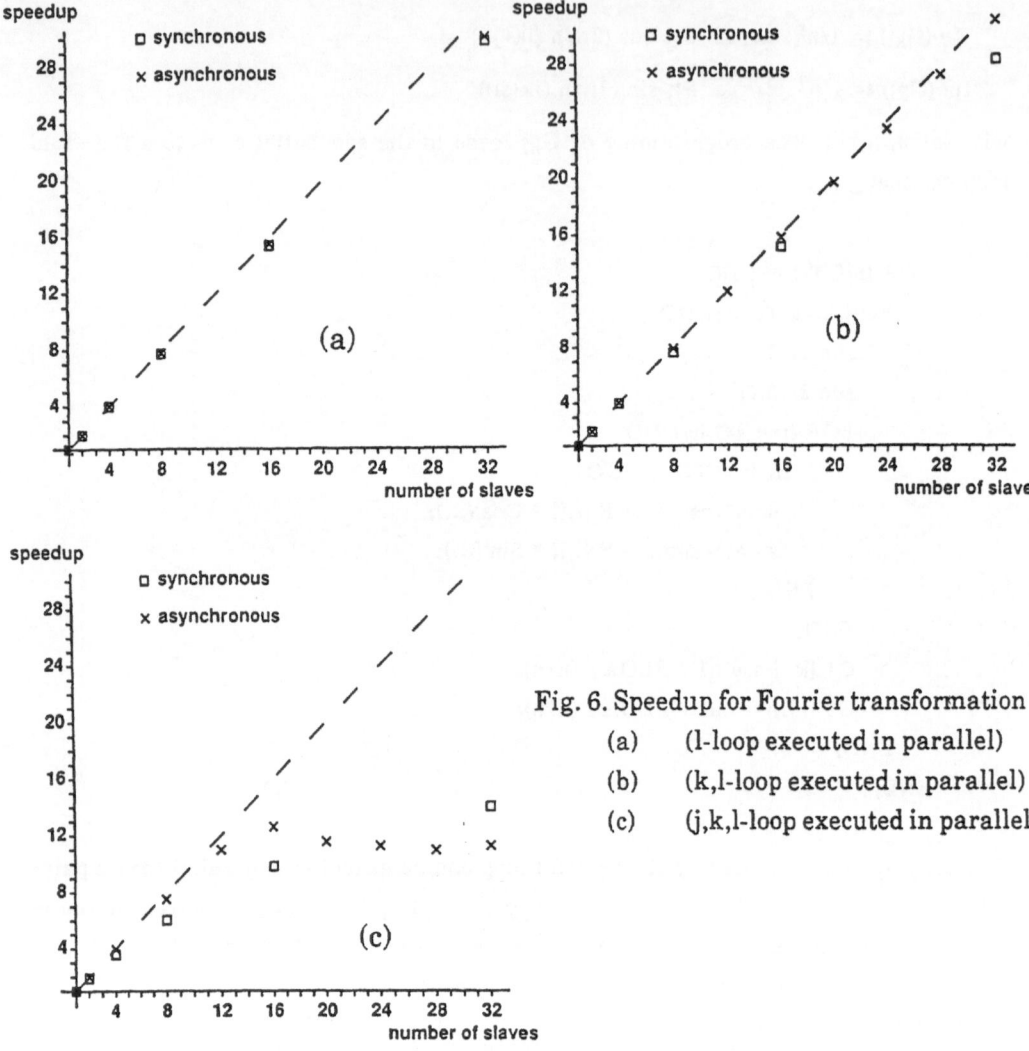

Fig. 6. Speedup for Fourier transformation
(a)      (l-loop executed in parallel)
(b)      (k,l-loop executed in parallel)
(c)      (j,k,l-loop executed in parallel)

synchronous method for 32 processes. In this case the non-overlapping execution of calculation and communication has an effect. If the level of parallelism is lowered by one more step (jkl-parallel) a further sequential addition has to be executed, as described above. This acts as a heavy brake on the speedup (Fig. 6c) from approximately 16 processors and upwards. In accordance with eq. (3) the additional communication overhead is remarkable which causes a hyperbolic drop in the speedup. Applying full ijkl-parallelism the effect would even more dramatically be visible. On the whole, this application shows that the Parawell needs the largest possible tasks in order to operate efficiently while the granularity of parallelism is more likely to be found in the lower two levels of this example when using vector computers.

## 5.3    Laplace equation

One of the most important partial differential equations is the Laplace equation. It states

$$\partial^2 u / \partial x^2 + \partial^2 u / \partial y^2 = 0$$

for a function u(x,y) in a two-dimensional area with fixed boundary values and is one of the basic equations in the theory of electricity and in thermodynamics. Since in this case the solving method used is also meant to be applicable to partial differential equations of a more complicated type (for example in fluid mechanics) the grid relaxation method was selected for it, more specifically, the checkerboard procedure (see Hockney and Jesshope 1981). As known from other parallel computer implementations, the definition area is discretized and mapped out into as many sections (strips) as there are processors available. Within the sections the iteration

$$u_{ij}{}^{(n+1)} := (u_{i+1,j}{}^{(n)} + u_{i-1,j}{}^{(n)} + u_{i,j+1}{}^{(n)} + u_{i,j-1}{}^{(n)}) / 4$$

for $u_{ij} := u(x_i, y_j)$ has to be carried out until convergence. Problems with the parallel implementation occur where these sections border each other and the data from the boundaries of neighbouring sections are required. In the course of computing therefore the function values of the boundaries must be communicated to the respective neighbouring sections. If this occurs after each iteration it is referred to as a synchronous relaxation. If the data is exchanged irregularly during processing so that data from different iterations may be combined in bordering strips, the relaxation is called asynchronous. (This should not be confused with synchronous or asynchronous execution of tasks on the Parawell!). An asynchronous iteration is quite difficult to implement using a master-slave programming concept. For this reason we shall limit our discussion to the synchronous procedure.

In contrast to the preceding example the time spent communicating is (for a given problem size) not independent of the number of sections but rather increases linearly when increasing the section number. In addition to the contribution ΔT caused by the operating system in eq. (3) there is an algorithmic contribution here too. For this reason when tasks are distributed synchronously a drop in the speedup can be expected into the hyperbolic region as illustrated in Fig. 5. The speedup measured (Fig. 7a-c for each one of the areas of $50^2$, $150^2$ and $300^2$ grid points) displays this behavior in the synchronous case, meaning a drop after attaining a maximum. Using asynchronous task creation the speedup increases - as in the ideal model - in an almost linear progression when the number of processors is low, but then breaks off drastically. The analysis of run times shows a sharp rise in the amount of communication time for processor numbers where the curve drops, so that the total computing time even starts

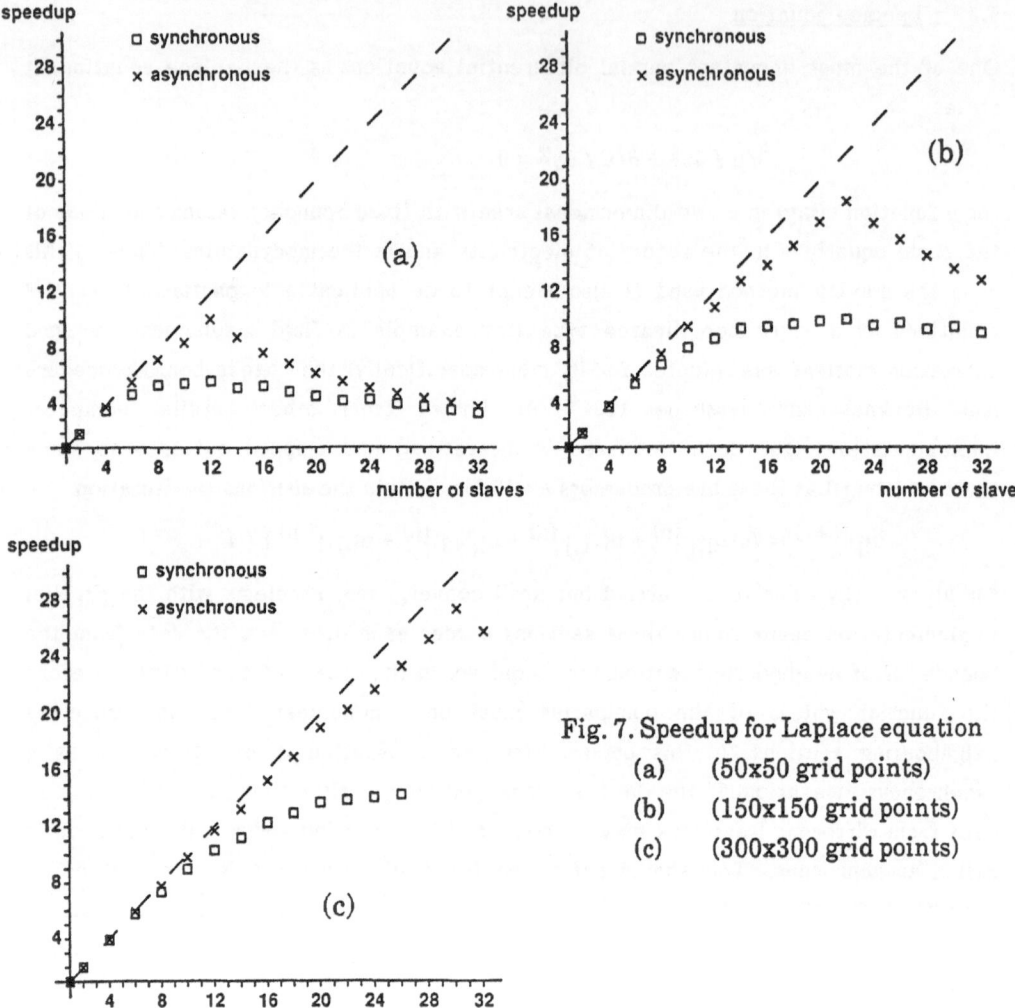

Fig. 7. Speedup for Laplace equation

(a) (50x50 grid points)

(b) (150x150 grid points)

(c) (300x300 grid points)

increasing again. This results in decreasing speedup. But if the discretization area is sufficiently fine-meshed the linear part can be exploited well as is the case for realistic applications.

## 5.4 N-body problem

In the field of molecular dynamics researchers are working on the simulation of a many-particle system over a period of time. Complicated physical processes can be simulated such as the diffusion of atoms or phase transitions in solids. The theoretical basis of this are the equations of motion in classical mechanics. This could deal with microscopic systems such as cluster formation of molecules on a surface, or with macroscopic systems such as the N-body problem of celestial mechanics. The advantage of the latter system is that no boundary effects need to be taken into consideration.

A body with a mass $m_j$ and the coordinates $x_j \in R^3$ is moving in the gravitational field

$$F_j = \sum_{i=1}^{N} \gamma \, m_i \, m_j \, (x_i - x_j) \, / \, |(x_i - x_j)|^3$$

of the other N-1 bodies according to the equation of motion:

$$m_j \, d^2x_j \, / \, dt^2 = F_j \, .$$

This system of ordinary differential equations can already no longer be solved analytically as soon as the number of bodies is three. In the most rudimentary (linear) approximation, the changes in position and velocity over a period of time can be calculated step-by-step using:

$$v_j^{(n+1)} = v_j^{(n)} + \Delta t * F_j^{(n)} \, / \, m_j \, ,$$

$$x_j^{(n+1)} = x_j^{(n)} + \Delta t * v_j^{(n+1)} \, ,$$

where the starting position and velocity have to be given. This is a procedure which must be executed sequentially for each time step. The calculation of $F_j$ is the most time-consuming part. The above equations can be computed in parallel for all j. To do this on the Parawell, all the slaves must have stored the coordinates of all the masses. Thus, after each iteration step, the new coordinates for each mass have to be communicated to all the slaves. The effort required (if N $\leq$ number of slaves) is proportional to $N^2$, as is the case for calculating $F_j$. But by using the broadcast mechanism (one slave $\rightarrow$ all slaves) the communication effort can be reduced to a factor proportional to N. If N exceeds the number of processors, this holds also for non-broadcasted data transfer, but the broadcast mechanism reduces the communication time by a constant factor. This being so, an increase in N improves the ratio of computation time to communication time. This can be seen directly from the speedups measured (Fig. 8, N=60 and N=210). Obviously the slave tasks are so large that all 32 processors of the Parawell can be efficiently put to work on this application.

## 5.5    Mandelbrot set

The calculation of the Mandelbrot set - in contrast to the examples discussed to this point - demands executing tasks of widely differing lengths. The Mandelbrot set is defined to be the set of all those points p in the complex number plane for which the sequence

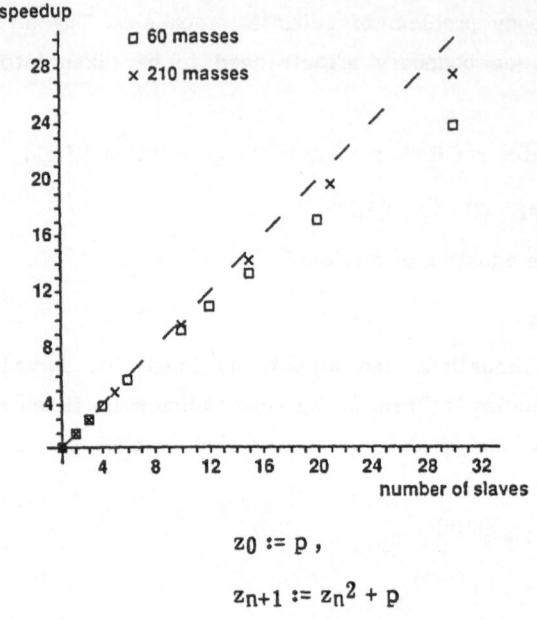

Fig. 8. Speedup for N-body problem

$$z_0 := p ,$$

$$z_{n+1} := z_n^2 + p$$

remains bounded. The non-divergent region shows up a fractal structure at its edges, which can be viewed as an example for deterministic chaos. In the computer implementation the membership of a point to the Mandelbrot set is determined when the sequence, after reaching a certain iteration number, has not diverged. Divergence is recognized by the fact that the absolut value of $z_n$ has surpassed a certain limit. The graphical output of the results on the screen is included in the speedup measurements. Convergent and divergent areas are displayed by means of different colors of the corresponding pixels. In the parallel implementation each individual task is given a pixel area on the screen, in the case studied a region of 9, 81 or 729 pixels.

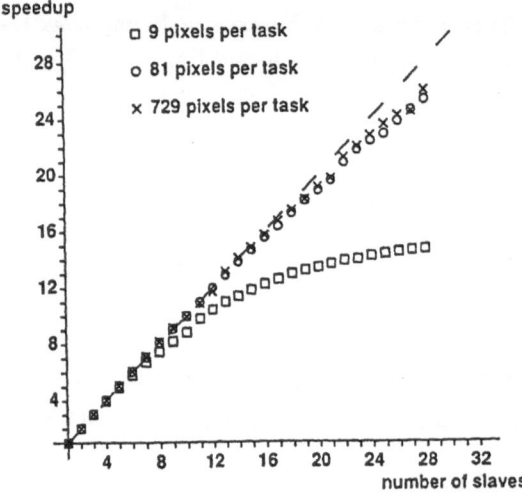

Fig. 9. Speedup for Mandelbrot set

The speedup curve (Fig. 9) shows a nearly linear increase for 81 and 729 pixels per task, whereas with each task getting 9 pixels the curve flattens out early. The measured ratio between slave computing time ($T_a$) and communication time ($T_c$) exploiting only one slave is shown in Table 2. In this case, output collisions from different slaves cannot occur. According to Fig. 9, for the two higher pixel numbers the ratio is sufficiently large in relation to the number of slaves. For 9 pixels per task the linear speedup is only given in the lower range of slave numbers. As the number of processors increases, the master can no longer manage the communication requests so that the slaves must remain idle waiting for output. While the computing effort per slave drops off as $1/k$, the number of communication collisions increases with $k^2$. Although the prerequisite of equal task size is not given here, the value of $T_a/T_c=12$ is comparable with the result of equation (4) according to which a linear speedup can be expected for $k \leq 11$, neglecting all overheads. From Fig. 9 it can be seen that the curve in fact does begin to deviate from its linear progress in the region of $k \approx 11$.

| pixels / task | $T_a$ / $T_c$ |
|:---:|:---:|
| 729 | 668 |
| 81 | 78 |
| 9 | 12 |

Table 2. $T_a$ /$T_c$ obtained by the Mandelbrot program

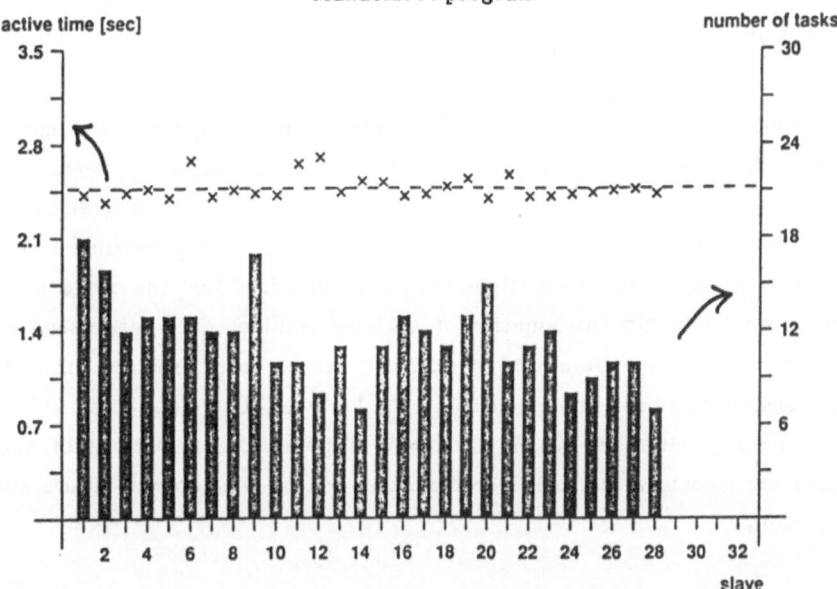

Fig. 10. Computing times and number of tasks per slave for the Mandelbrot set

In Fig. 10 the computing times of the slaves and the number of tasks they have executed (81 pixels per task) are listed. Although the number of tasks differs by approximately a factor of 2, the computing times are almost equal (The "break-outs" upwards are caused by the fact that towards the end of the computation not all the slaves can continue to be employed). The Parawell is clearly capable of dealing efficiently with tasks of varying size.

## 6.    CONCLUSIONS

The scientific algorithms studied in this paper for which versions for vector computers already exist, can be implemented well on a parallel computer. Already existing sequential programs must be rewritten for parallel processing. On the Parawell, synchronous task execution has proven to be relatively easy while for the asynchronous mode which is required for a theoretical linear speedup significantly more time and effort may be necessary. Thus, for example, when solving the Laplace equation asynchronously, several synchronization variables must be introduced.

The hierarchical system architecture of the Parawell leads in principle to a communication bottleneck but this does not come to pass during realistic applications. Instead of the speedup according to eq. (1) studied here we can also consider what is known as the "scaled speedup". Using this, one enlarges along with the number of processors the size of the problem at the same time, so that the computing work per task remains constant. But this does not avoid the arising of a communication bottleneck since this is caused by the central communication unit (master).

If the system is enlarged the bottleneck can be avoided by introducing additional levels of hierarchy. This requires in turn the use of other synchronization and communication concepts. These could be programmed explicitly for each application and the additional levels of hierarchy but this would make the writing of programs difficult to manage as a whole. A better solution would be to use mechanisms which are independent of specific applications (although specific for one system). Message passing protocols or the introduction of a global common data space such as in the parallel language Linda (Gelernter et al. 1985) can be considered for this. The latter concept has already been implemented on the Parawell (Borrmann et al. 1988). Also a message passing interface is at the user's disposal. A comparison of these two programming models with the hierarchical one will be the topic of future studies on the Parawell.

This work was supported in part within the ESPRIT project 1532. I would like to thank Mr E. Reyzl for permitting us to print the results of the Mandelbrot program. I am grateful to A. Klein, Dr. L. Borrmann, M. Herdieckerhoff and E. Reyzl for the many encouragements which contributed to the inception of this work.

## REFERENCES

Borrmann,L.; Herdieckerhoff,M.; Klein,A., 1988: Tuple space integrated into Modula2, implementation of the LINDA concept on a hierarchical multiprocessor. Conpar 1988, Manchester, to appear.

Detrich,J.H.; Falson,D.G.; Rosenzweig,L.J., 1988: ICAP/3090 at IBM Kingston: Evolution of software to support parallel execution. Third Int. Conf. on Supercomp., Boston.

Dongarra,J.; Hiromoto,R.E., 1984: A collection of parallel linear equations routines for the Denelcor HEP. Parall. Comp. 1, pp.133-142.

Gelernter,D. et al., 1985: Parallel Programming in Linda. Proc. 1985 Int. Conf. on Parallel Processing, Univ. Park, Pa., pp. 255-263.

Hey,A.J.G.; Pritchard,D.J., 1988: Parallel applications on the RTP Supernode machine. Conpar 1988, Manchester, Conf. paper B, pp. 160-167.

Hiromoto,R., 1986: Some issues in parallel processing as encountered on the Denelcor HEP. Parall. Comp. 3, pp.111-127.

Hockney, R.W.; Jesshope, C.R., 1981: Parallel Computers. Adam Hilger Ltd., Bristol.

Klein,A.; Eckardt,H.; Istavrinos,P., 1987: Parallelrechner-Architekturen: Eine Studie zum Stand der Technik. Siemens technical report.

Kober,R., 1977: The Multiprocessor System SMS 201 - Combining 128 Microprocessors to a Powerful Computer. Dig. of Papers, Compcon Fall 1977, pp. 225-229.

Kober,R. (ed.), 1988: Parallelrechner-Architekturen, Ansätze für imperative und deklarative Sprachen. Springer-Verlag, Berlin.

Larson,J.L., 1984: Multitasking on the Cray X-MP-2 multiprocessor. IEEE Computer 17, pp. 62-69.

Larson,J.L., 1988: Practical concerns in multitasking on the Cray X-MP. Multiprocessing in meteorological models (edts. G.-R. Hoffmann and D. F. Snelling), Springer-Verlag, Berlin, pp. 53-65.

# Numerical Solution of the Primitive Equations on the Connection Machine

JAMES J. TUCCILLO

NOAA/National Weather Service, Automation Division, National Meteorological Center, Washington, D.C. 20233, USA

## 1. INTRODUCTION

Since the late 1960's, the hydrostatic primitive equations have been the basis for operational Numerical Weather Prediction (NWP). These equations describe the time rate of change of the three dimensional atmospheric state variables: wind, temperature, moisture and pressure. Numerical time integration, from a set of initial conditions, for a period of 2 to 10 days provides guidance that has become indispensable to the operational forecaster. These numerical solutions require significant computer resources and operational weather centers have sought out the most advanced digital computers available. The most advanced systems, however, are often saturated shortly after installation as the NWP models increase in resolution and sophistication. The nature of the problem is such that a doubling of the spacial resolution, in 3 dimensions, increases the CPU requirements by a factor of 16 and the memory requirements by a factor of 8. The demand for increased memory and computational speed will most likely continue into the foreseeable future as modelers strive for increased accuracy through better spacial resolution and greater sophistication in the representation of physical processes.

The last 20 years has seen remarkable growth in the computational speed of computers and the size of random access memories. ( see Fig. 1 ). The development of vector processors capable of processing many operands in a pipelined manner has been a major development. The CRAY-1 and CDC CYBER 205 are the most popular examples of this architecture. As

Topics in Atmospheric and Oceanic Sciences
© Springer-Verlag Berlin Heidelberg 1990

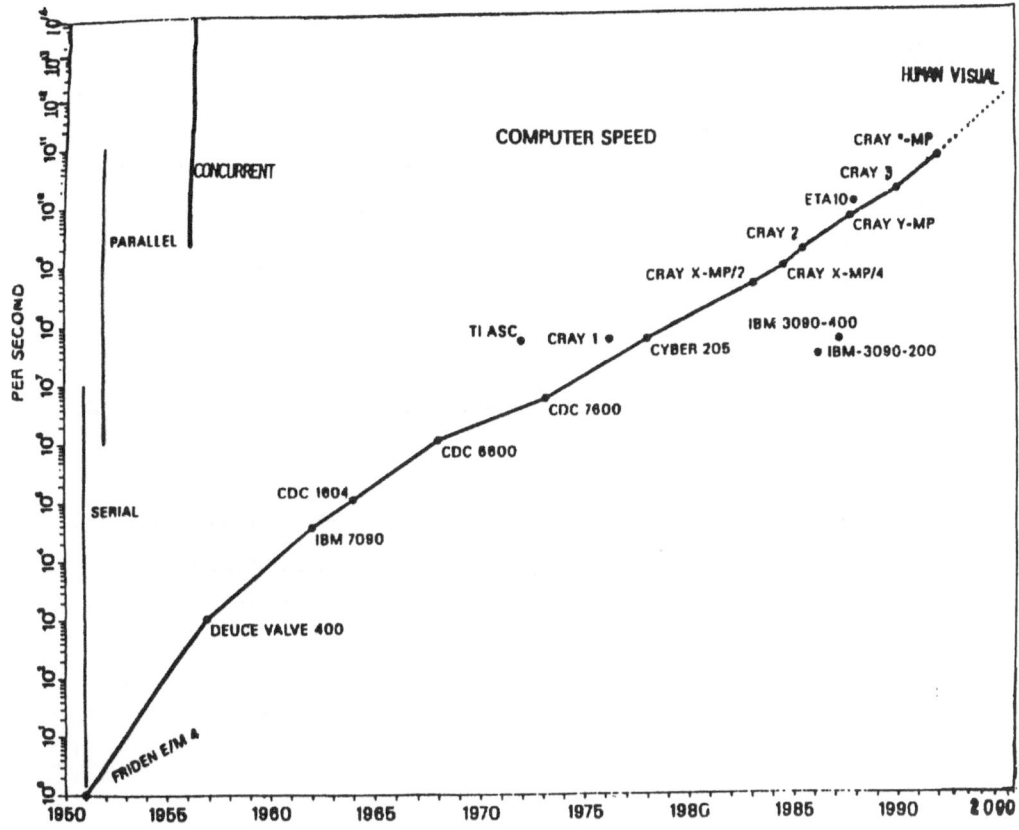

Fig. 1  History of peak supercomputer performance in Megaflops

pipelines machines approach a ceiling in performance do to a
limit on the speed of signal propagation in semiconductor
chips, the emphasis of the supercomputer industry has shifted
towards the development of parallel architectures. Many
computationally intensive tasks are inherently parallel and
architectures which can exploit that parallelism can be used
to solve problems once thought to be intractable.

Parallel processors have developed along at least three
paths. The first involves the interconnection of a few very
fast processors to a shared memory. Examples of this approach
include the current architectures of Cray Research and
ETA/CDC. The second path is represented by the so-called

Dataflow machines. In this architecture a detailed analysis
of the program is performed. Instructions are generated to
perform operations as operands become available thus avoiding
the von Neumann bottleneck. The Very Long Instruction Word (
VLIW ) architecture of Multiflow Computer is an example. The
third path involves the interconnection of thousands of
relatively slow processors. The Connection Machine ( CM ) of
Thinking Machines Corporation is an example of this type of
architecture.

This paper will be concerned with discussing the NWP problem
on the CM. Section 2 will present the formulation of an NWP
model including the finite difference operators and the
organization of calculations. In section 3 the CM will be
introduced and the hardware characteristics discussed.
Section 4 will describe how the NWP model was implemented on
the CM. Section 5 will present performance figures for the
model on the CM and several other architectures. Section 6
will be the summary and conclusions.


2.    MODEL FORMULATION

In this section a grid-point, primitive equation model will
be presented.


## 2.1    Primitive equations on a polar stereographic projection

The primitive equations on a polar stereographic projection
in the sigma coordinate ( Phillips, 1957 ) are presented
below.

$$\frac{\partial u}{\partial t} = - mu\frac{\partial u}{\partial x} - mv\frac{\partial u}{\partial y} - \dot{\sigma}\frac{\partial u}{\partial \sigma} - m C_p \theta_v\frac{\partial P}{\partial x} - m\frac{\partial \Phi}{\partial x} + fv + \frac{uy - vx}{2a^2} + \left(\frac{\partial u}{\partial t}\right)_{turbulence} \quad (1)$$

$$\frac{\partial v}{\partial t} = - mu\frac{\partial v}{\partial x} - mv\frac{\partial v}{\partial y} - \dot{\sigma}\frac{\partial v}{\partial \sigma} - m C_p \theta_v\frac{\partial P}{\partial y} - m\frac{\partial \Phi}{\partial y} - fu - \frac{uy - vx}{2a^2} + \left(\frac{\partial v}{\partial t}\right)_{turbulence} \quad (2)$$

$$\frac{\partial \theta}{\partial t} = - mu\frac{\partial \theta}{\partial x} - mv\frac{\partial \theta}{\partial y} - \dot{\sigma}\frac{\partial \theta}{\partial \sigma} + \left(\frac{\partial \theta}{\partial t}\right)_{turbulence} + \left(\frac{\partial \theta}{\partial t}\right)_{precipitation} + \left(\frac{\partial \theta}{\partial t}\right)_{radiation} \quad (3)$$

$$\frac{\partial q}{\partial t} = - mu\frac{\partial q}{\partial x} - mv\frac{\partial q}{\partial y} - \dot{\sigma}\frac{\partial q}{\partial \sigma} + \left(\frac{\partial q}{\partial t}\right)_{turbulence} + \left(\frac{\partial q}{\partial t}\right)_{precipitation} \quad (4)$$

$$\frac{\partial p_*}{\partial t} = - m^2 \int_0^1 \frac{\partial}{\partial x}\left(\frac{u\, p_*}{m}\right) + \frac{\partial}{\partial y}\left(\frac{v\, p_*}{m}\right) d\sigma \quad (5)$$

$$\frac{\partial \Phi}{\partial P} = - C_p \theta_v \quad (6)$$

$$p_*\frac{\dot{\sigma}}{\partial \sigma} = - m^2\left(\frac{\partial}{\partial x}\left(\frac{u\, p_*}{m}\right) + \frac{\partial}{\partial y}\left(\frac{v\, p_*}{m}\right)\right) - \frac{\partial p_*}{\partial t} \quad (7)$$

where

$$\sigma = \frac{p}{p_*}$$

$$P = \left(\frac{p}{1000}\right)^{0.286}$$

$$m = \frac{2}{1 + \sin\phi}$$

$$x = \frac{2\, a \cos\phi \cos\lambda}{1 + \sin\phi}$$

$$y = \frac{2\, a \cos\phi \sin\lambda}{1 + \sin\phi}$$

$$f = 2\,\Omega \sin\phi$$

and the other symbols have the usual meteorological meaning.

344

These equations describe the time rate of change of four
quantities, u, v, θ, q, which vary in 3-dimensions and one
quantity, p*, which varies in 2-dimensions. In terms of Fig.
2, u, v, θ, q, are defined over NX, NY, and NZ while p* is
defined over NX and NY.

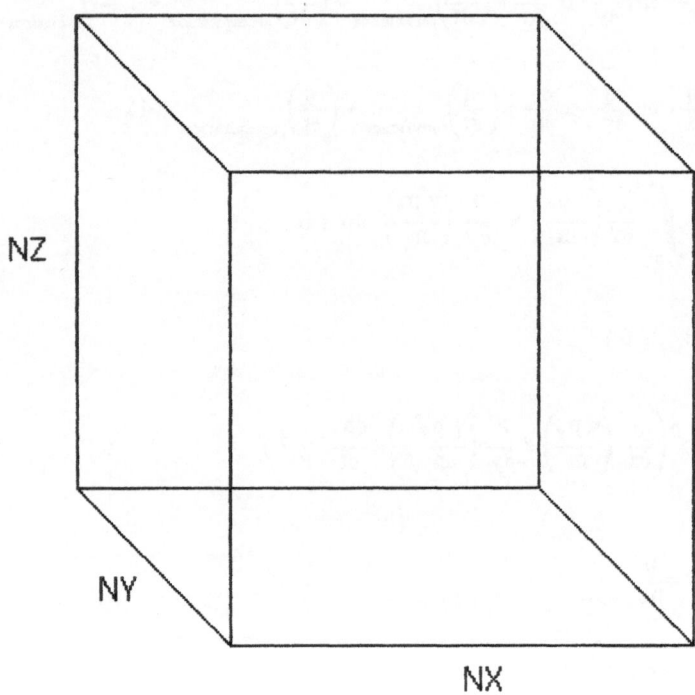

Fig. 2  Geometric definition of NX, NY and NZ.

## 2.2  Horizontal and Vertical Grid Structure

The horizontal grid is the "B" grid described by Arakawa (
1972 ) and illustrated in Fig. 3. This grid has very good
geostrophic adjustment properties and the placement of the
variables facilitates the programming effort. The vertical
structure is presented in Fig.4 and represents a standard
configuration found in many models.

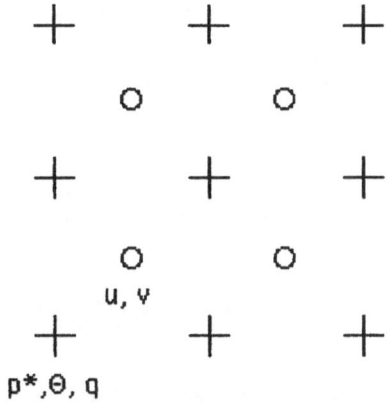

Fig. 3 The Arakawa "B" grid. The circle points represent the location of the u and v wind components and the plus points represent the location of p*, theta and specific humidity.

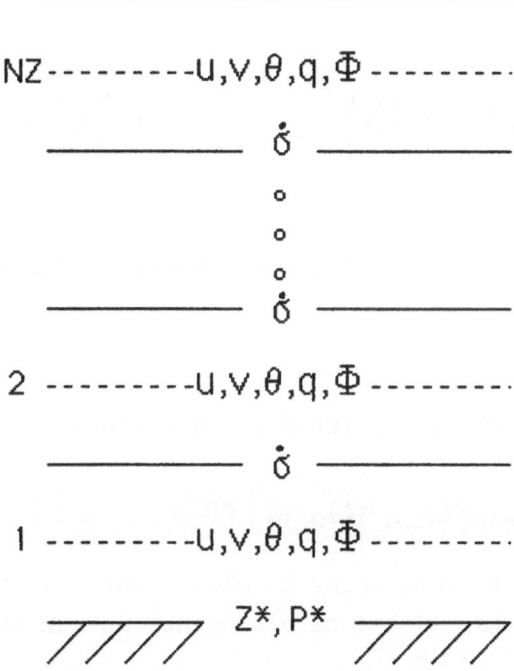

Fig. 4 Vertical sigma structure of the model. The solid lines represent the interfaces between sigma layers and the dashed lines represent the mid-points of the sigma layers.

## 2.3 Space differencing and averaging operators

The x and y horizontal derivative are approximated by the following formulas:

$$\frac{\partial()}{\partial x} \approx \overline{()_x}^{\,y} \equiv \left( ()_{i+1,j} - ()_{i,j} + ()_{i+1,j+1} - ()_{i,j+1} \right) * \left( \frac{0.5}{delx} \right)$$

$$\frac{\partial()}{\partial y} \approx \overline{()_y}^{\,x} \equiv \left( ()_{i,j} - ()_{i,j+1} + ()_{i+1,j} - ()_{i+1,j+1} \right) * \left( \frac{0.5}{dely} \right)$$

where the i and j indices refer to either the plus points for a derivative at the circle points or the circle points for a derivative at the plus points. Please note that i increases in the eastward direction and j increases in the southward direction.

The vertical advection terms are approximated by the following formula:

$$\dot{\sigma}\frac{\partial()}{\partial \sigma} \approx \left( \dot{\sigma}_{k+\frac{1}{2}} \left( \frac{()_{k+1} - ()_k}{\sigma_{k+1} - \sigma_k} \right) + \dot{\sigma}_{k-\frac{1}{2}} \left( \frac{()_k - ()_{k-1}}{\sigma_k - \sigma_{k+1}} \right) \right) * 0.5$$

where the k + 1/2 and k - 1/2 indices refer to the sigma layer interfaces and the k + 1 and k - 1 indices refer to the sigma layers.

The horizontal averaging operator is as follows:

$$\overline{()}^{\,xy} = \left( ()_{i,j} + ()_{i+1,j} + ()_{i,j+1} + ()_{i+1,j+1} \right) * 0.25$$

where the i and j indices refer to plus points for an average at the circle points and the circle points for an average at the plus points.

The formulation for the vertical integration of the vertical velocity and geopotential are standard and not presented here.

## 2.4  Time differencing

The time differencing scheme is the split-explicit method
described by Gadd ( 1978 ). In this scheme the terms
responsible for gravity-inertia oscillations are time
integrated seperately from the terms associated with
advection and physical processes. The advantage of this
approach is one of economy since the terms associated with
fast moving and meteorologically unimportant waves can be
solved with a small timestep for computational stability
while the remining terms are solved less frequently with a
longer time step. The net result is a considerable savings in
computer time over a scheme which solves all terms with a
timestep needed for stability of the gravity-inertia waves.
Details of the method and a stability analysis can be found
in Gadd's paper and the references within.

### 2.4.1  Gravity-Inertia terms

The gravity-inertia terms are integrated with the forward-
backward scheme as described by Gadd. The first step is a
forward time difference of the surface pressure tendency
equation.

$$p_*^{n+1} = p_*^{n} + F^n * \Delta t_{gw}$$

where F represents the RHS of (5).

Next the thermodynamic equation and moisture conservation
equations are forward time differenced with only the vertical
advection terms after the vertical velocity is computed with
(7).

$$\theta^{n+1} = \theta^{n} + VA^n * \Delta t_{gw}$$

$$q^{n+1} = q^{n} + VA^n * \Delta t_{gw}$$

where VA represents the vertical advection terms of (3) and
(4).

With the updated surface pressure, theta, and specific
humidity, the geopotential is vertically integrated using
(6). The equations of motion, (1) and (2), are then updated.

$$u^{n+1} = u^n + \left[ VA^n + PGF^{n+1} + MCT^n + f\left(\frac{v^n + v^{n+1}}{2}\right) \right] * \Delta t_{gw}$$

$$v^{n+1} = v^n + \left[ VA^n + PGF^{n+1} + MCT^n - f\left(\frac{u^n + u^{n+1}}{2}\right) \right] * \Delta t_{gw}$$

where VA represents the vertical advection, PGF represents
both pressure gradient terms, MCT represents the terms with
map coordinates and f is the coriolis parameter.

After two consecutive timesteps, the advective terms of (1),
(2), (3), and (4) are evaluated. This procedure is described
in section 2.4.2.

## 2.4.2   Horizontal advection terms

The horizontal advection terms are solved with the Lax-
Wendroff scheme ( Lax and Wendroff, 1960). In this scheme,
provisional values of the prognostic variables are computed
at time level n + 1/2. Using these provisional values, the
forcing is recomputed and the updated values of the
prognostic variables are obtained at time level n + 1. This
procedure is represented below.

$$u^{n+\frac{1}{2}} = \overline{\left(u^n\right)}^{xy} + HA^n * \frac{\Delta t_{adv}}{2}$$

$$v^{n+\frac{1}{2}} = \overline{\left(v^n\right)}^{xy} + HA^n * \frac{\Delta t_{adv}}{2}$$

$$\theta^{n+\frac{1}{2}} = \overline{\left(\theta^n\right)}^{xy} + HA^n * \frac{\Delta t_{adv}}{2}$$

$$q^{n+\frac{1}{2}} = \overline{\left(q^n\right)}^{xy} + HA^n * \frac{\Delta t_{adv}}{2}$$

$$u^{n+1} = u^n + HA^{n+\frac{1}{2}} * \Delta t_{adv}$$

$$v^{n+1} = v^n + HA^{n+\frac{1}{2}} * \Delta t_{adv}$$

$$\theta^{n+1} = \theta^n + HA^{n+\frac{1}{2}} * \Delta t_{adv}$$

$$q^{n+1} = q^n + HA^{n+\frac{1}{2}} * \Delta t_{adv}$$

where HA represents the horizontal advection terms of (1),
(2), (3), and (4). The n + 1/2 provisional values are
computed at the circle points for those prognostic variables
defined at the plus points. Similarly, the n + 1/2
provisional values are computed at the plus points for those
prognostic variables defined at the circle points.


2.5   Horizontal Boundary Conditions

For this study the model was configured with cyclic boundary
conditions in the x-direction and fixed boundary conditions
in the y-direction for all prognostic variables. In real data
applications tendencies from a larger scale model can be
applied to the boundaries in a manner similar to many
operational limited-area models.

## 2.6 <u>Sequence of calculations</u>

For each full timestep there are two gravity-inertia wave
timesteps and one advective timestep. The sequence of steps
to solve the equations for each full timestep are presented
in Fig. 5. For each step in the sequence the calculations
take place at all **NX** x **NY** horizontal grid points in parallel.

<u>GRAVITY-INERTIA WAVE TERMS</u>

FOR GW = 1 TO 2 DO

$$p*(n+1) = p*(n) + \ldots$$

FOR K = 1 TO NZ DO

$$\sigma(K) = \sigma(K-1) + \ldots$$

$$\theta(K)^{(n+1)} = \theta(K)^{(n)} + \ldots$$

$$q(K)^{(n+1)} = q(K)^{(n)} + \ldots$$

$$\Phi(K) = \Phi(K-1) + \ldots$$

$$u(K)^{(n+1)} = u(K)^{(n)} + \ldots$$

$$v(K)^{(n+1)} = v(K)^{(n)} + \ldots$$

END FOR

END FOR

<u>ADVECTIVE TERMS</u>

FOR K = 1 TO NZ

$$u(K)^{(n+1)} = u(K)^{(n)} + \ldots$$

$$v(K)^{(n+1)} = v(K)^{(n)} + \ldots$$

$$\theta(K)^{(n+1)} = \theta(K)^{(n)} + \ldots$$

$$q(K)^{(n+1)} = q(K)^{(n)} + \ldots$$

END FOR

Fig. 5 Sequence of calculations to advance the solution by
on full timestep. 'K' refers to the vertical layer and 'n'
indicates the time level.

The calculation of the vertical velocity and the geopotential couple each vertical layer to the one below it for the solution of the inertia-gravity wave terms. The horizontal advective terms computed in the second 'k' loop are not vertically coupled. The vertical coupling of the layers is of no consequence for the CM as the parallelism is across the horizontal domain. As we will see later, parallel processing of such a fine-grained algorithm is a problem on the CRAY Y-MP. The CRAY will function best if the code is setup so that multitasking is by vertical layer. The recursive propery of the vertical velocity and geopotential make a restructuring of the code necessary for optimum execution on the CRAY.

3.    THE CONNECTION MACHINE

The Connection Machine ( Hillis, 1986) is a single instruction/multiple data (SIMD) parallel computer with up to 65536 processors controlled by a conventional front-end computer (see Fig. 6). Each processor has 8K bytes of memory yielding a total memory capacity of 512 Megabytes or 128 million 32-bit floating point numbers. Problems requiring more physical processors than are available are supported through a virtual processor mechanism which is invisible to the user. Each physical processor can simulate several virtual processors with an associated decrease in memory and increase in execution time. For example, if each physical processor simulates two virtual processors then the execution time will double and the memory available for each virtual processor will be 4K bytes. Fig. 7 presents various virtual processor configurations and the associated memory.

Interprocessor communication is handled by a network built on a 12-dimensional hypercube. This hardware supports two mechanisms for communication. The router is the more general mechanism and allows for data to be sent from any processor directly to any other processor. The less general method is refered to as NEWS communication after the four directions on a two-dimensional grid: North, East, West and South. The NEWS mechanism allows for the efficient exchange of information between adjacent processors on a grid.

352

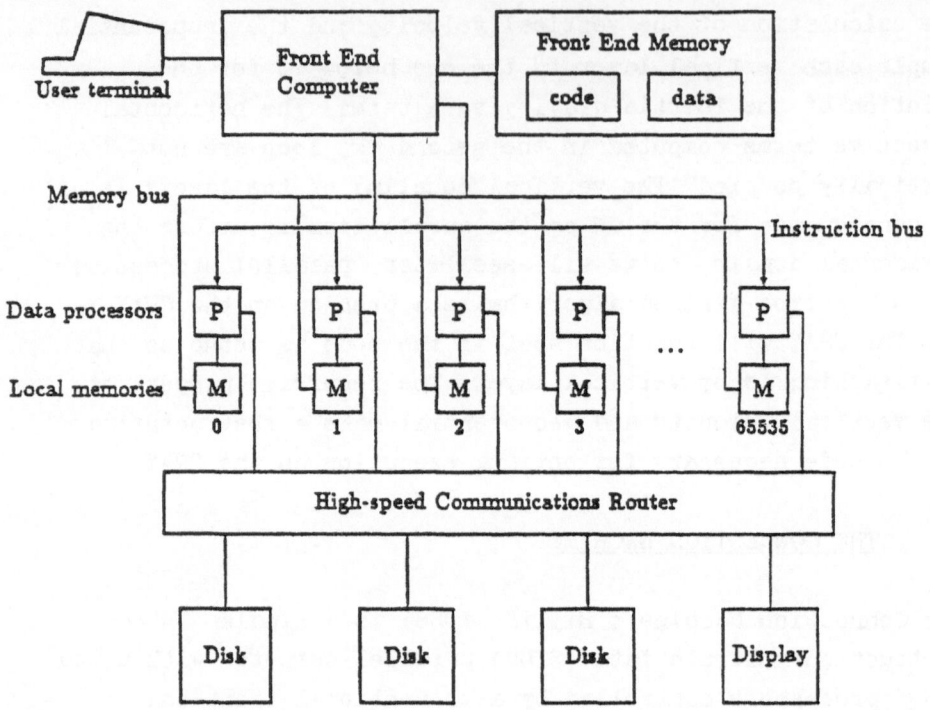

Fig. 6 Architecture of the Connection Machine. ( from
Thinking Machines Corporation Documentation of the Connection
Machine )

| Ratio n | Virtual processors | Memory each (CM-2) |
|---|---|---|
| 1 | 64K | 8K bytes |
| 2 | 128K | 4K bytes |
| 4 | 256K | 2K bytes |
| 8 | 512K | 1K bytes |
| 16 | 1M | 512 bytes |
| 32 | 2M | 256 bytes |
| 64 | 4M | 128 bytes |
| 128 | 8M | 64 bytes |
| 256 | 16M | 32 bytes |
| 512 | 32M | 16 bytes |
| 1K | 64M | 8 bytes |
| 2K | 128M | 4 bytes |
| 4K | 256M | 2 bytes |

Fig. 7 Various configurations of virtual processors and
memory on the Connection Machine. ( from Thinking Machines
Corporation Documentation of the Connection Machine )

# 4.    IMPLEMENTATION ON THE CONNECTION MACHINE

## 4.1    Data structure

The most straight forward implementation of the model on the
Connection Machine consist of assigning a virtual processor
to each column. The following C* code will define a data
structure called "state" which specifies the memory layout
within each processor and then creates a variable called
"points" which consists of NX by NY instances of the
structure.

```
#define   NX    256    /* number of grid points in x-dir */
#define   NY    256    /* number of grid points in y-dir */
#define   NZ     32    /* number of layers */

domain state {
        float u [ NZ ];    /* u wind                    */
        float v [ NZ ];    /* v wind                    */
        float t [ NZ ];    /* potential temperature */
        float q [ NZ ];    /* specific humidity      */
        float pstar   ;  /* surface pressure       */
        float zstar   ;}; /* terrain height          */

 domain state points [ NX * NY ];
```

The u and v wind components are staggered one half grid
distance in both the x and y directions from the potential
temperature points on the "B" grid. In terms of the processor
where they are stored, the wind components are colocated with
the mass point to the "northwest".

## 4.2   Finite difference and averaging operators

Interprocessor communication is slow compared to the floating
point performance for data within a processor. MACROS to
retrieve data from neighboring processors have been coded in
the Parallel Instruction Set ( PARIS ) of the Connection
Machine using the NEWS communication. These MACROS are called
XP1, XM1, YP1, and YM1 for x plus 1, x minus 1, etc. Using
these MACROS, the basic horizontal coupling operators can be
efficiently computed.

The C* code to compute the x-derivative, y-derivative and
four point average of u wind at the mass points in parallel
for some layer k is presented below.

```
temp1 = YM1 ( u [ k ] );
temp2 = temp1 + u [ k ];
temp3 = XM1 ( temp2 );

/* u bar */
ubar = ( temp2 + temp3 ) * 0.25;

/* dx of u */
dxu = ( temp2 - temp3 ) * 0.5;

/* dy of u */
temp3 = temp1 - u [ k];
dyu = ( temp3 + XM1 ( temp3 ) ) * 0.5;
```

The orientation of the points relative to each other for the
above code segment is shown in Fig. 8.

It can be shown that this code segment minimizes the amount
of interprocessor communication. Similar code exists to
compute the other horizontal coupling terms. The update of
the prognostic variables, once all the terms are computed,
will occur in the processor memories at the nominal
performance of the machine.

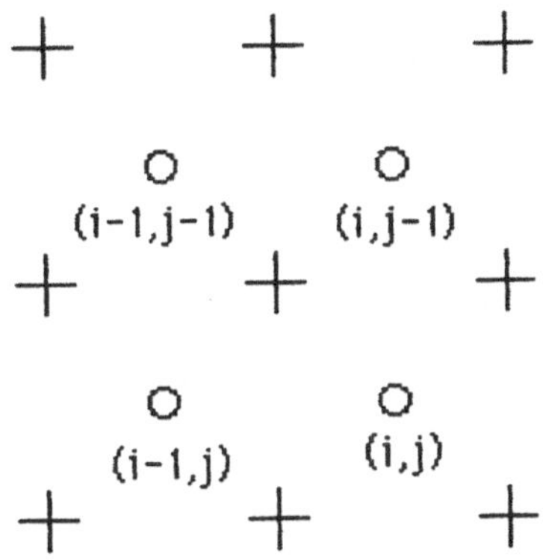

Fig. 8 Orientation of points for finite difference and averaging operators. I is the index in the x-direction and J is the index in the y-direction.

It is interesting to note that approximately 40% of the running time of the model is spent doing interprocessor communication. In other words, if data could be accesses from adjacent processors at the same speed as data within a processor the code would run 40% faster.

5.    PERFORMANCE FIGURES

Comparing the performance of algorithms on several different computer systems can be a difficult because the organization of the code may favor one system over the other. The requirement of contiguous long vectors for "good" vector performance on the CDC CYBER 205 is a well known example of this problem. For this study, every attempt was made to be fair to all systems. In this section, the performance of the Connection Machine and several other computer systems will be presented.

5.1    Model performance on different architectures

An ANSI 77 FORTRAN version of the model was designed which should execute efficiently on most systems. The code was

organized so that vectorizing compilers for vector
architectures would see a vector length equal to the
horizontal domain. The loops over the horizontal domain,
however, contains many instructions so that non-vector
architectures will see enough computational work to allow for
"instruction scheduling". The CYBER 205 compiler was able to
vectorize every horizontal loop. Chaining or linked triads
were encouraged through the liberal use of parenthesis to
help the compilers identify opportunities for these time
saving instructions.

All timings of the FORTRAN version of the code were done for
a 50 x 50 horizontal grid with 32 vertical layers. The
minimum grid distance on the image plane was set at 40 kms
and the appropriate time step for computational stability was
used. The forecast length was set for 24 hours or one
forecast day. This configuration represents a limited area
domain of about 2000 kms on a side. This domain size is
unrealistically small and was chosen to so that CPU timings
from a variety of systems with much different performance
characteristics could be obtained. A realistic configuration
would be a 256 x 256 horizontal grid corresponding to a
domain of approximately 10000 km on a side. Since the amount
of computational work is linear with the number of grid
points, the timings obtained with the 50 x 50 grid can be
scaled to arrive at timings for a 256 x 256 grid. For the
most restrictive architecture, the CYBER 205, this is valid
because a vector length of 2500 ( 50 X 50 ) is long enough
for vector efficiency of over 90%. I am assuming that
sufficient memory is available to hold the larger domain and
that memory conflicts are not significantly changed. Assuming
64-bit floating point precision, this problem will need about
80 Megabytes of memory for a 256 x 256 grid with 32 layers.
The introduction of physical parameterization for turbulence,
radiation, and precipitation will increase the memory
requirements.

Table 1 shows the CPU timings for various systems. The NAS
9050 is an IBM 370 plug compatible system featuring scalar
processing with a 38ns clock. It performance is comparable to
the CDC 205 using the scalar processor only. The Very Long
Instruction Word Multiflow system does very well compared to
the NAS and 205 when you consider that its price tag in about
$500K. The 205, using its vector processor, achieves a
speedup of about 10 to 1 for 64 bit and 20 to 1 for 32 bit
over the scalar processor. These are typical values for very
well vectorized code. The speed difference between the CRAY X-
MP and Y-MP reflects, almost exactly, the difference in the
cycle time ( 8.5ns for the X-MP vs.6.2ns for an early YMP ).

| SYSTEM | PRECISION | COMPILER | CPUTIME |
|--------|-----------|----------|---------|
| NAS 9050 | 32 bit | IBM VS-FORTRAN | 55417 s |
| CDC 205(scalar) | 32 bit | CDC FTN200 | 46784 s |
| CDC 205(scalar) | 64 bit | CDC FTN200 | 40186 s |
| Multiflow 14/200 | 32 bit | TRACE FORTRAN | 34524 s |
| CDC 205(vector) | 64 bit | CDC FTN200 | 3382 s |
| CRAY X-MP(1 proc) | 64 bit | CFT77 | 2439 s |
| CDC 205(vector) | 32 bit | CDC FTN200 | 1861 s |
| CRAY Y-MP(1 proc) | 64 bit | CFT77 | 1730 s |

TABLE 1. CPU times for the FORTRAN version of the model. All
times were computed for a 50 x 50 horizontal grid and scaled
to a 256 x 256 grid. The forecast length is 24 hours, there
are 32 vertical layers and the grid distance is 40 kms.

## 5.2    Model performance on the Connection Machine

CPU timings for the C* version of the model on the Connection
machine are presented in table 2. The domain size considered
is again 256 x 256 with 32 layers. The numbers presented are
a combination of measured and computed results using the 16K
Connection Machine at NRL. Since this problem is completely
parallel and the Connection Machine scales linearly, the
computed results are accurate.

The best wall time performance is achieved when one processor
is assigned to each vertical column, a VP ratio of 1. A 64K
processor machine is required and it will solve the problem
in a little less than half of the time required by the CRAY
Y-MP using one processor. The best use of the processors in
terms of speed per physical processor is obtained when
several virtual processors are assigned to each physical
processor. Pipelining of instructions across several virtual
processors within a physical processor results in a better
utilization of the hardware as shown by the CPU/VP ratios in
table 2. A VP ratio of 4 is probably best because anything
higher results in too little memory per processor to solve
the problem once physical parameterizations are added. For a
64K machine, a 512 X 512 grid would yield a VP ratio of 4 and
would take 2438 secs to solve.

| PHYSICAL PROCESSORS | VP RATIO | CPUTIME | CPU/VP RATIO |
|---------------------|----------|---------|--------------|
| 64K                 | 1        | 768 s   | 768 s        |
| 16K                 | 4        | 2438 s  | 610 s        |
| 8K                  | 8        | 4656 s  | 582 s        |

TABLE 2. CPU times for the C* version of the model with a
256 x 256 grid, 40km grid spacing, 32 layers and a forecast
length of 24 hours for different numbers of virtual
processors per physical processors. The CPU time normalized
by the VP ratio is also presented.

## 5.3    Multitasking on the CRAY Y-MP

The FORTRAN version of the model was also run on the CRAY
Y-MP using multiple processors. The code was multitasked
using the recently available autotasking software. This
software features a preprocessing step which analyzes the
data dependencies within the code and inserts microtasking
compiler directives into the source prior to the actual
compilation process. The processed source is available for
inspection and may be further modified prior to compilation.
This preprocessing step automates what was previously a
manual task. As with all source preprocessors, one should
expect to go through several iterations before obtaining an
optimized version of the code.

The source output from the autotasking software was modified
by including additional compiler directives. The FORTRAN
statements remained unchanged. The timing results on the CRAY
Y-MP using 4 and 8 processors is shown in Table 3. The
results, indicated in terms of speedup over a uniprocessed
run, indicate a point of diminishing returns with 4
processors.

| NUMBER OF PROCESSORS | SPEEDUP OVER 1 PROCESSOR |
| --- | --- |
| 4 | 2.6 |
| 8 | 2.8 |

TABLE 3.    Speedup over 1 processor for multitasked versions
of the code on the CRAY Y-MP.

The explanation of why an additional four processors failed
to speedup the code significantly is as follows. Refering
back to Fig. 5, the first 'k' loop contains a recursive
computation for the vertical velocity and geopotential. This
dependency prevents the 'k' loop from being microtasked. The

microtasking is then applied to the NX x NY dimension. For
the NX = NY = 50 grid dimension used in the run, the
synchonization of processors at the end of each horizontal
loop becomes a bottleneck. In other words, the calculations
are too fine grained. The second 'k' loop contains no
vertical dependencies and was microtasked. Since a majority
of the work is contained in the first 'k' loop, a point was
reached where additional processors could not be effectively
used.

The solution, fortunately, is straight forward. An increase
in the horizontal domain to the desired size of 256 x 256
should result in significantly less synchonization overhead
at the end of each horizontal loop. Alternatively, a seperate
'k' loop could be constructed to compute the vertically
coupled portions of the inertia-gravity wave calculations.
The first 'k' loop would not contain any vertical
dependencies and could then be microtasked. This additional
'k' loop would obviously microtask over the horizontal loop
but would result in less code being subjected to a fine-
grained bottleneck. A significant increase in temporary
storage, however, would be required. Unfortunately, as of
this time neither of these opportunities has been pursued.


6.    SUMMARY AND CONCLUSION


Solution of the primitive equations on the CM has been found
to be straight forward and very efficient. The execution time
for the dynamics is comparable to vector supercomputers. This
paper did not consider physical parameterizations, however,
some general conclusions can be reached. Since a processor
was assigned to each column and physical parameterizations
generally do not involve horizontal information exchange, it
can be anticipated that physics can be computed very quickly
on the CM. The nominal floating point performance of the
machine, 1 Gigaflop, should be achievable as all calculations
will involve data already in the processor memories. Perhaps
more importantly, physical parameterization routines can be

coded as serial code with looping over 'k'; a very natural way of thinking. The projection of the code so as to execute on all processors simultaneously is straighforward.


7.    ACKNOWLEDGEMENTS

I would like to thank John Church of NRL for helping me get started on the Connection Machine and for being available to answer many questions. Robert Whaley of Thinking Machines Corporation wrote the MACROS used for interprocessor communications and offered many useful suggestions. Jim Abeles, Mic Talian and Steve Perry of Cray Research generously volunteered to run the model on the Cray systems and discuss the results with me. Chuck Aston and Louis Hackerman of Multiflow Computer also volunteered to run the model on their system and supply me with the results. I would like to especially thank Fran Balint for giving me the time necessary to investigate the Connection Machine.


8.    REFERENCES

Arakawa, A., 1972: Design of the UCLA general circulation model, Numerical Simulation of Weather and Climate, Dept. of Meteorology, Univ. of California, Los Angeles, Tech. Rept. No. 7.

Gadd, A.J., 1974: An economical explicit integration scheme. Meteor. Office Tech. Note 44, 7 pp.

Hillis, D.W., 1986: The Connection Machine. MIT Press, Cambridge, 1986.

Lax, P., and B. Wendroff, 1960: Systems of conservation laws. Comm. Pure and Appl. Math., 13, 217-237.

Phillips, N.A., 1957: A coordinate system having some special advantages for numerical forecasting. J. Meteor., 14, 184-185.

# A Distributed Memory Implementation of the Shallow Water Equations

DAVID A. TANQUERAY[1] and DAVID F. SNELLING[2]

[1] FPS Computing, Apex House, London Road, Bracknell, Berks. RG12 2TE, U.K.
[2] Department of Computer Studies, University of Leicester, Leicester, LE1 7RH, U.K.

## Abstract

The shallow water equations have been used to test the behaviour of a variety of parallel computer systems in numerical weather prediction. The algorithm is fully vectorisable, but in its parallel form, requires a high degree of synchronisation and data motion. This paper describes an implementation on the FPS T Series which is a distributed memory parallel computer in which each node is a vector processor. The strategies for data partitioning and the overlap of communications with computation are discussed and some comparative performance results presented.

## 1. Introduction

The emerging technology of massively parallel computers offers the potential of performing numerical weather prediction with greater speed and accuracy than can be achieved with conventional machines.

The Shallow Water equations (see Hoffmann et. al. [3] and Sadourney [5]) are the basic set of equations used in full scale weather prediction and as such are representative of the type of calculation involved. They are fully vectorisable, require a large volume of data and high degree of communication and have been studied in the past with respect to parallel computing. They thus make an ideal vehicle for evaluating a distributed memory parallel computer system for this application.

In this paper the method of implementing the shallow water equations on the FPS T Series is described. It is shown that the communication can be overlapped with the computation in a fairly straightforward way. The timings given were obtained on a 16 processor system.

Topics in Atmospheric and Oceanic Sciences
© Springer-Verlag Berlin Heidelberg 1990

## 2. The Shallow Water Equations

In this section we provide a brief summary of the shallow water equations. A more detailed discussion of the formulation of these can be found in Hoffmann [3] or Sadourney [5]. The shallow water equations model a single layer of fluid within a Cartesian domain. Although they are a primitive model of the atmosphere, their structure is sufficiently representative to be of interest in evaluating the behaviour of parallel computer systems.

The primary variables in the formulation of the following shallow water equations are u and v, the velocities in the x and y directions; P, the pressure; and H, which is a value related to the height of the fluid; U and V, the mass fluxes; and Z, the potential velocity. The vorticity term, normally present in these equations, has been rewritten in terms of u and v. The final formulation of the equations is given below. A more detailed discussion of these equations and the discretisation technique may be found in Hoffmann [3].

$$\partial u/\partial t - ZV + \partial H/\partial x = 0$$
$$\partial v/\partial t + ZU + \partial H/\partial y = 0$$
$$\partial P/\partial t + \partial U/\partial x + \partial V/\partial y = 0$$

In this implementation the boundaries are assumed to be periodic in both the east/west and north/south directions (see Figure 1). This was done in the original presentation of this model to ease the explanation of the concepts. In a more accurate model the boundary conditions would be continuous only in the east/west direction. This means that the problem is being solved on a two dimensional torus, a simplification adopted as a straightforward simulation of an infinite two dimensional field. This was implemented in the original code by allocating an additional row and column (the shaded areas) for each of the variables. Periodic continuation consists of copying the row and column from the opposite edges into them as indicated by the arrows.

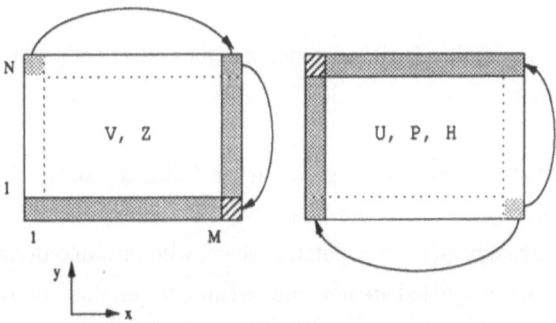

Cartesian domain with periodic boundary conditions.

Figure 1. The solution domain

## 3. Distributed Memory Implementation

In a distributed memory system the data arrays must be partitioned across the processors. In this implementation of the shallow water equations a horizontal slice of the domain for each variable is stored in each processor (see Figure 2). This is in contrast to all previous implementations, which have assumed that all variables were stored entirely in a global shared memory.

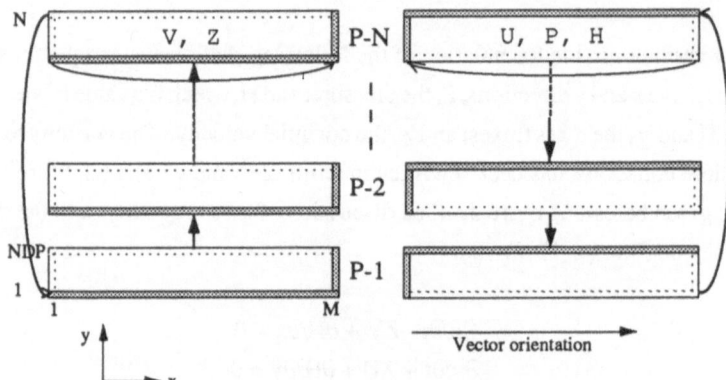

Periodic continuation is generalised to slices of the domain.

Figure 2. Domain partition on the T Series

The number of slices is simply the number of processors, represented by P-1 to P-N in Figure 2, and the size of each slice is determined by the size of the total problem divided by the number of processors. The processors may therefore be configured as a one dimensional torus (or simple ring of processors). In this case the east/west transfers remain as simple copies within each processor and only the north/south periodic continuation requires data to be transferred between processors. Because of the structure of the algorithm, the periodic continuation for some of the variables is performed north to south while for other variables it is south to north (see Figure 2). The T Series is able to perform transfers in both directions at once.

This implementation is simply a logical extension of the concept of periodic continuation to the the interface between processors.

The general structure of the algorithm given in the following pseudo code implies that the periodic continuation is synchronous and occurs twice in the loop. However, these transfers can also run in parallel with some of the computation since, with careful ordering of the calculations, the results of each variable updated are not required until the computation of others is complete. The way this overlap operates is illustrated in Figure 3. Each of the transfers illustrated was implemented very simply by a single call to an asynchronous routine followed by a Wait for Completion call at the appropriate place (indicated by the corresponding arrow head).

Initialisation

While (nstep <= max_steps) Do

Compute U, V, Z, & H from u, v, & P.
**Periodic continuation (U, V, Z, & H).**

Compute unew, vnew, & Pnew from uold, vold, Pold, U, V, Z, & H.
**Periodic continuation (unew, vnew, Pnew).**

Update old values from old, current, & new values.
Copy new values to current values.

End_While.

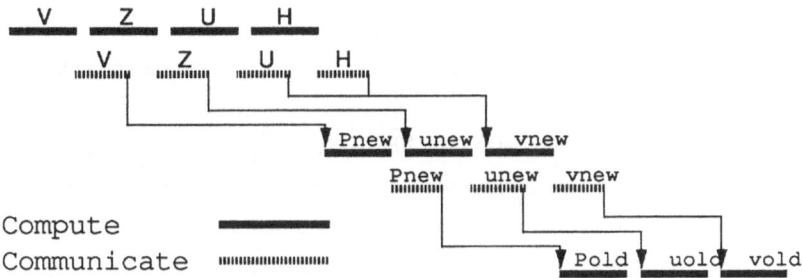

Figure 3. Communications overlap strategy

## 4. The FPS T Series Implementation

The FPS T Series is a Parallel Vector Supercomputer with a distributed memory in which the processing nodes are connected as a binary n-cube, or hypercube.

Each processor consists of a Transputer control unit, with interprocessor links, 1 Mbyte vector memory organised in two banks and a full 64-bit floating point vector unit with a peak speed of 12 Mflops. The basic system with 16 nodes may be replicated to build larger configurations. For a detailed description of the T Series architecture see Hawkinson [2] or Tanqueray [6].

### 4.1. Programming approach

A single Fortran 77 or C program is normally loaded into all processing nodes of the T Series to operate on a part of the data in each node. There is thus no requirement for explicit Fork or Taskstart operations, these being implicit in the architecture.

Furthermore there is no explicit synchronisation of the processors required other than that implied by data flow considerations. This results from the message passing nature of the distributed memory architecture.

Normally one processor will be used to read in the parameters for a run and broadcast them to all the others. All processors then proceed independently, only stopping to wait for data from a neighbour when needed. At the end of a run the results are then collected onto a single node to be formatted for output.

## 4.2. Vectorisation Strategy

The data allocation strategy for the shallow water equations has the advantage that all calculations within each processor can be performed on long vectors with unit stride. This is achieved by computing spurious results at the additional boundary locations which are subsequently updated during periodic communication. Because of the structure of the T Series vector memory architecture, unit stride is a requirement for high performance. Note however, that most vector machines also achieve higher performance when the stride is carefully managed.

Greater performance can be achieved on the FPS T Series by allocating particular arrays to specified memory banks. In this case all the variables were stored in bank B, the larger of the two. A single additional array was allocated in bank A to hold intermediate vector results and provide the second source for most of the vector operations without the need for copying the data. A typical example is shown in Figure 4, in which P must first be copied to the scratch area in bank A, but thereafter the operation runs at full speed. Note that the intermediate vector result of the VASM (vector add scalar multiply) operation is written to the scratch area ready for the next vector multiply. The result of a vector operation may be stored into both memory banks simultaneously with no additional penalty.

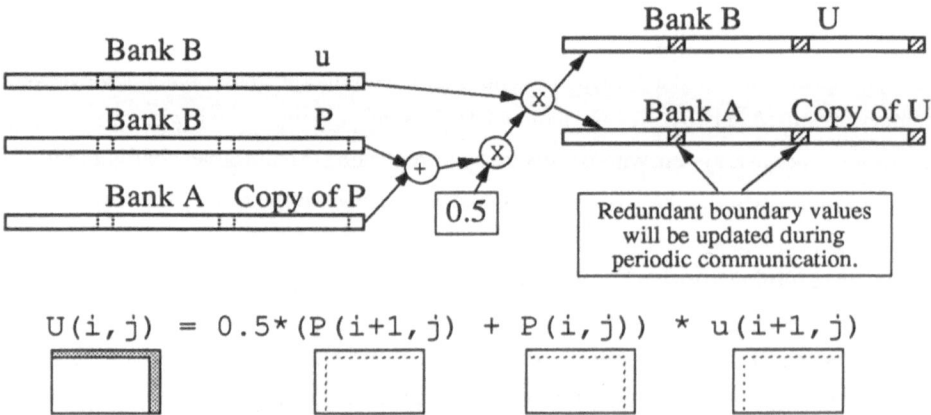

$$U(i,j) = 0.5*(P(i+1,j) + P(i,j)) * u(i+1,j)$$

- All rows are processed in a single very long vector operation.
- Results may be stored in two memory banks, without penalty.

Figure 4. Vectorisation strategy

Because the T Series compiler does not incorporate an automatic vectoriser, the vector unit is accessed on each node by calls to the T Series Math Library routines which take care of all issues associated with data alignment within and across memory banks. In many cases this representation was more intuitive than the lists of array indices used in standard FORTRAN 77. However, greater performance can be achieved by optimising the alignment of data in memory and direct use of the low level vector operations. Results for compiled Fortran, and both levels of vector usage are given in the results section below.

## 4.3. Communications Strategy

Interprocessor communication is accomplished by calls to high level communication routines from Fortran or C. These routines, supplied as part of the FPS parallel processing library, allow the user to regard the system as an arbitrary collection of processors which can be configured to suit the application topology (see Figure 5), and perform such high level functions as data broadcast and collection, rotation or shifting of data within one dimension of a torus, and basic one to one transfers or swaps. Communication can be synchronous or asynchronous with multiple transfers active simultaneously. These routines free the programmer from the details of hardware interconnect and thus simplify the process of parallel implementation. For a description of these routines and their characteristics see Carlile [1].

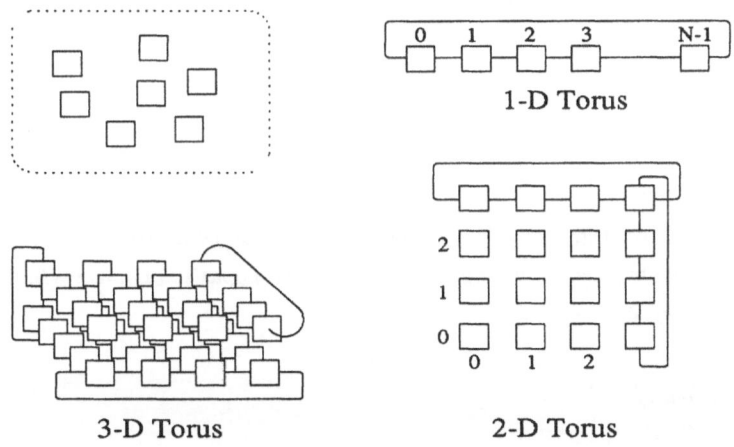

Figure 5. T Series programmable topologies

Having defined a topology, one or more logical links (analogous to files or Unix pipes) can be opened for parallel communications. A logical link bears no relation to any interprocessor path but serves solely to establish a queue for communication requests. The length of the queue determines the number of asynchronous calls that may be outstanding. If the queue is of zero length, communication is synchronous. Communications over multiple links take place independently and in parallel. Waiting on a link waits for all outstanding I/O on that link.

In this implementation of the shallow water equations, three links were used in the main loop corresponding to the three points that must wait for data at each stage of the loop (i.e. the arrow heads in Figure 3).

The following is an example of one of the high level communication calls used:

```
ISTAT=ROTAT_1D(LINK,0,-1,CU(1,1),CU(1,NDPP1),MP1B)
```

This routine rotates data around the one-dimensional torus (ring) thus implementing the periodic continuation and distributed data movement in a single call. In this example, the source address for each MP1B bytes transfer in each processor is CU(1,1) and the corresponding destination address is CU(1,NDPP1). The -1 indicates the direction of the transfer (north to south) and the distance (number of processors) to move each set of data. The LINK is a logical path or queue to be used.

The following pseudo code gives the sequence of operations within each part of the main loop and indicates in italics where the communication takes place.

Initialisation: Configure 1-D torus; and Open control link
If (processor 0) Read control parameters
Broadcast parameters from processor 0
Initialise data; and Open data links (Sync/Async)

Main Loop:
  Loop 100: Compute V
      Vector stride copy West column to East within each processor
      *Rotate North row around the torus to neighbour's South row  (Initiate)*
      Repeat for U, Z and H

  Loop 200: *Wait for V*
      Compute Pnew
      Copy West column to East within each processor
      *Rotate South row around the torus (in reverse) to neighbour's North (Initiate)*
      Wait for Z;      and repeat for unew
      Wait for U and H;  and repeat for vnew

  Loop 300: *Wait for Pnew:*    and update Pold and P
      Wait for unew;    and update uold and u
      Wait for vnew;    and update vold and v

## 5.    Results

The results of running the shallow water equations model on the FPS T Series are shown in Table 1. This gives the fully optimised timings of the three main loops over a domain of 64 x 64 grid points. For comparison purposes the timings given in Hoffmann [3] for the Cray-1 are also shown. The reason that this particular comparison is made is that the theoretical peak speed of the Cray-1 (160 Mflops) most closely matches that of the T Series model measured (192 Mflops) which was the basic 16 node system.

Figure 6 shows the method of measuring the times of the three principal loops taking into account the communication overhead. Due to the heavily overlapped nature of the implementation the effect of using synchronous versus asynchronous communications moves the overhead from one loop to another. Consequently it is not meaningful to make direct comparisons between these results for any individual loop. The whole cycle results however do give a good measure of the various effects.

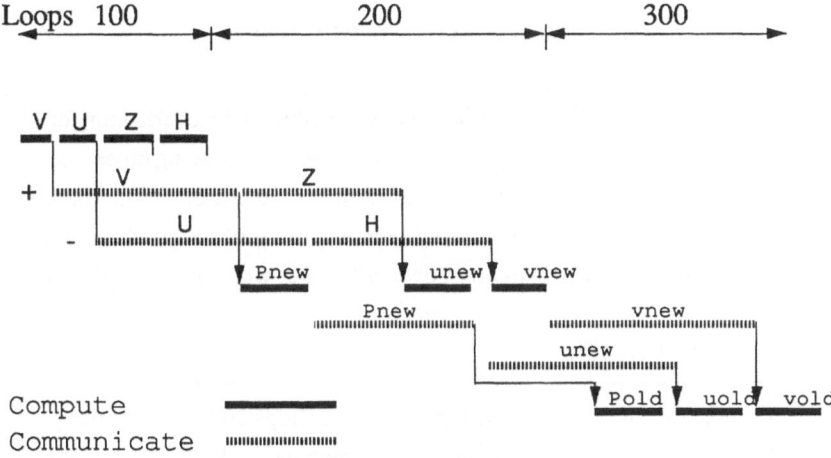

Figure 6. Communications Timing

It should be noted that all the timings given in [3] exclude the periodic continuation time since all the machines tested had a common shared memory. However, when evaluating distributed memory systems such as the T Series, this now becomes an essential aspect of the performance and should of course be included. Consequently, three figures are given for the 16 node T Series system to show this effect.

The first, labelled compute, excludes the periodic and interprocessor communication, and measures only the speed of the arithmetic.

370

The second, labelled Sync I/O, includes all periodic communication, which is performed synchronously with no overlap. In this case the periodic communications time is included in the loop that initiates it.

The last column gives the Asynchronous timings in which the data transfers are initiated to run in parallel with one another and arithmetic computation. In this case the communications overhead is split between the initiation overhead and the loop that waits for it.

| Computational Section | Mflops Cray-1 | Mflops 16 Node T Series | | |
|---|---|---|---|---|
| | | Compute | Sync I/O | Async |
| U, V, Z, H | 70.0 | 19.2 | 7.8 | 11.1 |
| unew, vnew, Pnew | 63.8 | 28.2 | 10.3 | 12.6 |
| uold, vold, Pold | 47.9 | 41.7 | 34.3 | 20.9 |
| Whole cycle | - | 25.7 | 10.7 | 13.1 |

16 Nodes = 4 + 1 Rows per Processor
Table 1. Results for 64 x 64 grid

While the latter results do not show much improvement due to the overlapping of I/O, the size of the problem is such that only four horizontal rows of the grid are being processed in each node for a full row of data transmission. In order to show the benefit of a better balance between compute and communications, the problem was rerun with a grid size of 256 x 256 points. The corresponding results from these runs are shown in Table 2, which indicate that it is possible to hide all or most of the data movement time in large scale problems.

| Computational Section | Mflops Cray-1 | Mflops 16 Node T Series | | |
|---|---|---|---|---|
| | | Compute | Sync I/O | Async |
| U, V, Z, H | (70.0) | 33.0 | 25.8 | 29.8 |
| unew, vnew, Pnew | (63.8) | 52.3 | 40.3 | 44.5 |
| uold, vold, Pold | (47.9) | 85.3 | 83.9 | 77.2 |
| Whole cycle | - | 46.4 | 37.1 | 41.1 |

16 Nodes = 16 + 1 Rows per Processor
Table 2. Results for 256 x 256 grid

Implementing this shallow water model on the T Series in parallel took one man-day of effort. This included entering and testing the code (240 lines) and the communication structure. The further time spent vectorising the code and then optimising it are shown in Table 3. Most of this time was spent in conventional single processor optimisation.

| Development Stage | Effort | Performance (Mflops) | |
|---|---|---|---|
| | | 64x64 | 256x256 |
| Parallel communication | 1 day | .11 | .12 |
| Generic Vectorisation | 2 days | 7.7 | 19.7 |
| Optimised Vector | 10 days | 13.1 | 41.1 |

Table 3. Development effort

## 6. Conclusions

In this paper we have shown that a standard numerical weather prediction program can be implemented on a distributed memory vector computer, with good performance. Parallel efficiency of 89 percent was achieved on a large data set.

The program was parallelised and the necessary data transfers implemented with only one day's effort using the high level communications library available on the FPS T Series. Since these routines directly express the desired operations in a single call, they greatly simplify the implementation and debugging of parallel programs.

In earlier studies on this code the implicit assumption had been made that parallel processors suitable for meteorological applications would share a common memory. Consequently the timings measured taskstart rather than data movement overhead.

## References

[1] B.R. Carlile and D. Miles, Structured Asynchronous Communication Routines for the FPS T Series, submitted to the *3rd Conference on Hypercube Concurrent Computers and Applications,* (ACM 1988).

[2] S. Hawkinson, The FPS T Series: A Parallel Vector Supercomputer, Floating Point Systems Inc. (1986).

[3] G.-R. Hoffmann, P.N. Swarztrauber, and R.A.Sweet, Aspects of Using Multiprocessors for Meteorological Modelling, *Multiprocessing in Meteorological Models,* Hoffmann & Snelling eds. (Springer-Verlag, Berlin, 1988).

[4] M. Ikeda, Multitasking with a Memory Hierarchy, *Multiprocessing in Meteorological Models,* Hoffmann & Snelling eds. (Springer-Verlag, Berlin, 1988).

[5] R. Sadourney, The Dynamics of Finite Difference Models of the Shallow Water Equations, *Journal of Atmospheric Sciences* **32** (1975).

[6] D. A. Tanqueray, The Floating Point Systems T Series, *Multiprocessing in Meteorological Models,* Hoffmann & Snelling eds. (Springer-Verlag, Berlin, 1988).

# Highly Parallel Architectures in Meteorological Applications (Summary of the Discussion)

D. K. MARETIS

European Centre for Medium-Range Weather Forecasts, Shinfield Park, Reading, Berkshire RG2 9AX, U.K.

## 1. Introduction

Great efforts have been made in the last few years to develop parallel computers that can deliver high-performance computer power at a reduced hardware cost compared with conventional mainframes or vector supercomputers. The ICL DAP almost a decade ago, and now TM's Connection Machine and the AMT DAP have shown that these machines can provide both a tolerable programming environment and parallel implementations of a wide range of problems.

Similarly, multi-processor hypercube machines in the USA and reconfigurable transputer networks in Europe have shown that the potential of highly-parallel MIMD computers can be realized.

The economic advantages of these massively parallel computers open up new opportunities both for research and industrial applications.

Numerical weather prediction is not only a time-critical application, but is also demanding of both CPU resources and memory, due to the large amount of data involved. Therefore, the questions that immediately arise are whether such highly-parallel architectures are well suited for running NWP models, and if so, whether the very nature of these applications pose special requirements in architectures, topologies, memories and software.

The following discussion report gives a brief summary of the user's computational needs and expectations when running meteorological models, and their experiences to date with parallel machines, as well as an assessment of expected future developments based both on today's knowledge and state-of-the art.

## 2. Architectural Aspects

### 2.1 Number of processors

There are two mainstreams of computer development at present: The coarsely-parallel architectures like the well-known vector computers with 2-16 powerful processors, and a trend towards further increase in the number of processors (up to 64), and the highly-parallel machines built from large numbers of small processors (up to 64000 in the case of the Connection Machine).

Topics in Atmospheric and Oceanic Sciences
© Springer-Verlag Berlin Heidelberg 1990

Conventional vector systems with shared memory and containing a small number of powerful processors still seem to best meet the requirements of the meteorological community. However, the non-scalability of those architectures, together with possible semiconductor technology constraints limiting the performance of conventional processors, could reinforce the drive to highly-parallel distributed memory systems.

However, there is still no agreement on the number of processors which would provide the best possible configuration for meteorological applications, since this seems to be application dependent.

It is also not possible to give a straight answer to the question whether SIMD or MIMD architectures are more suitable forodels, becauy depends on the model used. As an example, the physics of the present ECMWFR spectral model would lead to a strong imbalance if it were run on a SIMD machine.

## 2.2   Topology

Most of the highly parallel architectures offer the possibility of increasing the number of processors used for a certain application, and also of adapting the topology to the problem (and scale of the problem) to be solved. The reconfigurability of these systems is one of their strong points. However, it is still not clear which topology should be given preference for use for meteorological applications. Furthermore, there are great problems when porting software from one configuration to another, or to a different topology. Rewriting of great parts of the code is required if the program is to run on a different configuration/topology.

## 2.3   Memory and Communications

But there are some more major problems to be solved before highly parallel architectures can replace the conventional vector processors in meteorological applications: the need for large memories and the communication between processors.

It is a well-known fact that meteorological applications need a large amount of data, i.e. they have very large memory requirements. Many of the conventional processors provide large shared memories, even when they also have some sort of local ones. In contrast to this, almost all new highly-parallel architectures have only varying amounts of distributed memories. Access to needed data could, therefore, become a major bottleneck when models run on such novel architectures.

Furthermore, the processors of a highly-parallel machine must communicate with one another, and in addition to this data must be trasferred between them and from some kind of globaly accessible storage onto their local (distributed) memories. The problems of communication between the processors and of transport of data have been tackled in different ways by different manufacturers. Acceptable solutions, however, have not yet been developed, and a great deal of research and development efforts is presently going into improving existing and/or developing new ways of communications. So, e.g., there is still much work to be done on detailed aspects of message-passing networks. Research on, and extensive simulations of, routing strategies is also required, as well as full-scale performance evaluation.

3.    Meteorological Applications

3.1   Grid/Spectral Models

It seems that both grid and spectral methods could be used for meteorological applications on highly parallel architectures.  Both can be implemented without major difficulties.

Grid point methods are more flexible with respect to variation in strong local  nonlinearities or complicated geometry.  On the other hand spectral techniques are more accurate in simulating smooth flows.  Spectral models perform quite effectively and efficiently at low resolutions, and they are thus well suited to establishing the climatology of a model.  The decision in favour of a spectral model at ECMWF was taken because at that point it produced the best forecast for a given computational cost.  This still remains one of the main requirements of meteorological centers to the performance (possibilities) of parallel computers.  It is also highly desirable to use the same source for operational and research activities, and with varying resolutions without great effort or changes.  This is becoming increasingly important since future operational activities may well make use of ensembles of lower resolution forecasts.

3.2   Analysis

The analysis also has high memory requirements, but here, I/O  and data handling problems make the introduction of highly parallel systems difficult.

4.    User interface

4.1   Languages

The shortcomings of Fortran are well known. But it is still widely used for meteorological applications because programmers do not see another viable alternative, and Fortran code is relatively portable over a wide range of machines and architectures.  Also it is usually the language where most optimization effort has gone both by manufactureres and scientists, and hence it often produces the best code.

The wide range of applications written in Fortran, and the existence of sophisticated compilers that can efficiently vectorize such code, are the main reasons for the widespread resistance in accepting alternative massively parallel computer architecturarge Fortran codes cannot be routinely ported to the wide variety of parallel hardware now available, nor do compilers exist to take advantage of the parallelism in the hardware architectures.

With the marked reluctance to restructure and rewrite codes, only the provision of a parallel programming support environment able to provide meteorological users with the compilers and software tools necessary to ease the program 'migration' problem, might persuade users to accept the new architectures and migrate their codes.

## 4.2 Autotasking

New ways of supporting concurrent programming (multiprocessing) on vector processors have evolved over recent years. They provide an automatic mechanism for exploiting parallelism without programmer intervention, while still allowing for manual fine-tuning of the application. Such 'autotasking' facilities must be also provided on highly parallel computers if users are to be able to make full use of the potential (parallelism) of those architectuers. This is even more necessary since the great number of processors involved makes it more difficult to distribute the work among them in the most effective way, and keep the load on the communication links at an acceptable level.

## 4.3 Portability

One of the main drawbacks of the highly parallel systems is the great difficulty in porting applications from the one architecture to another, and even between different topologies of the same machine. Software tools are therefore required which will facilitate the portability of software across a wide range of architectures anpoloes. A first step in this direction could be the development of software aids for migrating and adapting Fortran codes to the new architectures.

## 4.4. Debugging

Although the vendors of vector processors have now put some more effort into the development of debugging aids, there is still a long way to go before one can debug a multiprocessor program as easily as a uniprocessor program.

So, it is not at all surprising that - due to the much greater number of processors involved - debugging programs which run on highly parallel systems is a far more difficult task than if they were running on conventional vector processors.

Effective and user-friendly debugging aids, therefore, are indispensable, before real meteorological applications can be developed and run on parallel architectures.

## 4.5   Visualization

Over the last few years, it has been realized that visualization tools are not only very useful for checking the results of running programs, but that they could also provide a very powerful tool during the development and debugging of programs. A facility to graphically display intermediate results while the program is running could help detect errors and/or inconsistencies at an early stage, thus providing invaluable help during debugging. Although some manufacturers seem to put considerable effort into developing such tools, it is quite unlikely that they could be made available on highly parallel architectures in the near future.

## 5.   CONCLUSIONS

Meteorological models will increasingly require high computing power, as well as large and fast memories. The increasing number of CPUs available to one application and the difficulties in distributing the work in an efficient way among the available CPUs make the development of automatic parallelization tools (like Cray's autotasking) indispensable. This is even more important in the case of highly parallel machines due to the great numbers of processors involved.

Whatever the architecture, it is imperative to have good software tools as part of a parallel programming environment. Vectorizing and parallelizing compilers, debuggers, performance analyzers, visualization tools, expert systems to aid the algorithmic selection etc. are needed if the computer architectures available at present, or under development are to be used effectively.

The economic and other advantages of massively parallel computers can be exploited in meteorological applications only if the problems of communication between processors and of data transfer are solved, if comprehensive software tools for migration of existing codes, distribution of work among the available processors and debugging are made available to users, and, finally, if the research into parallel algorithms can provide substantial help in the choice of the numerical methods and the distribution of work among the processors.

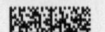